Commutative Semigroup Rings

Chicago Lectures in Mathematics

commutative semigroup rings

Robert Gilmer

The University of Chicago Press
Chicago and London

To the memory of Tom Parker,
who kindled my initial interest in
commutative semigroup rings.

Chicago Lectures in Mathematics Series
Irving Kaplansky, *Editor*

The Theory of Sheaves, by Richard G. Swan (1964)
Topics in Ring Theory, by I. N. Herstein (1969)
Fields and Rings, by Irving Kaplansky (1969; 2d ed. 1972)
Infinite Abelian Group Theory, by Phillip A. Griffith (1970)
Topics in Operator Theory, by Richard Beals (1971)
Lie Algebras and Locally Compact Groups, by Irving Kaplansky (1971)
Several Complex Variables, by Raghavan Narasimhan (1971)
Torsion-Free Modules, by Eben Matlis (1973)
The Theory of Bernoulli Shifts, by Paul C. Shields (1973)
Stable Homotopy and Generalized Homology, by J. F. Adams (1974)
Commutative Rings, by Irving Kaplansky (1974)
Banach Algebras, by Richard Mosak (1975)
Rings with Involution, by I. N. Herstein (1976)
Theory of Unitary Group Representation, by George W. Mackey (1976)
Infinite-Dimensional Optimization and Convexity, by Ivar Ekeland and Thomas Turnbull (1983)

The University of Chicago Press, Chicago 60637
The University of Chicago Press, Ltd., London

Printed and bound by CPI Group (UK) Ltd, Croydon, CR0 4YY

93 92 91 90 89 88 87 86 85 84 1 2 3 4 5

1980 Mathematics Subject Classification: 13, 20M14

ISBN: 978-0-226-29391-2 (cloth) 978-0-226-29392-9 (paper)
LCN: 83-51596

CONTENTS

PREFACE

This book contains an account of the current state of knowledge on some topics in the area of commutative semigroup rings. Much of the material in the book comes from literature of the last ten years, and in this regard, a bit of historical information may be in order. Work on group rings goes back at least forty years to Higman's paper [76]. Original work in the area focused on two related problems: the isomorphism problem for groups over a fixed coefficient ring, and the problem of determining the structure of the unit group of a group ring. In this work, the cofficient ring was usually taken to be a field (in which case the term group algebra, rather than group ring, was frequently used) or the ring Z of integers or (occasionally) the ring of algebraic integers in a finite algebraic number field — hence a commutative ring in any case. But restriction to the case of abelian groups was relatively uncommon, and from a ring–theoretic point of view, most of the work in group rings before 1970 was in noncommutative ring theory.

In commutative algebra, semigroup rings arise more naturally than group rings; for example, each polynomial ring over R is a semigroup ring over R, subrings of $R[X_1,\ldots,X_n]$ generated over R by pure monomials are semigroup rings, and residue class rings $R[\{X_\lambda\}]/I$, where I

is generated by "pure difference binomials", are also semigroup rings. Thus, semigroup rings were seen in the early 1970's as a general concept under which several classes of important examples in commutative algebra could be considered. And with the presentation in [52] of an example of a group algebra K[G] that is two-dimensional, non-Noetherian, and factorial, a new problem area in commutative ring theory was stimulated — that of determining, for a commutative ring R and commutative semigroup S, conditions under which the semigroup ring of S over R has a given ring-theoretic property E. In order to solve problems of this type, detailed results about other aspects of commutative semigroup rings — for example, nilpotents, units, integral closure, and dimension — were sometimes required. To a large degree, Chapters II–V recount work done in the problem areas just described.

The book is written for a person with a fair background in commutative ring theory, who may know little about the theory of commutative semigroups. Thus, Chapter I develops almost from scratch most of the semigroup theory that is subsequently used. The reader is expected to be familiar with the material in Chapters 1–3 of Kaplansky's book [83] and with the material related to primary decomposition in Noetherian rings found in Sections 4–7 of Chapter IV of [136]. The material just cited is a sufficient background in ring theory for Chapters I and II of this book. In Chapters III–V, proofs of some results from my book [51] are repeated here, while other results from [51] are merely cited. Whether such a result is proved or cited is usually determined by its importance to the topic in semigroup rings being developed. Except for one result from the paper [41] referred to in Section 20, Chapter V, all the ring theory needed is either developed in this book or it can be found in [83], [136], or [51].

As indicated above, Chapter I is devoted to the development of a basic theory of commutative semigroups. Section 4 of this chapter deals with congruences, and is used throughout the rest of the book. Section 5 treats Noetherian semigroups, and is essential for the proof in Section 7 that a monoid ring is Noetherian if and only if the coefficient ring is Noetherian and the monoid in question is finitely generated.

After defining a semigroup ring in Section 7 and presenting some basic properties of such rings, the rest of Chapter II is devoted to an investigation of zero divisors, nilpotent elements, idempotents, and units of semigroup rings. Known results for polynomial rings serve as a model for the development in Sections 8–11. Thus, the symbol $R[X;S]$ is used in Chapters I and II for the semigroup ring of S over R, and elements of $R[X;S]$ are written as "polynomials" $f = \Sigma_{i=1}^{n} f_i X^{s_i}$, with "exponents" s_i in S. The general aim in §§8–11 is to determine, under various hypotheses on R and S, equivalent conditions in terms of $\{f_i\}_1^n$ and $\{s_i\}_1^n$ in order that f should be a zero divisor, nilpotent, etc. of $R[X;S]$.

Chapters III–V consider problems about semigroup rings that are more global in nature than those discussed in the first two chapters. In these three chapters, $R[S]$ is the symbol used to denote the semigroup ring of S over R. Chapter III treats questions of the form "determine necessary and sufficient conditions on the unitary ring R and the monoid S in order that the monoid ring R[S] should be an integral domain with property E". More specifically, Sections 12–15 handle the cases where property E is the property of being integrally closed, a Prüfer domain, a factorial domain, or a Krull domain, respectively. In particular, the example in [52] referred to in the second paragraph above followed resolution of the problem of determining conditions under which R[S] is factorial, as described in Section 14. Chapter IV has the same basic theme as Chapter III, but properties E that are of interest in rings with zero divisors are examined in Chapter IV. For example, E is the property of being (von Neumann) regular in Section 17, while it's the property of satisfying a specified chain condition in Section 20.

Various topics are included in Chapter V. In Section 21, the (Krull) dimension of R[S], for S a cancellative monoid, is determined in terms of the dimension of an appropriate polynomial ring over R. The R–automorphisms of the group ring $R[Z^n]$ are determined in Section 22. Sections 23 and 24 treat two different cases of the question of whether isomorphism of $R_1[S]$ and $R_2[S]$ implies isomorphism of R_1 and R_2. Specifically, Section 23 provides an affirmative answer in the case where R_1 and R_2 are fields and S contains a periodic element, and Section 24 treats the case where S is Z^n for some n. Finally, Section 25 is a survey of some known results concerning the isomorphism problem for groups mentioned in the first paragraph of this preface — that is, does $R[G_1] \simeq R[G_2]$ imply $G_1 \simeq G_2$?

The book ends with a selected bibliography, an index of main notation, an index of some main results, and a topic index. While the bibliography contains 137 items, I did not attempt to give a complete bibliography. All but about 25 of the references are cited somewhere in the text, usually in the remarks that conclude each section; I consulted these 25 items in writing the book, but had no occasion to refer to them specifically. The index of main results is designed to help the reader locate them and related results, for I have followed Kaplansky's practice in [83] of calling no result a lemma or a proposition; each result in [83] is called a theorem, but I've included some labeled as corollaries as well. More than any other aspect of [51], I've heard complaints about (and experienced) the inadequacy of the topic index; I hope that this index is more helpful in tracking items down.

Much stimulation for research has come from errors or gaps published in the mathematical literature. In spite of four proof readings, I'm sure that <u>Commutative Semigroup Rings</u> will contain its share of errors, and I can only hope that some of these will stimulate, rather than frustrate, the reader.

Most of this book was written during the Fall semester of 1982 while I was on sabbatical leave from Florida State University at the University of Connecticut. I hereby express my appreciation to FSU for its financial support and to UConn for its hospitality and a good atmosphere for work. In particular, Gene Spiegel of UConn helped me work out kinks in the material and in its presentation on a number of occasions; I'm grateful for these and for his other deeds of help and kindness. I finished the book during the summer of 1983, when I received partial support from National Science Fonda-tion Grant MCS 8122095. Again I acknowledge with gratitude this financial support from NSF.

Chapter I

COMMUTATIVE SEMIGROUPS

This chapter contains a portion of the basic theory of commutative semigroups. Other parts of the theory are scattered through the remaining four chapters of the text, closer to the material where applied. Many ring theorists are unfamiliar with the theory of semigroups, and hence the topic, insofar as we subsequently need to use it, is developed in this chapter almost from scratch. Section 1–5 are fundamental for all subsequent chapters, while the material in Section 6 concerning factorization in commutative monoids is not referred to again until Section 14.

§1. Basic Notions

This section introduces basic terminology and notation concerning commutative semigroups that are used in the remainder of the text. Only two results are stated formally in the section, and the proof of the second of these is given in outline form.

A semigroup is a nonempty set closed under an associative binary operation. If $(S,\circ)$ is a semigroup, then S is commutative (or abelian) if it is commutative under the operation $\circ$, and S has an identity element if there exists an identity element with respect to $\circ$; a semigroup with identity is called a monoid. Except for a brief foray in Section 23, we deal exclusively with commutative semigroups in the text, but comments about the noncommutative case are included in remarks at the end of some sections. The semigroup operation is denoted by $+$ unless there is a strong tradition favoring a different (usually multiplicative) notation. Each commutative semigroup S can be imbedded as a subsemigroup of a commutative monoid. One way to accomplish such an imbedding is to take a symbol t not in S and extend the semigroup operation $+$ on S to the set $S^0 = S \cup \{t\}$ by defining $x + t = t + x = x$ for each $x \in S^0$. Most of the semigroups considered in the sequel are assumed to be monoids.

If A and B are nonempty subsets of the semigroup S, then $A + B$ is defined to be $\{a + b | a \in A, b \in A\}$. A subset T of S is a subsemigroup of S if T itself is a semigroup under the operation on S, and a subsemigroup T is a subgroup of S if T itself is a group. We remark that if S is a monoid, then a subgroup of S need not contain the identity element of S. If $\{S_\alpha\}$ is a family of subsemigroups of S, then $\cap S_\alpha$ may be empty; if $\cap S_\alpha$ is nonempty, then it is a subsemigroup of S. Thus, if B is a nonempty subset of S, then B generates a subsemigroup $\langle B \rangle$ of S consisting of all finite sums $\Sigma_{i=1}^{n} b_i$, with $b_i \in B$ for each i. An <u>ideal</u> of S is a nonempty subset I of S such that $I \supseteq s + I = \{s + i \mid i \in I\}$ for each $s \in S$. The intersection of a family of ideals of S is an ideal, provided it is nonempty. If B is a nonempty subset of S, then $B \cup (B + S)$ is the ideal of S generated by B; if S is a monoid, then $B \subseteq B + S$ and $B + S$ is the ideal of S generated by B. An ideal I of S is <u>proper</u> if $I \neq S$, and I is <u>prime</u> if $x + y \in I$ implies $x \in I$ or $y \in I$ for $x,y \in S$. A proper ideal I of S is prime if and only if $S \backslash I$, the complement of I in S, is a subsemigroup of S. If I is an ideal of S, then the <u>radical</u> of I, denoted $\text{rad}(I)$ or $\sqrt{I}$, is defined to be $\{s \in S \mid$ there exists $n \in Z^+$ such that $ns \in I\}$; it is easy to show that $\text{rad}(I)$ is an ideal of S. The first result lists some basic properties of semigroup ideals.

THEOREM 1.1. Let S be a semigroup.

(1) S is a group if and only if S is the only ideal of S.

(2) If I is an ideal of S and T is a subsemigroup of S such that $I \cap T = \phi$, then there exists a prime ideal P of S containing I such that $P \cap T = \phi$.

(3) If I is an ideal of S, then $\text{rad}(I)$ is the

intersection of the family of prime ideals of S containing I.

Proof. The proof of (1) is elementary, and (3) follows from (2). We prove (2). By Zorn's Lemma, it suffices to show that I is prime if I is maximal with respect to failure to meet T. Thus, assume that $a,b \in S\backslash I$. Each of the ideals $I \cup \{a\} \cup (a + S)$ and $I \cup \{b\} \cup (b + S)$ meets T, and hence there exist $s_1, s_2 \in S$ such that $a + s_1 \in T$ and $b + s_2 \in T$. Consequently, $a + s_1 + b + s_2 \in T$. Since $I \cap T = \phi$, it follows that $a + b \notin I$, and hence I is prime.

If S is an additive monoid with identity (zero) element 0 and if $s \in S$, then s is said to be invertible if there exists $x \in S$ such that $s + x = 0$; the set G of invertible elements of S forms a subgroup of S, and G is the largest subgroup of S containing 0. A finite sum $\Sigma_{i=1}^{n} s_i$ of elements of S is invertible if and only if each s_i is invertible. Thus, $S\backslash G$ is a prime ideal of S if $G \neq S$. If H is any subgroup of S containing 0, then as in the case of groups, H induces a partition of S into disjoint cosets $s + H$ of H. In fact, if the relation $\sim$ on S is defined by $a \sim b$ if $a = b + h$ for some $h \in H$, then $\sim$ is an equivalence relation on S and $s + H$ is the equivalence class of $s \in S$.

An element s of a semigroup S is said to be cancellative if $s + a = s + b$ implies $a = b$ for $a, b \in S$. The set C of cancellative elements of S may be empty; if $C \neq \phi$, then C is a subsemigroup of S, a a finite sum $\Sigma_{i=1}^{n} s_i$ is in C if and only if each $s_i \in C$ and hence $S \backslash C$ is a prime ideal of S if $S \neq C$. If $S = C$, we say that S is a cancellative semigroup. A familiar result from elementary group theory states that a finite cancellative semigroup is a group. It is clear that a subsemigroup of a group is cancellative. Theorem 1.2 proves a form of the converse.

THEOREM 1.2. If C is a subsemigroup of an additive semigroup S and if each element of C is cancellative in S, then there exists an imbedding f of S into an abelian monoid T such that (1) $f(c)$ has an inverse $-f(c)$ in T for each $c \in C$, and (2) $T = \{f(s) - f(c) | s \in S \text{ and } c \in C\}$. The monoid T is determined, up to semigroup isomorphism, by properties (1) and (2). If S is cancellative and $S = C$, then T is a group.

Outline of a proof. The construction of T is like the familiar one used in constructing the ring Z of integers from the set Z_0 of nonnegative integers, starting from Peano's postulates. To wit, let $A = S \times C$ and let $\sim$ be the equivalence relation on A defined by $(s_1,c_1) \sim (s_2,c_2)$ if $s_1 + c_2 = s_2 + c_1$ (the cancellative property of C is used in proving transitivity of $\sim$). Denote by $[s,c]$ the equivalence class of (s,c) under $\sim$ and let T be the set of equivalence classes $[s,c]$. The set T is an abelian monoid with zero element $[c,c]$ under the operation $[s_1,c_1] + [s_2,c_2] = [s_1 + s_2, c_1 + c_2]$, and the mapping $f : S \longrightarrow T$ defined by $f(s) = [s + c, c]$ is an imbedding of S in T. If $c \in C$, then $f(c) = [2c, c]$ has inverse $[c, 2c]$ in T, and an arbitrary element $[s, c]$ of T can be written as $[s + c, c] + [c, 2c] = f(s) - f(c)$. It is clear that T is determined up to semigroup isomorphism by the properties (1) and (2), and if $S = C$, then an arbitrary element $[s, c]$ of T has inverse $[c, s]$ in T so that T is a group. This completes the proof of Theorem 1.1.

The monoid T constructed as in the proof of Theorem 1.2 is called the <u>quotient monoid of</u> S <u>with respect to</u> C. By abuse of notation, we write the elements of T in the form $s - c$ instead of $f(s) - f(c)$ and we consider S to be a subset of T. If S is cancellative, then the group T of Theorem 1.2 is called the <u>quotient group</u> of S; to within isomorphism, it is the smallest group in which S can be imbedded. One way of classifying cancellative semigroups is in terms of properties of their quotient groups. For example, abelian groups are frequently classified as being either torsion-free, mixed, or torsion groups. We recall the definitions. An abelian group G is <u>torsion-free</u> if 0 is the only element of G of finite order; G is a <u>torsion group</u> if each element of G has finite order, and G is <u>mixed</u> if it contains elements of infinite order and nonzero elements of finite order. If S is a cancellative semigroup with quotient group G, then the condition that G be torsion-free, for example, can be stated in terms of S; in fact, G is torsion-free if and only if S satisfies the following condition.

(1.3) For any positive integer n and any $x,y \in S$, the equality $nx = ny$ implies that $x = y$.

We note that the condition in (1.3) makes sense whether S is assumed to be cancellative or not; it is used as the definition of a torsion-free semigroup. We will encounter torsion-free semigroups in Section 3 in determining the class of (totally) ordered semigroups. We avoid the use of the term torsion semigroup, preferring the term periodic semigroup instead. The definition is given in Section 2; we remark that for a cancellative semigroup S, the definition is such that S is periodic if and only if its quotient group is a torsion group.

Section 1 Remarks

The material in Section 1 is both basic and elementary, and hence complete proofs have not been given in the section. Some notation and terminology have been used without explanation. For example, $\subseteq$ denotes inclusion, $<$ proper inclusion, and $A\backslash B$ denotes the complement of B in A, where B is a subset of the set A; the notions of semigroup isomorphism, isomorphic semigroups, and an imbedding of one semigroup in another are defined in the expected ways.

Many of the notions in Section 1 are meaningful, and have been studied, for noncommutative semigroups. But Theorem 1.2 does not extend to the noncommutative case. Malcev in [94] constructed a noncommutative cancellative semigroup S that is not imbeddable in a group; he also showed that the semigroup ring of S over the field Q of rational numbers is a noncommutative ring without zero divisors that is not imbeddable in a field.

Clifford and Preston's book [32] is a standard reference on the theory of (not necessarily commutative) semigroups. In addition, Redei's book [122] treats finitely generated commutative semigroups, and especially the theory of congruences on such semigroups. The papers [31], [39] and [73] are primarily concerned with the basic theory of commutative semigroups.

§2. Cyclic Semigroups and Numerical Monoids

Two concrete classes of semigroups frequently encountered in the general theory are the semigroups $\langle s\rangle$ generated by a single element and submonoids of the monoid Z_0 of nonnegative integers. This section contains basic results concerning semigroups in these two classes.

An additive semigroup S is said to be cyclic (or monogenic) if there exists $s \in S$ such that $S = \langle s\rangle = \{s, 2s, \ldots, ns, \ldots\}$. We determine the structure of such semigroups. Clearly there are two cases to consider. First, the elements ms and ns may be distinct for distinct integers m and n; in this case S is isomorphic to the additive semigroup Z^+ of positive integers, and S has infinite order. The second case, where $ms = ns$ for some $m \neq n$, is covered in Theorem 2.1.

THEOREM 2.1. Assume that $S = \langle s\rangle$ is a cyclic semigroup such that $ms = ns$ for some $m \neq n$. Let k be the smallest integer such that $ks = rs$ for some $r < k$, and let $m = k - r$.

(1) For $u \geq v \geq r$, the equality $us = vs$ holds if and only if m divides $u - v$.

(2) $S = \{s, 2s, \ldots, (k-1)s\}$ has cardinality $k - 1$.

(3) $G = \{rs, (r+1)s, \ldots, (k-1)s\}$ is a subgroup of S; its identity element is hs, where h is the integer between r and $k - 1 = r + m - 1$ that is divisible by m.

Proof. (1) Assume that $u - v = ma$. Since $rs = rs + ms$, it follows easily that $rs = rs + mas$. Hence $vs = rs + (v-r)s = rs + mas + (v-r)s = (v+ma)s = us$.

For the converse, assume that $us = vs$. Since $\{r, r+1, \ldots, k-1\}$ is a set of m consecutive integers, there exist unique integers u_1, v_1 in this set so that $u \equiv u_1 \pmod{m}$ and $v \equiv v_1 \pmod{m}$. The part of (1) already proved shows that $u_1 s = us = vs = v_1 s$, and the choice of k implies that $u_1 = v_1$. Therefore $u \equiv v \pmod{m}$.

The choice of k also implies that the set $\{s, 2s, \ldots, (k-1)s\}$ has cardinality $k - 1$, and (1) shows that it is equal to S . Moreover, (1) also shows that the mapping $as \longrightarrow a + (m)$ of G into $Z/(m)$ is a semigroup isomorphism; since $Z/(m)$ is a group, then G is also a group and its identity is hs , where $h + (m) = 0 + (m)$.

If s, r and m are as in Theorem 2.1, then r is called the _index_ of s , and m is the _period_. We also refer to r and m as the _index_ and _period_ of the semigroup $\langle s \rangle$; the integer $r + m - 1$ is called the _order_ of s , and also the order of $\langle s \rangle$. As in the case of groups, the order of $\langle s \rangle$ is the number of elements in $\langle s \rangle$. If $\langle s \rangle$ is finite, then s is said to be _periodic_, and a semigroup S is _periodic_ if each element of S is periodic. If $\langle s \rangle$ is infinite, then s is _aperiodic_, and S is _aperiodic_ if each non-identity element of S is aperiodic. The index and period of a cyclic semigroup determine the cyclic semigroup up to isomorphism. Moreover, for any positive integers r and m , there exists a cyclic semigroup $C(r,m)$ of index r and period m . One such semigroup is the multiplicative semigroup generated by the class of X in $Z[X]/(X^{r+m} - X^r)$; another is the multiplicative semigroup in $Z/(2^m - 1)Z \oplus Z/q^rZ$ generated by $(\bar{2},\bar{q})$, where q is a prime greater than $2^m - 1$. We note that $C(r,m)$ is a group iff $r = 1$ iff $C(r,m)$ is a monoid; $C(1,m)$ is, of course, the cyclic group of order m . On the other hand, each $C(r,m)$ contains a unique idempotent element (in additive notation, an element t is idempotent if $t = t + t = 2t$) .

The additive semigroup Z^+ of positive integers is the infinite cyclic semigroup and $(Z_0,+)$ is the monoid obtained by adding an identity element to Z^+ as in Section 1. On the other hand, the fundamental theorem of arithmetic shows that the multiplicative semigroup Z^+ is the internal weak direct sum of countably many copies of $(Z_0,+)$, in the sense of the following definition. Let S be an additive monoid with zero element 0 . If $\{S_\alpha\}_{\alpha\in A}$ is a family of submonoids of S , each containing 0 , then we say that S is the internal weak direct sum of the family $\{S_\alpha\}$ (and we write $S = \Sigma^w_{\alpha\in A}S_\alpha$) if each element of S is expressible in the form $\Sigma^n_{i=1}s_{\alpha_i}$, with $s_{\alpha_i} \in S_{\alpha_i}$ for each i , and if an equality $\Sigma^n_{i=1}s_{\alpha_i} = \Sigma^n_{i=1}t_{\alpha_i}$ implies that $s_{\alpha_i} = t_{\alpha_i}$ for each i . The definition implies that $S = \Sigma_\alpha S_\alpha$ and that $S_\alpha \cap \Sigma_{\beta\neq\alpha}S_\beta = (0)$ for each $\alpha \in A$, but is not equivalent to these two conditions since elements of S_α need not be invertible in S . In case the set $A = \{1,2,\ldots,n\}$ is finite, we write $S = S_1\oplus\ldots\oplus S_n$. To return to the monoid $(Z^+,\cdot)$, let $p_1 < p_2 < \ldots$ be the set of prime integers and let S_i be the submonoid $\{p_i^j\}_{j=0}^\infty$ of Z^+ for each i . Unique factorization in Z^+ translates precisely to the statement that Z^+ is the weak direct sum (or product) of the family $\{S_i\}_{i=1}^{\infty}$; we return to these notions in Section 6 in the definition of a factorial monoid.

We turn to a consideration of submonoids of the additive monoid Z_0 . Such a monoid is called a numerical monoid. If S is a nonzero numerical monoid, let d be the greatest common divisor (g.c.d) of the elements of S . Then $S = dt$, where T is isomorphic to S and the g.c.d of elements of T is 1 ; we call such a numerical monoid T primitive. To determine all numerical monoids, it suffices to determine the primitive ones. Theorem 2.2 foreshadows what is proved in Theorem 2.4 concerning such monoids.

THEOREM 2.2. Let a and b be relatively prime positive integers. If $n \geq (a - 1)(b - 1)$, then there exist nonnegative integers x and y such that $n = xa + yb$.

Proof. The case where a or b is 1 is trivial, so we assume that $1 < a < b$. Write $ib = q_i a + r_i$, where $0 \leq r_i < a$ and $0 \leq i \leq a - 1$. Since b is a unit modulo a , the set $\{r_i\}_0^{a-1}$ is a complete set of residues modulo a . If $n \geq (a - 1)(b - 1)$, write $n = ta + r$, where $0 \leq r < a$, and hence $r \in \{r_i\}_0^{a-1}$. If $r = r_j$, then $t \geq q_j$, for $t < q_j$ implies that $ta + rj = n < q_j a + r_j = jb$. Therefore $jb - n$ is divisible by a . But $jb - (a - 1)(b - 1) \leq (a - 1)$, a contradiction. Therefore $t \geq q_j$ and $n = (t - q_j)a + q_j a + r_j = (t - q_j)a + jb$.

Theorem 2.2 implies that a numerical monoid containing two relatively prime integers contains an ideal $K + Z_0 =$

$\{x \in Z_0 | x \geq K\}$ of Z_0 for some $K \in Z^+$. The next result shows that each nonzero primitive numerical monoid contains a pair of relatively prime integers.

THEOREM 2.3. If S is a nonzero primitive numerical monoid, then there exist positive integers $a,b \in S$ such that $\gcd\{a,b\} = 1$.

Proof. There exists a finite subset $\{a_1,a_2,\ldots,a_n\}$ of $S \cap Z^+$ with g.c.d. 1 . Choose integers $x_1,\ldots,x_n$ such that $1 = x_1a_1+\ldots+x_na_n$. Then $1 + [k(a_1+\ldots+a_{n-1}) - x_n]a_n$ $(x_1 + ka_n)a_1+\ldots+(x_{n-1} + ka_n)a_{n-1}$ for each integer k , and for k sufficiently large, each $x_i + kx_n$ is positive so that $\Sigma_{i=1}^{n-1}(x_i + ka_n)a_i = 1 + [k(a_1+\ldots+a_{n-1}) - x_n]a_n$ is in S and is relatively prime to $a_n \in S$.

THEOREM 2.4. Let S be a nonzero numerical monoid and let d be the g.c.d. of the elements of S .

(1) There exists a positive integer K such that S contains kd for each $k \geq K$.

(2) S is finitely generated. Hence Z_0 satisfies the ascending chain condition (a.c.c.) on submonoids.

(3) There exists a finite set B of monoid generators S such that if C is any set of generators for S as a monoid, then $B \subseteq C$.

Proof. Statement (1) follows at once from Theorems 2.2 and 2.3. To prove (2) , we note that the set $\{s_i\}_{i=1}^t$ of elements of S that are less than Kd is finite, and S is generated by $\{s_i\}_{i=1}^t \cup \{Kd,(K + 1)d,\ldots,(2K - 1)d\}$. The fact that each submonoid of Z_0 is finitely generated implies, in the usual way, that a.c.c. on submonoids is satisfied in Z_0 .

To prove (3) , let b_1 be the smallest positive element of S . Clearly b_1 belongs to each generating set C for S . If $\{b_1\}$ generates S as a monoid, then we take $B = \{b_1\}$. Otherwise, let b_2 be the smallest element of S not in the monoid $\langle b_1 \rangle^0$ generated by b_1 . Again it is clear that $\{b_1, b_2\} \subseteq C$ for any generating set C of S . Continuing this process, we obtain $\langle b_1, b_2, \ldots, b_n \rangle^0 = S$ for some n since Z_0 satisfies a.c.c. on submonoids, and we take $B = \{b_1, b_2, \ldots, b_n\}$.

We have already observed that each nonzero numerical monoid is isomorphic to a primitive one. The next result shows that distinct primitive numerical monoids are not isomorphic. If S_1 and S_2 are semigroups, then as usual, a <u>homomorphism</u> of S_1 into S_2 is a mapping of S_1 into S_2 that preserves the semigroup operation.

THEOREM 2.5. Assume that S and T are nonzero numerical monoids. If $h : S \longrightarrow T$ is a homomorphism, then there exists a nonnegative rational number q so that $h(s) = qs$ for each $s \in S$. Thus h is either injective or it maps each element of S to 0 . If S and T are primitive and h is an isomorphism of S onto T , then $S = T$.

<u>Proof</u>. Let s be a nonzero element of S , let $t = h(s)$, and let $q = t/s$. Take $s' \in S$. If $s' = 0$, it is clear that $h(s') = 0 = qs'$. If $s' \neq 0$, then $h(ss') = sh(s') = s'h(s)$ so that $h(s') = qs'$. Thus

h is injective or the zero mapping, according as $q \neq 0$ or $q = 0$. If S and T are primitive and h is surjective, then primitivity of S and the inclusion $qS \subseteq Z_0$ imply that q is an integer, and primitivity of T and the equality $T = qS$ show that $q = 1$. Thus, isomorphism of S and T implies equality, for S and T primitive.

If S and B are as in part (3) of Theorem 2.4, then the cardinality of B is called the rank of S. The submonoid $\langle k, k+1, \ldots, 2k-1\rangle^0$ of Z_0 has rank k for any $k \in Z^+$. In contrast with Z_0, we note that submonoids of $Z_0 \oplus Z_0$, the external direct sum of Z_0 with itself, need not be finitely generated. For example, $(1,k) \notin \langle (1,0),(1,1),\ldots,(1,k-1)\rangle^0$ for any k.

It is an easy step to pass from submonoids of Z_0 to submonoids of the additive group of integers.

THEOREM 2.6. If S is a submonoid of Z containing both positive and negative integers, then S is a subgroup of Z.

Proof. Let $S_1 = S \cap Z_0$ and $S_2 = S \cap (-Z_0)$. If d_1 and d_2 are the greatest common divisors of the elements of S_1 and S_2, respectively, then since Z_0 and $-Z_0$ are isomorphic, Theorem 2.4 implies that there exist positive integers K_1 and K_2 such that $kd_1 \in S_1$ for each $k \geq K_1$ and $-md_2 \in S_2$ for each $m \geq K_2$. Let $d = \text{g.c.d.}\{d_1, d_2\}$. We show that $S = dZ$, the inclusion $S \subseteq dZ$ being clear. There exist integers x and y such that $d = xd_1 + yd_2$, and since $xd_1 + yd_2 = (x + rd_2)d_1 + (y - rd_1)d_2$ for any integer r, we can assume without loss of generality that

$x \geq K_1$ and $-y \geq K_2$. Thus $d \in S$ and a similar argument shows that $-d \in S$ as well. Therefore $dZ \subseteq S$ and $S = dZ$.

Theorem 2.6 can be extended to submonoids of the additive group Q . Before stating this result (Theorem 2.9), however, we review some of the basic theory of subgroups of Q ; this material will be referred to again in Chapter 3. First, the mapping $x \longrightarrow tx$ is an automorphism of Q for each nonzero element $t \in Q$. Thus, each nonzero subgroup G of Q is isomorphic to a subgroup containing Z , and in considering such subgroups, we frequently assume that G contains Z . Recall that a group H is said to have property **E** locally if each finitely generated subgroup of H has property E . In particular, H is locally cyclic if each finitely generated subgroup of H is cyclic.

THEOREM 2.7. Let $\{a_i/b_i\}_{i=1}^n$ be a finite subset of $Q\backslash\{0\}$, where $b_i > 0$ for each i and each a_i/b_i is in lowest terms. If $m = \text{lcm}\{b_1,\ldots,b_n\}$, then the subgroup of Q generated by $\{1\} \cup \{a_i/b_i\}_1^n$ is cyclic and is generated by $1/m$. In particular, Q is locally cyclic.

Proof. In this proof, we use (Y) to denote the subgroup of Q generated by the subset Y of Q . The inclusion $(1,a_1/b_1,\ldots,a_n/b_n) \subseteq (1/m)$ is clear. To prove the reverse inclusion, we use induction on n . If $n = 1$, write $1 = a_1x + b_1y$ for integers x and y . Then $1/b_1 = x(a_1/b_1) + y \in (1,a_1/b_1)$. If $n = 2$, let $d = \gcd\{b_1,b_2\}$ and write $d = b_1x + b_2y$. Then $1/m = d/b_1b_2 = x(1/b_2) + y(1/b_1) \in (1,a_1/b_1,a_2/b_2)$, so the equality $(1/m) = (1,a_1/b_1,\ldots,a_n/b_n)$ holds for $n = 1$ or 2..

Assume the result for $n = k$, and in the case where $n = k + 1$, let $m' = \text{lcm}\{b_1,\ldots,b_k\}$ so that $m = \text{lcm}\{m', b_{k+1}\}$. The cases $n = k$ and $n = 2$ yield the desired conclusion:

$$1/m \in (1/m', a_{k+1}/b_{k+1}) \subseteq (1, a_1/b_1, \ldots, a_{k+1}/b_{k+1}) .$$

An arbitrary finitely generated subgroup (Y) of Q is contained in $(\{1\} \cup Y)$. Since the latter group is cyclic and since a subgroup of a cyclic group is cyclic, it follows that (Y) is cyclic and Q is locally cyclic.

COROLLARY 2.8. Assume that H is a subgroup of Q.

(1) H is the union of an ascending sequence of cyclic groups.

(2) If H contains Z, then H is generated by the set $\{1/p^k \mid p$ is prime, $k \geq 0$, and $1/p^k \in H\}$.

Proof. (1): Let $\{h_i\}_1^\infty$ be an enumeration of the elements of H. If $H_i = (h_1,\ldots,h_i)$ for each i, then each H_i is cyclic, $H_1 \subseteq H_2 \subseteq \ldots$, and $H = \cup_1^\infty H_i$.

Statement (2) follows immediately from Theorem 2.7.

THEOREM 2.9. If S is a submonoid of Q containing both positive and negative rationals, then S is a subgroup of Q.

Proof. It suffices to show that $-s \in S$ for each nonzero $s \in S$. Choose $t \in S$ such that $st < 0$. The subgroup G of Q generated by $\{s,t\}$ is cyclic; assume that g is a generator of G. If $s = mg$ and $t = ng$, then $mn < 0$. Thus, the subsemigroup $\langle s,t \rangle$ of G is a group (we're using the fact that G admits only two total orders — one in which $g > 0$ and a second in which $-g > 0$.) It follows that $-s \in \langle s,t \rangle \subseteq S$.

Because of its connection with numerical semigroups, we conclude this section with a theorem concerning the multiplicative semigroup of a commutative ring. If A and B are semigroups, then the <u>external direct sum</u> of A and B , denoted $A \dot{\oplus} B$, is the set $A \times B$, with operation defined by the coordinatewise operations on A and B .

THEOREM 2.10. Let S be the direct sum of the semigroups $(Z^+,\cdot)$ and $(Z^+,+)$. If h is a homomorphism of S into a finitely generated commutative semigroup T , then there exist $m,n,a \in Z^+$ with $m \neq n$ such that $h(m,a) = h(n,a)$.

<u>Proof</u>. Assume that $T = \langle t_1,\ldots,t_k \rangle$. For each of the first $k + 2$ primes p_j , we write $h(p_j,1) = \Sigma_1^k a_{ji}t_i$. For $1 \leq j \leq k + 2$, let v_j be the $(k + 1)$–tuple $(1,a_{j1},\ldots,a_{jk})$. Thus, the set $\{v_j\}_{j=1}^{k+2}$ is linearly dependent over the rational field Q . An equation of linear depencence of $\{v_j\}$ over Q can be written in the form $\Sigma b_jv_j = \Sigma c_jv_j$, where b_j and c_j are positive integers and $b_j \neq c_j$ for some j . The equality $\Sigma b_jv_j = \Sigma c_jv_j$ implies that $\Sigma b_j = \Sigma c_j = a$. Let $m = \Pi_{j=1}^{k+2}p_j^{b_j}$ and let $n = \Pi_{j=1}^{k+2}p_j^{c_j}$. Then $h(m,a) = \Sigma_{j=1}^{k+2}h(p_j^{b_j},b_j) = \Sigma_1^{k+2}b_jh(p_j,1) = \Sigma b_jv_j = \Sigma c_jv_j = h(n,a)$; moreover, $m \neq n$ since $b_j \neq c_j$ for some j .

THEOREM 2.11. If the multiplicative semigroup of the commutative ring R is finitely generated, then R is

finite.

Proof. Consider first the case where R has prime characteristic p. In this case we show that $\{x^i\}_{i=1}^{\infty}$ is finite for each $x \in R$; this is sufficient since $(R,\cdot)$ is finitely generated. The ring R can be considered as a vector space over Z_p, the field with p elements. The assumption that $\{x^i\}_{i=1}^{\infty}$ is linearly independent over F_p leads to a contradiction, for in that case the subspace V of R spanned by $\{x^i\}_1^{\infty}$ is isomorphic as a ring to the ring $XZ_p[X]$, where X is an indeterminate over Z_p. But if $\{P_i\}_{i=1}^{\infty}$ is a complete set of nonassociate prime elements of $Z_p[X]$ distinct from X, then the mapping $(p_1^{e_1}\cdots p_v^{e_v},a) \longrightarrow x^a[P_1(x)]^{e_1}\cdots[P_v(x)]^{e_v}$ is an imbedding of S (as in Theorem 2.10) in $(R,\cdot)$, contrary to that result. Thus $\{x^i\}_1^{\infty}$ is linearly dependent, so some x^u is a linear combination of $x,x^2,\ldots,x^{u-1}$. It follows easily that V is spanned by $\{x^i\}_1^{u-1}$, so V is finite and $\{x^i\}_1^{\infty}$ is also finite. This completes the proof in the case where R has prime characteristic. The rest of the proof amounts to a reduction to that case.

In the general case, let $\{x_i\}_{i=1}^{k}$ be a finite set of generators for $(R,\cdot)$ and let S be as above. For any $x \in R$, the mapping $(m,a) \longrightarrow mx^a$ is a homomorphism of S into $(R,\cdot)$, and hence Theorem 2.10 shows that for some $m,n,a \in Z^+$, $mx^a = nx^a$ and $m \neq n$. Thus, a power of x has finite additive order. We choose the positive integers K and b so that $Kx_i^b = 0$ for $1 \leq i \leq k$. Let I be the ideal of R consisting of all elements $r \in R$ such that

$Kr = 0$. The residue class ring R/I is finite since only finitely many monomials $x_1^{h_1}\ldots x_k^{h_k}$ in $x_1,\ldots,x_k$ are such that $h_i < b$ for each i . Therefore R/I has nonzero characteristic, and the definition of I implies that R also has nonzero characteristic c . Let $c = p_1^{e_1}\ldots p_v^{e_v}$ be the prime factorization of c . If $R_j = \{r \in R \mid p_j^{e_j} r = 0\}$ for $1 \le j \le v$, then R is the direct sum of the ideals $R_1, R_2, \ldots, R_v$, and R_j has characteristic $p_j^{e_j}$ for each j As R_j is a homomorphic image of R , its multiplicative semigroup is finitely generated. Thus, to prove that R is finite, it suffices to prove that each R_i is finite. We therefore assume without loss of generality that the characteristic of R is a prime power p^e . Then $R > pR > \ldots > p^{e-1}R > p^eR = (0)$, and the additive group of $p^iR/p^{i+1}R$ is a homomorphic image of that of R/pR for each i . Thus, if R/pR is finite, then R is finite. Now R/pR is a ring of characteristic p which is a homomorphic image of R . By the first case of the theorem considered, R/pR is finite and this completes the proof of the theorem.

Section 2 Remarks

If B is a subset of a monoid S , then the definition of the <u>submonoid of</u> S <u>generated by</u> B appears implicitly in the proof of Theorem 2.4. The term is defined, of course, as the intersection of the family of submonoids of S containing $B \cup \{0\}$. It is denoted by $\langle B\rangle^0$, for it amounts to adjoining the zero element of S to the subsemigroup of S generated by B .

Additional material on numerical semigroups can be found

in [112, Sect. 38], [75], and [71]. In particular, there has been some work on determining the smallest positive integer K such that $K + Z_0$ is contained in a given primitive numerical monoid S. For S of rank 2 generated by a and b, the value $K = (a - 1)(b - 1)$ given in Theorem 2.2 cannot, in general, be improved. Much of the interest in numerical monoids in commutative algebra stems from the study of subrings of the polynomial or power series ring in an indeterminate t over R that are generated over R by a set $\{t^{n_i}\}_{i=1}^{k}$ of pure monomials. The ring in question is just the set of all polynomials or power series over R whose exponents come from the monoid $\langle n_1,\ldots,n_k\rangle^0$; in the case of polynomials, the ring $R[t^{n_1},\ldots,t^{n_k}]$ is the semigroup ring of $\langle n_1,\ldots,n_k\rangle^0$ over R. Two recent papers dealing with rings of this kind are [22] and [49]; polynomial rings are treated in [22], while [49] considers power series.

Theorem 2.8 was proved by Isbell, and our proof is merely an elaboration of the proof given in [79]. The question of whether Theorem 2.8 extends to noncommutative rings apparently remains open.

§3. Ordered Semigroups

An examination of the proofs in Section 2 shows that the order relation on Z plays a significant role in several of the results. In this section we determine the class of abelian semigroups that admit a relation of total order compatible with the semigroup operation. First we need some definitions.

A relation $\sim$ on a semigroup $(S,+)$ is said to be compatible with the semigroup operation if $a \sim b$ implies $a + x \sim b + x$ for $a,b,x \in S$. A relation $\sim$ on an arbitrary set X is called a partial order if it is reflexive, transitive, and asymmetric ($\sim$ is asymmetric if $a \sim b$ and $b \sim a$ imply that $a = b$), and a partial order $\sim$ is a total order on X if for distinct elements $a,b \in X$, either $a \sim b$ or $b \sim a$. A partial order is frequently denoted by $\leq$ and the notation $a > b$ or $b < a$ is used to mean $b \leq a$ and $b \neq a$. To say that a semigroup S is partially ordered (respectively, totally ordered) under a relation $\leq$ means that $\leq$ is a partial order (respectively, total order) on S and that $\leq$ is compatible with the semigroup operation on S.

If S admits a total order compatible with the semigroup operation, then it is easy to show that S is torsion-free and cancellative. To see this, take distinct elements $a,b \in S$ — say $a < b$. Then $a + x < b + x$ for each $x \in S$, so S is cancellative. Moreover, $2a < a + b < 2b$ and by induction, $na < nb$ for each $n \in Z^+$, implying that S is also torsion-free. We prove presently that conversely, each torsion-free cancellative semigroup can be totally ordered. The next result shows that to do so, it suffices to consider the case of a torsion-free group.

THEOREM 3.1. Let S be a torsion-free cancellative semigroup with quotient group G. Then S admits a total order compatible with its operation if and only if G has this property.

Proof. If G is totally ordered under $\leq$, then the relation $\leq$ induces a total order on S compatible with the semigroup operation. Conversely, if S is totally ordered under $\leq$, then we define a relation $\sim$ on G as follows. Each element of G is expressible in the form $s - t$ for some $s,t \in S$. For $g_1 = s_1 - t_1$ and $g_2 = s_2 - t_2$ in G, define $g_1 \sim g_2$ to mean that $s_1 + t_2 \leq s_2 + t_1$. Then $\sim$ is a well defined relation of total order on G that is consistent with the group operation on G and extends the relation $\leq$ on S; we only verify that $\sim$ is well defined and that it agrees with the relation $\leq$ on S. Thus, if $g_1 = s_1 - t_1 = s_1' - t_1'$ and $g_2 = s_2 - t_2 = s_2' - t_2'$, where $s_1 + t_2 \leq s_2 + t_1$, then

$$s_1 + t_2 + (s_2' + t_1') \leq s_2 + t_1 + (s_2' + t_1')$$

$$s_1' + t_1 + s_2 + t_2' \leq s_2 + t_1 + s_2' + t_1'$$

$$s_1' + t_2' \leq s_2' + t_1' .$$

This proves that $\sim$ is well defined. The imbedding of S in G is $s \longrightarrow (s + c) - c$, where $c \in S$. For $s,t \in S$, we have $s \sim t$ if and only if $(s + c) + c \leq (t + c) + c$, hence if and only if $s \leq t$.

Theorem 3.1 leaves us with the problem of proving that a torsion-free abelian group can be totally ordered. A bit more can be proved — namely, that any partial order on a torsion-free abelian group G can be extended to a total order. Some additional terminology facilitates the proof. Thus, assume that $\leq$ is a partial order on G and define $P = \{x \in G | x \geq 0\}$. Then P is called the positive cone of $\leq$, and P satisfied the following conditions (i) and (ii):

(i) $P + P \subseteq P$; that is, P is a subsemigroup of G.

(ii) $P \cap (-P) = \{0\}$.

Moreover, if $\leq$ is a total order on G, then condition (iii) is satisfied.

(iii) $P \cup (-P) = G$.

A subset L of G satisfying (i) and (ii) is called a positive subset of G. Such a subset induces a partial order $\sim$ on G that is compatible with the group operation. The relation $\sim$ is defined by $g \sim h$ if $g - h \in L$, and $\sim$ is a total order if and only if L satisfies (iii). These correspondences between partial orders on G and positive subsets of G are inverses of each other, and their respective restrictions to the total orders on G and the set of positive subsets of G satisfying (iii) are also inverses. The set of positive subsets of G is partially ordered under $\subseteq$, and the inclusion $L_1 \subseteq L_2$ is equivalent to the condition that $\sim_2$, the partial order on G induced by L_2, is an extension of the partial order $\sim_1$ induced by L_1. Theorem 3.2 is stated in this vein.

THEOREM 3.2. If P_0 is a positive subset of a torsion-free abelian group G, then P_0 can be imbedded in P, a positive subset of G satisfying condition (iii).

<u>Proof</u>. If $\{P_\alpha\}$ is a linearly ordered family of positive subsets of G, then it is clear that $\cup P_\alpha$ is also positive. Hence the set of positive subsets of G is inductive under $\subseteq$, and P_0 is contained in a maximal positive subset P. To complete the proof, we show that P satisfies condition (iii) — that is, $G \subseteq P \cup (-P)$. Thus, take $g \in G$ and consider the following three cases: (1) $n_0 g \in P$ for some $n_0 \in Z^+$; (2) $n_0(-g) \in P$ for some $n_0 \in Z^+$; (3) for all $n \in Z^+$, neither ng nor $-ng$ is in P.

In Case 1, let $P^* = P + \langle g\rangle^0 = \{p + kg \mid p \in P, k \in Z_0\}$. That P^* satisfies (i) is clear. Take $x \in P^* \cap (-P^*)$, say $x = p_1 + k_1 g = -(p_2 + k_2 g)$. Then $n_0 x = n_0 p_1 + k_1(n_0 g) = -(n_0 p_0 + k_2 n_0 g)$, and since $n_0 g \in P$, it follows that $n_0 x \in P \cap (-P) = \{0\}$. Therefore $x = 0$ since G is torsion-free. Thus P^* satisfies (ii), and maximality of P implies that $g \in P = P^*$. It follows from Case 1 that $-g \in P$ in Case 2.

In Case 3 we again let $P^* = P + \langle g\rangle^0$. If $x = p_1 + k_1 g = -(p_2 + k_2 g) \in P^* \cap (-P^*)$ in this case, then $p_1 + p_2 = (k_1 + k_2)(-g) \in P$. By assumption, $k_1 + k_2 = 0$ so that $k_1 = k_2 = 0$, and hence $x = p_1 = -p_2 = 0$. It follows that $P^* \cap (-P^*) = \{0\}$, so again $g \in P$ (which implies that Case 3 does not arise). This completes the proof.

If $\{Z_\alpha\}_{\alpha \in A}$ is a family of groups, each isomorphic to Z, and if $G = \Sigma^c_{\alpha \in A} Z_\alpha$ is the complete direct sum of the family $\{Z_\alpha\}$, then $P = \{\{n_\alpha\} \mid n_\alpha \geq 0 \text{ for each } \alpha\}$ is a positive subset of G. The induced partial order, defined by $\{a_\alpha\} \geq \{b_\alpha\}$ if $a_\alpha \geq b_\alpha$ for each α, is called the <u>cardinal order</u> on G. One extension of the cardinal order on G to a total order is the <u>lexicographic order</u> on G, defined by well-ordering the set A, then setting $\{a_\alpha\} > \{b_\alpha\}$ if $a_{\alpha_0} > b_{\alpha_0}$ for the first element α_0 of A for which $a_\alpha \neq b_\alpha$. The external weak direct sum $G_0 = \Sigma^w_{\alpha \in A} Z_\alpha$ of the family $\{Z_\alpha\}$ is a subgroup of G, and the orders on G_0 induced by the cardinal and lexicographic orders, respectively, are called the cardinal and lexicographic orders on G_0. Another total order on G_0 in frequent use is the <u>reverse lexicographic order</u>, whose positive cone consists of 0 and the nonzero elements of G_0 whose last nonzero coordinate is positive. We note that the assumption that each Z_α is isomorphic to Z plays only a minor role in the definitions above; the definitions are meaningful for any family $\{Z_\alpha\}$ of totally ordered semigroups. It is the case where each Z_α is isomorphic to Z, however, that is used in the next corollary.

COROLLARY 3.3. If $\{g_\alpha\}_{\alpha \in A}$ is a free subset of a torsion-free abelian group G, then there exists a total order $\leq$ on G such that $0 \leq g_\alpha$ for each α.

<u>Proof</u>. The subgroup $\Sigma_{\alpha \in A}(g_\alpha)$ of G generated by $\{g_\alpha\}$ is isomorphic to $\Sigma^w_{\alpha \in A} Z_\alpha$, where $Z_\alpha \simeq Z$ for each α. Extending the cardinal order on $\Sigma_{\alpha \in A}(g_\alpha)$ to a total order yields the desired conclusion.

For the sake of future reference, we state formally the

result proved concerning orderability of a commutative semigroup.

COROLLARY 3.4. The abelian semigroup S admits a total order compatible with its semigroup operation if and only if S is torsion-free and cancellative.

Also for the sake of reference, we record one consequence of the proof of Theorem 3.2.

COROLLARY 3.5. Assume that P_0 is a positive subset of the torsion-free group G and that $x \in G$ is such that $nx \notin P_0$ for each $n \in Z^+$. Then P_0 can be extended to a positive subset P of G such that P satisfies (iii) and $-x \in P$.

Section 3 Remarks

The result that a torsion-free abelian group can be totally ordered is due to Levi [92], but the method used in the proof of Theorem 3.2 is closer to that of [80]; the class of lattice-ordered abelian groups is also discussed in [80]. For additional information concerning the question of what nonabelian groups can be totally ordered, see Chapter 3 of [46]. In particular, a (multiplicative) group with the property that $x \neq y$ implies $x^n \neq y^n$ for all $n \in Z^+$ need not admit a total ordering compatible with the group operation [93].

§4. Congruences

If S is a semigroup, the homomorphisms defined on S and the homomorphic images of S play an important role in determining the structure of S. An equivalent and convenient way of considering homomorphisms on S is through the notion of a congruence on S, defined as follows. A congruence on S is an equivalence relation on S that is compatible with the semigroup operation. Theorem 4.1 states the basic relationships between homomorphisms and congruences. The proof of Theorem 4.1 is routine and will be omitted.

THEOREM 4.1. Let S be an additive semigroup.

(1) If $\sim$ is a congruence on S, then for $s \in S$, denote by $[s]$ the equivalence class of s under $\sim$ for each $s \in S$, and let $S/\sim = \{[s] \mid s \in S\}$. Then $S/\sim$ is an abelian semigroup under the operation $[a] + [b] = [a + b]$, and the mapping $f : S \longrightarrow S/\sim$ defined by $f(s) = [s]$ is a homomorphism of S onto $S/\sim$; moreover, $f(s_1) = f(s_2)$ if and only if $s_1 \sim s_2$.

(2) Conversely, if $h : S \longrightarrow T$ is a homomorphism of S onto T, define the relation ρ on S by $a \rho b$ if $h(a) = h(b)$. Then ρ is a congruence on S and the semigroups S/ρ and T are isomorphic under the mapping $[s] \longrightarrow h(s)$, where $[s]$ denotes the equivalence class of $s \in S$ under ρ.

The semigroup $S/\sim$ defined in (1) of Theorem 4.1 is called the factor semigroup of S with respect to $\sim$. Theorem 4.1 shows that, to within isomorphism, these factor semigroups represent all homomorphic images of S.

If $\sim_1$ and $\sim_2$ are congruences, then we write $\sim_1 \leq \sim_2$ if $a \sim_1 b$ implies $a \sim_2 b$ for $a,b \in S$. Considering $\sim_1$ and $\sim_2$ as subsets of S, the relation $\sim_1 \leq \sim_2$ holds if and only of $\sim_1$ is contained in $\sim_2$. Thus, $\leq$ is a partial order on the set of congruences on S. Another equivalent form of the relation $\sim_1 \leq \sim_2$ is the assertion that the natural mapping $[s]_1 \longrightarrow [s]_2$ of $S/\sim_1$ onto $S/\sim_2$ is well defined (if the mapping is well defined, it is a homomorphism). The largest congruence on S is $S \times S$, the universal congruence under which any two elements of S are related, and the smallest congruence on S is the identity congruence $i = \{(s,s) \mid s \in S\}$. The intersection of any family of congruences on S is again a congruence on S; hence any family $\{\sim_\alpha\}_{\alpha \in A}$ of congruences has both a <u>least upper bound</u> (denoted $\text{l.u.b.}\{\sim_\alpha\}$ or $\text{lub}\{\sim_\alpha\}$) and a <u>greatest lower bound</u> (denoted $\text{g.l.b.}\{\sim_\alpha\}$ or $\text{glb}\{\sim_\alpha\}$) in the set of congruences on S. The relation $\sim = \text{glb}\{\sim_\alpha\}$ is easy to describe: $a \sim b$ means that $a \sim_\alpha b$ for each α. To describe $\text{lub}\{\sim_\alpha\}$, we note that any subset ρ of $S \times S$ generates a congruence on S, namely, the intersection of all congruences on S that contain ρ. We proceed to describe this congruence; since $\text{lub}\{\sim_\alpha\}$ is the congruence generated by $\bigcup_\alpha \sim_\alpha$, a description of $\text{lub}\{\sim_\alpha\}$ follows. The congruence $\sim$ generated by ρ is constructed in three steps.

1. Set $\rho_0 = \rho \cup \rho^{-1} \cup i$, where $\rho^{-1} = \{(b,a) \mid (a,b) \in \rho\}$ and i is the identity congruences on S. Then ρ_0 contains ρ and is both reflexive and symmetric. Clearly ρ and ρ_0 generate the same congruece on S.

2. Set $\rho_1 = \rho_0 \cup \{(a + c, b + c) \mid (a,b) \in \rho_0 \text{ and } c \in S\}$.

Then ρ_1 is reflexive, symmetric, and is compatible with the semigroup operation on S. If S is a monoid, then $\rho_1 = \{(a + c, b + c) \mid (a,b) \in \rho_0$ and $c \in S\}$ since ρ_0 is contained in this set. Again it is clear that ρ_0 and ρ_1 generate the same congruence.

3. Let $\sim = \{(a,b) \mid$ there exist $a_0, a_1, \ldots, a_t \in S$ such that $a = a_0$, $b = a_t$, and $(a_i, a_{i+1}) \in \rho_1$ for $i = 0,1,\ldots,t-1\}$. Then $\sim$ is the congruence on S generated by ρ.

In case $\rho = \bigcup_\alpha \sim_\alpha$ is the union of a family of congruences on S, then $\rho = \rho_0 = \rho_1$, and only Step 3 above need be applied to ρ in order to obtain $\mathrm{lub}\{\sim_\alpha\}$.

A congruence $\sim$ on S is said to be finitely generated if there exists a finite subset ρ of $S \times S$ such that $\sim$ is the congruence generated by ρ, and S is Noetherian if each congruence on S is finitely generated. As usual, each congruence on S is finitely generated if and only if S satisfies a.c.c. on congruences, which is true if an only if each strictly increasing set $\sim_1 < \sim_2 < \ldots$ of congruences on S is finite. In Section 7 we prove that a monoid S is Noetherian if and only if S is finitely generated as a semigroup. We return briefly to a consideration of least upper bounds and greatest lower bounds of congruences in Section 5. The rest of this section is devoted to an examination of some examples of congruences that naturally arise in the remainder of the text.

THEOREM 4.2. Assume that T is a subsemigroup of the semigroup S. Define a relation $\sim$ on S by $a \sim b$ if $a + t = b + t$ for some $t \in T$. Then $\sim$ is a congruence on S and $[t]$ is cancellative in $S/\sim$ for each $t \in T$. If ρ is any congruence on S such that the image of T in S/ρ consists of cancellative elements of S/ρ, then $\rho \geq \sim$.

<u>Proof</u>. Clearly $\sim$ is reflexive and symmetric. If $a \sim b$ and $b \sim c$, then $a + t_1 = b + t_1$ and $b + t_2 = c + t_2$ for some $t_1, t_2 \in T$, so $a + (t_2 + t_1) = c + (t_2 + t_1)$ and $a \sim c$. Moreover, $a + t_1 = b + t_1$ implies $a + s + t_1 = b + s + t_1$ for each $s \in S$ so that $\sim$ is a congruence on S. If $[a] + [t] = [b] + [t]$ for some $a,b \in S$ and $t \in T$, then $a + t + t_1 = b + t + t_1$ for some $t_1 \in T$, and hence $a \sim b$ and $[a] = [b]$. Thus, $[t]$ is cancellative in $S/\sim$ for each $t \in T$. The assertion that $\sim \leq \rho$ for ρ as described is clear.

In the case where $T = S$, the congruence $\sim$ defined on S as in Theorem 4.2 is called the <u>cancellative congruence</u> on S. For a subsemigroup T of S such that each element of T is cancellative in S, the quotient monoid of S with respect to T was defined in Section 1. For a general subsemigroup T, the quotient monoid of S with respect to T is defined to be the quotient monoid of $S/\sim$ with respect to the subsemigroup $\{[t] \mid t \in T\}$, where $\sim$ is the congruence defined in Theorem 4.2. The definition is reminiscent of the definition of the quotient ring R_N of a commutative ring R with respect to a multiplicative system N, and indeed there is a connection between the two, as the next result shows.

THEOREM 4.3. Assume that N is a multiplicative system in the commutative ring R. Regarding R as a semigroup under multiplication, the quotient monoid of R with respect to its subsemigroup N is the multiplicative semigroup of the ring R_N .

Proof. We recall the definition of R_N . Let I be the ideal of R consisting of all $r \in R$ such that $rn = 0$ for some $n \in N$ and let ϕ be the natural homomorphism from R onto R/I . Then $\phi(N)$ is a regular multiplicative system in $\phi(R)$ and R_N is defined to be $\phi(R)_{\phi(N)} = \{\phi(r)/\phi(n) \mid r \in R, n \in N\}$. The elements of the quotient monoid of R with respect to N can be represented in the form $[r]/[n]$, where $[x]$ denotes the class of $x \in R$ with repect to $\sim$, and $\sim$ is defined on R by $a \sim b$ if $an = bn$ for some $n \in N$. To complete the proof, we show that $[x] = \phi(x)$ for each $x \in R$. Thus, $y \in [x]$ if and only if $yn = xn$ for some $n \in N$. But $yn = xn$ iff $(y - x)n = 0$ iff $y - x \in I$ iff $y \in x + I$. Therefore $[x] = x + I = \phi(x)$, and this completes the proof.

The next result is essentially contained in Section 1. We restate it here in the language of congruences. The proof should be familiar from group theory, and hence it is omitted.

THEOREM 4.4. Assume that S is a monoid and that H is a subgroup of S containing 0 . For $a,b \in S$, define $a \sim b$ to mean that $a = b + h$ for some $h \in H$. Then $\sim$ is a congruence on S , $[a] = a + H$, and if H is the group of all invertible elements of S , then $[0]$ is the only invertible element of $S/\sim$.

We remark that if S is a group, then the only congruences

on S arise as in Theorem 4.4 — that is, as congruences with respect to a subgroup of S. To see this, let $\sim$ be a congruence on the group S and let $H = \{s \in S \mid s \sim 0\}$. Then H is a subgroup of S, for $a \sim 0$, $b \sim 0$ imply that $a + b \sim b$ so $a + b \sim 0$; also, $a - a \sim 0 - a$ so $0 \sim -a$. Thus H is a subgroup of S and $a \sim b$ iff $a - b \sim 0$ iff $a - b \in H$ iff $a \in b + H$.

There is an analogue, for torsion-freeness, of the cancellative property in Theorem 4.2. We can kill more than the torsion-free bird, however, with a single stone.

THEOREM 4.5. Assume that S is an additive semigroup and that M is a multiplicative semigroup of positive integers. For $a,b \in S$, define $a \sim b$ to mean that $ma = mb$ for some $m \in M$. Then $\sim$ is a congruence on S, and if $m[a] = m[b]$ for some $[a],[b] \in S/{\sim}$ and some $m \in M$, then $[a] = [b]$.

Proof. The reflexive and symmetric properties of $\sim$ are immediate. If $m_1 a = m_1 b$ and $m_2 b = m_2 c$, then $m_1 m_2 a = m_1 m_2 c$ and $m_1 m_2 \in M$, so $\sim$ is also transitive. Moreover, $m_1 a = m_1 b$ implies that $m_1(a + x) = m_1(b + x)$ for $x \in S$; hence $\sim$ is a congruence on S. The equality $m[a] = m[b]$ means that $nma = nmb$ for some $n \in M$. Since $nm \in M$, it follows that $a \sim b$ and $[a] = [b]$.

In case $M = Z^+$ in Theorem 4.5, the semigroup $S/{\sim}$ is torsion-free and $\sim$ is the smallest congruence ρ on S such that S/ρ is torsion-free. In considering nilpotent elements of a semigroup ring over a ring of prime characteristic p in Section 9, we shall encounter the case of

Theorem 4.5 where $M = \{p^i\}_{i=0}^{\infty}$ is the set of powers of p ; the congruence is denoted by $\underset{p}{\sim}$ in this case, and is referred to as p-equivalence. If $\underset{p}{\sim}$ is the identity congruence on S , we say that S is p-torsion-free. Another congruence encountered in Section 9 is that of asymptotic equivalence, defined by setting $a \sim b$ if there exists $K \in Z^+$ such that $ka = kb$ for each $k \geq K$. Theorem 4.6 provides an alternate way of viewing this relation, and Theorem 4.7 is the analogue, for asymptotic equivalence, of Theorem 4.5. We say that S is free of asymptotic torsion if distinct elements of S are not asymptotically equivalent.

THEOREM 4.6. Assume that $a,b \in S$ are such that $na = nb$ for some $n \in Z^+$. Let M be the set of all such positive integers n . Then M is a subsemigroup of the additive semigroup Z^+ , and the following conditions are equivalent.

(1) a and b are asymptotically equivalent.

(2) M is primitive.

(3) There exist $n_1, n_2, \ldots, n_k \in M$ such that g.c.d.$\{n_1, \ldots n_k\} = 1$.

(4) There exist $m,n \in M$ such that m and n are relatively prime.

Proof. That M is closed under addition is immediate. Conditions (1) and (2) are equivalent by the definition of asymptotic equivalence, and the equivalence of (2) , (3) , and (4) follows from Theorems 2.2 and 2.3 and from the definition of primitivity.

THEOREM 4.7. Assume that S is a semigroup and that $\sim$ is the relation of asymptotic equivalence on S. Then $\sim$ is a congruence on S, and the semigroup $S/\sim$ is free of asymptotic torsion. If ρ is any congruence on S such that S/ρ is free of asymptotic torsion, then $\rho \geq \sim$.

Proof. We prove only that $S/\sim$ is free of asymptotic torsion. Thus, assume that $[a]$, $[b]$ in $S/\sim$ are such that $[a]$ and $[b]$ are asymptotically equivalent. Then there exist relatively prime integers m,n so that $ma \sim mb$ and $na \sim nb$. Theorem 4.6 shows that there exists $k_1 \in Z^+$ such that $k_1ma = k_1mb$ and $\text{g.c.d.}\{k_1,n\} = 1$. By the same token, there exists $k_2 \in Z^+$, with k_2 relatively prime to k_1m, such that $k_2na = k_2nb$. Since k_1m and k_2n are relatively prime, it follows form (4.6) that $a \sim b$. Therefore $[a] = [b]$ and $S/\sim$ is free of asymptotic torsion.

Does there exist a minimal congruence Δ on S such that S/Δ is torsion-free and cancellative? The answer is affirmative, and the congruence in question is, in this case, the least upper bound of the cancellative congruence on S and the minimal congruence ρ on S such that S/ρ is torsion-free. Rather than regard Δ as this least upper bound, however, it's easier to define it directly. The congruence Δ will reappear in Section 8, where zero divisors of semigroup rings are considered.

THEOREM 4.8. Define the relation Δ on the semigroup S as follows: $a \Delta b$ if there exists $n \in Z^+$ and $x \in S$ such that $na + x = nb + x$. Then Δ is a congruence on S and is the smallest congruence on S such that S/Δ is torsion-free and cancellative.

Proof. We verify all the details for a change. Thus $a \Delta a$ since $a + x = a + x$ for each $x \in S$, and $na + x = nb + x$ implies $nb + x = na + x$. If $n_1a + x_1 = n_1b + x_1$ and $n_2b + x_2 = n_2c + x_2$, then it is easy to check that $n_1n_2a + n_2x_1 + n_1x_2 = n_1n_2c + n_2x_1 + n_1x_2$. Moreover, $na + x = nb + x$ implies that $n(a + y) + x = n(b + y) + x$ for each $y \in S$. Therefore Δ is a congruence on S. If $n[a] = n[b]$ in S/Δ, then $na \Delta nb$ so $mna + x = mnb + x$ for some m and x, implying that $a \Delta b$ and $[a] = [b]$. Similarly, $[a] + [c] = [b] + [c]$ means that $m(a + c) + x = m(b + c) + x$ for some m and x; again this implies that $a \Delta b$ and $[a] = [b]$. Therefore S/Δ is torsion-free and cancellative. Finally, if S/ρ is torsion-free and cancellative and if $a \Delta b$, then $na + x = nb + x$ for some n and x. Hence $n[a]_\rho + [x]_\rho = n[b]_\rho + [x]_\rho$ in S/ρ. Since S/ρ is cancellative, $n[a]_\rho = n[b]_\rho$, and since it is also torsion-free, $[a]_\rho = [b]_\rho$. Therefore $a \rho b$ and this completes the proof.

We consider one other class of congruences on a semigroup S before ending this section. Let I be an ideal of S. The relation $\sim$ on S defined by setting $a \sim b$ if either $a = b$ or $a,b \in I$ is a congruence on S; it is called the Rees congruence modulo I. The factor semigroup is sometimes denoted by S/I instead of $S/\sim$. Setwise, S/I can be thought of as $S \setminus I$, together with a symbol ∞, where $x + \infty = \infty + x = \infty$ for each $x \in S/I$ and where, for $x,y \in S \setminus I$, their sum in S/I is either $x + y$ or ∞, depending upon whether $x + y \in S\backslash I$ or $x + y \in I$. The factor semigroups S/I have some of the properties that one would anticipate of them from ring-theoretic notation. For example, if $\{J_\alpha\}$ is the set of ideals of S containing I , then $J_\alpha \longrightarrow J_\alpha/I$ is an inclusion-preserving bijection from $\{J_\alpha\}$ onto the set of ideals of S/I , and $S/J_\alpha \simeq (S/I)/(J_\alpha/I)$ for each α . This result can be used to show that a Noetherian semigroup satisfies the ascending chain condition for ideals, but we give a proof based on first principles in the next section.

A straightforward description of $\text{lub}\{\rho,\sim\}$, in the case where $\sim$ is a Rees congruence, is contained in the next result.

THEOREM 4.9. Assume that ρ and $\sim$ are congruences on the semigroup S , where $\sim$ is the Rees congruence modulo the ideal I of S . Let μ be the relation $\{(a,b) \mid$ there exist $c,d \in S$ such that $a \rho c$, $c \sim d$, and $d \rho b\}$. Then $\mu = \text{lub}\{\rho,\sim\}$.

Proof. It is clear that μ is reflexive, symmetric, and that it is compatible with the semigroup operation. If $a \mu b$ and $b \mu c$, then there exist $w,x,y,z \in S$ so that $a \rho w$, $w \sim x$, $x \rho b$, $b \rho y$, $y \sim z$, and $z \rho c$. If $w \rho x$ or $y \rho z$, then the transitive property of ρ and the definition of μ imply that $a \mu c$. In the other case, $w \neq x$ and $y \neq z$ so $w \sim x$ and $y \sim z$ imply that $w,x,y,z \in I$; therefore $w \sim z$ and again $a \mu c$. Thus μ is a congruence on S . Since the relations $p \leq \mu$, $\sim \leq \mu$, and $\mu \leq \text{lub}\{\rho,\sim\}$ are clear, it follows that $\mu = \text{lub}\{\rho,\sim\}$.

Section 4 Remarks

It is traditional to blur the distinction between consideration of a relation ρ on a set A as a subset of $A \times A$ or as a rule by which certain pairs of elements are related. Considered as a subset of $A \times A$, we write $\rho \subseteq A \times A$; considered as a rule, we write $a \rho b$. We have maintained this dual consideration of congruences in Section 4 and will continue to do so. If $\{\rho_i\}_1^n$ is a finite set of relations on A, then $\mathrm{glb}\{\rho_i\}_1^n$ is frequently denoted by $\wedge_{i=1}^n \rho_i$; considered as a subset of $A \times A$, we have $\wedge_{i=1}^n \rho_i = \cap_{i=1}^n \rho_i$. The corresponding lattice-theoretic notation for $\mathrm{lub}\{\rho_i\}_1^n$ is $\vee_{i=1}^n \rho_i$. For relations, $\vee_1^n \rho_i$ is $\cup_1^n \rho_i$ as a subset of $A \times A$, but whereas the finite intersection of congruences on a semigroup is a congruence, the set-theoretic union of finitely many congruences need not be transitive.

§5. Noetherian Semigroups

Recall that a semigroup S is Noetherian if the ascending chain condition for congruences holds in S. While the main purpose of this section is to prove Theorem 5.10, which shows that a Noetherian monoid is finitely generated, the material divides rather naturally into three related parts. Some basic consequences of the a.c.c. on congruences are developed in the first part. The second part includes several new concepts and is devoted to proving Theorem 5.8, a form of Primary Decomposition Theorem for congruences on a Noetherian semigroup. Theorem 5.10 and its proof constitute the third part of the section.

THEOREM 5.1. Let S be a Noetherian semigroup.

(1) The a.c.c. for ideals of S is satisfied. Hence each ideal of S is finitely generated.

(2) If S is a monoid and if G is the group of invertible elements of S, then the a.c.c. for subgroups of G is satisfied.

(3) If P is a proper prime ideal of S, then $P^c = S \setminus P$ is Noetherian.

(4) If A is an ideal of S such that A is a monoid, then A is Noetherian.

<u>Proof</u>. Let $I_1 \subseteq I_2 \subseteq \ldots$ be an ascending chain of ideals of S. Let $\sim_j$ be the Rees congruence modulo I_j for each j. Then $\sim_1 \leq \sim_2 \leq \ldots$, so there exists k such that $\sim_k = \sim_{k+1} = \ldots$. This implies that $I_k = I_{k+1} = \ldots$, for if $I_k < I_{k+1}$, then for $y \in I_{k+1} \setminus I_k$ and $x \in I_k$, we have $x \sim_{k+1} y$ and $x \not\sim_k y$, contrary to assumption.

The proof of (2) is similar to that of (1): if $H_1 \subseteq H_2 \subseteq \ldots$ is an ascending sequence of subgroups of G, then define $\sim_i$ by $a \sim_i b$ if $a \in b + H_i$. Theorem 4.4 shows that $\sim_i$ is a congruence on S, and it is clear that $\sim_1 \leq \sim_2 \leq \ldots$. Moreover, an equality $\sim_k = \sim_{k+1}$ implies that $H_k = H_{k+1}$, and this establishes (2).

(3) If $\sim$ is a congruence on P^c, then it is easy to verify that $\sim^* = \sim \cup (P \times P)$ is a congruence on S (the assumption that P^c is prime is used only to assert that P^c is a semigroup). Moreover, $\sim^*$ agrees with $\sim$ on P^c. Thus, if $\sim_1 \leq \sim_2 \leq \ldots$ is an ascending sequence of congruences on P^c, then $\sim_1^* \leq \sim_2^* \leq \ldots$ is an ascending sequence on S. Since $\sim_k^* = \sim_{k+1}^*$ implies that $\sim_k = \sim_{k+1}$, this completes the proof.

(4) Let z be the identity element of A. For a congruence $\sim$ on A, define the relation $\sim^*$ on S as follows. For $s,t \in S$, $s \sim^* t$ means that $(s + z) \sim (t + z)$. We verify that $\sim^*$ is a congruence on S and that $\sim^* \cap (A \times A) = \sim$. If $s \in S$, then $s + z \in A$ and $(s + z) \sim (s + z)$; therefore $s \sim^* s$. Similarly, $\sim^*$ is symmetric (transitive) because $\sim$ is symmetric (transitive). If $s \sim^* t$ and if $x \in S$, then $(s + z) \sim (t + z)$ and $x + z \in A$ imply that $s + x + z = [(s + z) + (x + z)] \sim [(t + z) + (x + z)] = t + x + z$. Hence $(s + x) \sim^* (t + x)$ and $\sim^*$ is a congruence on S. If $a,b \in A$, then $a \sim^* b$ iff $(a + z) \sim (b + z)$ iff $a \sim b$. Thus, $\sim^* \cap (A \times A) = \sim$. It then follows as in the proof of (3) that a.c.c. on congruences of S implies a.c.c. on congruences of A.

With regard to (2) of Theorem 5.1, we state the following result concerning Noetherian groups as a corollary.

COROLLARY 5.2. The following conditions are equivalent in an abelian group G .

(1) G is Noetherian.

(2) G satisfies a.c.c. on subgroups.

(3) G is finitely generated.

Proof. Since congruences on G are in one-to-one order-preserving correspondence with subgroups of G , conditions (1) and (2) are equivalent. Clearly (2) implies (3) , and it is well known that each subgroup of a finitely generated abelian group is finitely generated, so (3) implies (2) .

THEOREM 5.3. Assume that S is a monoid satisfying a.c.c. for ideals and that A is an ideal of S generated by the finite set $\{a_i\}_{i=1}^n$. Then either $S = \langle a_1, \ldots, a_n, A^c \rangle$ or there exist elements $a, y \in A$ such that $a = a + y$.

Proof. Set $T = \langle a_1, \ldots, a_n, A^c \rangle$ and assume that $T \neq S$. Choose $b \in S \setminus T$. Then clearly $b \in A$. Write $b = a_{i_1} + s_1$, where $1 \le i_1 \le n$ and $s_1 \in S$. Since $b \notin T$, then $s_1 \notin T$ so $s_1 \in A$ and $s_1 = a_{i_2} + s_2$. By induction, there exists $\{i_j\}_1^\infty \subseteq \{1, 2, \ldots, n\}$ and $\{s_j\}_1^\infty \subseteq A$ such that $s_j = a_{i_{j+1}} + s_{j+1}$ for each j . The ascending chain $b + S \subseteq s_1 + S \subseteq s_2 + S \subseteq \ldots$ of principal ideals of S stabilizes; say $s_k + S = s_{k+1} + S = \ldots$. Then $s_{k+1} = s_k + t = (s_{k+1} + a_{i_{k+1}}) + t$.

Thus, $a = s_{k+1}$ and $y = t + a_{i_{k+1}}$ are in A and $a = a + y$.

COROLLARY 5.4. A cancellative Noetherian monoid is finitely generated.

Proof. Let S be such a monoid and let G be the group of invertible elements of S. If $G = S$, then Corollary 5.2 shows that S admits a finite set $\{g_i\}_{i=1}^n$ of generators as a group, so $\{g_i\}_1^n \cup \{-g_i\}_1^n$ is a finite set of generators for S as a semigroup. If $G \neq S$, then let A be the ideal G^c of S. Theorem 5.1 shows that A has a finite set $\{a_i\}_1^m$ of generators as an ideal of S and that G is finitely generated both as a group and as a semigroup. By Theorem 5.3, either $S = \langle a_1, \ldots, a_m, G\rangle$ or there exist elements a, y of A such that $a = a + y$. An equality $a = a + y$ is impossible in this case since the cancellative property of S would imply that y is the identity element of S, contrary to the fact that $y \in A = G^c$. Therefore $S = \langle a_1, \ldots, a_m, G\rangle$ and S is finitely generated.

We turn toward a proof of Theorem 5.8, a form of Primary Decomposition Theorem. The statement and proof of this result use several new concepts. Let S be a semigroup. We denote by U the universal congruence $S \times S$ on S. If ρ is a congruence on S and $s \in S$, we define the relation $\rho(s)$ on S to be $\{(a,b) \mid (a + s)\ \rho\ (b+s)\}$. Properties of ρ transfer easily to $\rho(s)$ so that, in particular, $\rho(s)$ is a congruence on S. We denote by $[\rho]^\circ$ the set $\{s \in S \mid \rho(s) = U\} = \{s \in S \mid (s + a)\ \rho\ (s + b)$ for all $a,b \in S\}$, and $[\rho]$ is defined to be $\{s \in S \mid ns \in [\rho]^\circ$ for some $n \in Z^+\} = \{s \in S \mid$ there exists $n \in Z^+$ such that $(ns + a)\ \rho\ (ns + b)$ for all $a,b \in S\}$, the radical of $[\rho]^\circ$. Theorem 5.5 states some of the basic properties of $\rho(s)$, $[\rho]^\circ$, and $[\rho]$.

THEOREM 5.5. Assume that S is a semigroup and that $\rho, \rho_1, \ldots, \rho_n$ are congruences on S .

(1) If $a,b \in S$, then $\rho \leq \rho(a) \leq \rho(a)(b) = \rho(a + b)$. If S is Noetherian, then there exists $n \in Z^+$ such that $\rho(na) = \rho(n + 1)a) = \ldots$.

(2) If $[\rho]^\circ$ is nonempty, then $[\rho]^\circ$ and $[\rho]$ are ideals of S and $[\rho]^\circ \subseteq [\rho]$.

(3) If $\rho = \text{glb}\{\rho_i\}_1^n$ and $a \in S$, then $\rho(a) = \text{glb}\{\rho_i(a)\}_1^n$; moreover, $[\rho]^\circ = \cap_{i=1}^n [\rho_i]^\circ$ and $[\rho] = \cap_{i=1}^n [\rho_i]$.

Proof. (1): The relation $\rho \leq \rho(a)$ holds since ρ is compatible with $+$. Thus, $\rho \leq \rho(a) \leq \rho(a)(b)$. Moreover, for $s,t \in S$,

$$(s,t) \in \rho(a)(b) \text{ iff } (s + b)\ \rho(a)\ (t + b) \text{ iff }$$
$$(s + b + a)\rho(t + b + a) \text{ iff } (s,t) \in \rho(a + b) .$$

Hence $\rho(a)(b) = \rho(a + b)$.

(2): To show that $[\rho]^\circ$ is an ideal of S , take $a \in [\rho]^\circ$ and $s \in S$. Then $\rho(a + s) = \rho(a)(s) = U(s) = U$, so $a + s \in [\rho]^\circ$ and $[\rho]^\circ$ is an ideal. We observed in Section 1 that $\text{rad}([\rho]^\circ) = [\rho]$ is an ideal of S containing $[\rho]^\circ$.

(3): For $s,t \in S$, we have

$s\rho(a)t$ iff $(s + a)\ \rho\ (t + a)$ iff

$(s + a)\rho_i(t + a)$ for each i iff

$s\rho_i(a)t$ for each i .

Therefore $\rho(a) = \text{glb}\{\rho_i(a)\}_{i=1}^n$. This equality shows that $\rho(a) = U$ if and only if $\rho_i(a) = U$ for each i — that is, $a \in [\rho]^\circ$ if and only if $a \in \cap_1^n[\rho_i]^\circ$, so $[\rho]^\circ = \cap_1^n[\rho_i]^\circ$. To prove that $[\rho] = \cap_1^n[\rho_i]$, it then suffices, by induction, to show that if A and B are ideals of S such that $A \cap B \neq \phi$, then $\text{rad}(A \cap B) = \text{rad}(A) \cap \text{rad}(B)$; this we proceed to do. The inclusion $\text{rad}(A \cap B) \subseteq \text{rad}(A) \cap \text{rad}(B)$ is clear. Moreover, if $mx \in A$ and $nx \in B$, then $mnx \in A \cap B$, so $\text{rad}(A) \cap \text{rad}(B) \subseteq \text{rad}(A \cap B)$. This completes the proof of Theorem 5.5.

A congruence ρ on a semigroup S is said to be primary if, for all $s \in S$, $\rho \neq \rho(s)$ implies that $s \in [\rho]$; ρ is prime if $\rho \neq \rho(s)$ implies $s \in [\rho]^\circ$ for $s \in S$, and ρ is cancellative if $\rho = \rho(s)$ for each $s \in S$. It is clear that a cancellative congruence is prime and that a prime congruence is primary. A prime congruence ρ is cancellative if and only if $[\rho]^\circ$ is empty. In the case where $\rho = i$ is the identity congruence, i is cancellative if and only if S is a cancellative semigroup, and the definitions of i primary and prime closely parallel the way that the notions of primary and prime ideals are extended from rings to submodules of a module. If T is a subsemigroup of S , then by the restriction of ρ to T , denoted $\rho|_T$, we mean the relation $\rho \cap (T \times T)$ on T ; clearly $\rho|_T$ is a congruence on T . Theorem 5.6 catalogues some fundamental properties of primary, prime, and cancellative congruences.

THEOREM 5.6. Assume that S is a semigroup, that $\mu, \rho_1, \ldots, \rho_n$ are congruences on S, and that $\rho = \text{glb}\{\rho_i\}_1^n$. Let T be a subsemigroup of S.

(1) If μ is primary, then $[\mu]$ is a prime ideal of S.

(2) If each ρ_i is primary with $[\rho_1] = \ldots = [\rho_n]$, then ρ is primary and $[\rho] = [\rho_1]$. If each ρ_i is cancellative, then ρ is also cancellative.

(3) If μ is primary, prime, or cancellative, then $\mu|_T$ has the same property.

(4) If μ is primary, then the restriction of μ to $[\mu]^c$ is cancellative.

Proof. (1) Assume that μ is primary and that $a,b \in S$ are such that $a + b \in [\mu]$. Then there exists $n \in Z^+$ so that $(n(a + b) + x)\ \mu\ (n(a + b) + y)$ for all $x,y \in S$. If $(nb + x)\ \mu\ (nb + y)$ for all $x,y \in S$, then it follows from the definition of $[\mu]$ that $b \in [\mu]$. In the other case, there exist $x_1, y_1 \in S$ such that $(nb + x_1, nb + y_1) \notin \mu$. Since $(na + nb + x_1)\ \mu\ (na + nb + y_1)$ and since μ is primary, it follows that $mna \in [\mu]^\circ$ for some $m \in Z^+$, so $a \in [\mu]$. Thus $[\mu]$ is prime in S.

(2) Part (3) of Theorem 5.5 shows that $[\rho] = \cap_{i=1}^n [\rho_i] = [\rho_1]$. To see that ρ is primary, take $s \in S$ such that $\rho(s) \neq \rho$. Since $\rho(s) = \text{glb}\{\rho_i(s)\}$, it follows that $\rho_i \neq \rho_i(s)$ for some i. Therefore $s \in [\rho_i] = [\rho]$, and ρ is primary.

The verification of (3) is routine.

To prove (4), take $a,b,x \in [\mu]^c$ such that $(a + x)\ \mu\ (b + x)$. We must have $a\ \mu\ b$, for if not, then the definition of μ primary implies that $x \in [\mu]$, contrary to assumption. Therefore $a \mu b$ and the restriction of μ to $[\mu]^c$ is cancellative.

The proof that each congruence on a Noetherian semigroup can be represented as a finite intersection (greatest lower bound) of finitely many primary congruences now follows a path that, at least in terminology, is familiar from commutative ring theory: we define an irreducible congruence, prove that each congruence on a Noetherian semigroup is a finite intersection of irreducible ones, and that an irreducible congruence is primary. The first step is to define a congruence ρ on a semigroup S to be <u>irreducible</u> if cannot be expressed as the intersection of two strictly greater congruences on S .

THEOREM 5.7. Let S be a Noetherian semigroup.

(1) Each congruence on S is a finite intersection of irreducible congruences on S .

(2) Each irreducible congruence on S is primary.

<u>Proof</u>. If (1) fails, then among the congruences on S that are not expressible as a finite intersection of irreducible congruences, we can choose a maximal one ρ . Clearly ρ is not irreducible, so $\rho = \rho_1 \cap \rho_2$, where $\rho < \rho_1$ and $\rho < \rho_2$. Maximality of ρ implies that each ρ_i is a finite intersection of irreducibles, hence so is ρ , a contradiction. This establishes (1) .

To prove (2) , take an irreducible congruence ρ on S and an element $x \in S$ such that $\rho \neq \rho(x)$. We prove that $x \in [\rho]$. By assumption, there exist $u,v \in S$ such that

$(u,v) \notin \rho$ but $(u + x)\ \rho\ (v + x)$. Let ρ_1 be the congruence on S generated by $\rho \cup \{(u,v)\}$; we note that $\rho_1 \leq \rho(x)$. Let the integer n be such that $\rho(nx) = \rho((n + 1)x) = \ldots$, let μ be the Rees congruence modulo the ideal $nx + S$ of S, and let $\rho_2 = \mathrm{lub}\{\rho,\mu\}$. By Theorem 4.9, $\rho_2 = \{(a,b) \mid \text{there exist } c,d \in S \text{ such that } a\ \rho\ c,\ c\ \mu\ d, \text{ and } d\ \rho\ b\}$. Let $\rho_3 = \rho_1 \cap \rho_2$; we show that $\rho = \rho_3$. It is clear that $\rho \leq \rho_3$. Let $y,z \in S$ be such that $y\ \rho_3\ z$. Then $y\ \rho_1\ z$ implies $y\ \rho(x)\ z$ so $(y + x)\ \rho\ (z + x)$. Moreover, there exist $c,d \in S$ such that $y\ \rho\ c$, $c\ \mu\ d$, and $d\ \rho\ z$. If $c = d$, then $y\ \rho\ z$ and we're finished. On the other hand, if $c,d \in nx + S$, then $c = nx + c_1$ and $d = nx + d_1$ for some $c_1, d_1 \in S$. We have $c\ \rho(x)\ y$, $y\ \rho(x)\ z$, and $z\ \rho(x)\ d$. Therefore $c\ \rho(x)\ d$ or $(c + x)\ \rho\ (d + x)$. Consequently, $(c_1 + (n + 1)x)\ \rho\ (d_1 + (n + 1)x)$ so that $c_1\ \rho((n+1)x)\ d_1$. By choice of n, this means that $c_1\ \rho(nx)\ d_1$, $c = (c_1 + nx)\ \rho\ (d_1 + nx) = d$, and hence $y\ \rho\ z$ as we wished to show. We have proved that $\rho = \rho_3 = \rho_1 \cap \rho_2$. The relation $\rho_1 > \rho$ holds by definition of ρ_1. Since ρ is irreducible, $\rho = \rho_2 = \mathrm{lub}\{\rho,\mu\}$ so that $\mu \leq \rho$. Now $(n + 1)x \in nx + S$, so $((n + 1)x + s)\ \mu\ ((n + 1)x + t)$ for all $s,t \in S$. Consequently, $((n + 1)x + s)\ \rho\ ((n + 1)x + t)$ for all $s,t \in S$. This implies that $x \in [\rho]$, completing the proof of (2).

If ρ is a congruence on a semigroup S, then a representation $\rho = \rho_1 \cap \cdots \cap \rho_n$ of ρ as a finite intersection of congruences on S is said to be <u>canonical</u> if the following conditions are satisfied.

1) each ρ_i is primary,

2) $[\rho_i] \neq [\rho_j]$ for $i \neq j$,

and 3) $\rho_i \not\supseteq \cap_{j \neq i} \rho_j$ for each i .

The next result follows from Theorem 5.7 and from part (2) of Theorem 5.6.

THEOREM 5.8. Each congruence on a Noetherian semigroup admits a canonical representation.

The proof of Theorem 5.10 uses the fact that the identity congruence on a Noetherian semigroup is a finite intersection of primary congruences. Before embarking on the proof, we need to strengthen the conclusion of Theorem 5.3 in the case where the ideal A in the statement of that result is of the form $[\rho]$.

THEOREM 5.9. Let S be a monoid satisfying a.c.c. for ideals, let ρ be a congruence on S , and let $\{a_i\}_{i=1}^{n}$ be a finite set of generators for the ideal $A = [\rho]$ of S . Then $S = \langle [\rho]^{\circ}, a_1, \ldots, a_n, A^c \rangle$.

Proof. Let $T = \langle a_1, \ldots, a_n, A^c \rangle$. If $T = S$, we're finished; if not, then an examination of the proof of Theorem 5.3 shows that for $b \in S \backslash T$, there exists $y \in A$ such that $b = b + y$. Thus $b = b + my$ for each $m \in Z^+$, and for m suffiently large, $my \in [\rho]^{\circ}$. Hence $S \backslash T$ is contained in $[\rho]^{\circ}$ and $S = \langle [\rho]^{\circ}, a_1, \ldots, a_m, [\rho]^c \rangle$ in either case.

Before proving Theorem 5.10, we make the trivial observation that if ρ is a congruence on a monoid S , then $0 \in [\rho]^c$ if $\rho \neq U$. Since the congruence U is redundant in any intersection representation with more than one component, we tacitly assume in Theorem 5.10 that U is not one of the components in the representation of the identity congruence i ($i = U$ only if $S = \{0\}$) .

THEOREM 5.10. A Noetherian monoid is finitely generated.

Proof. Let S be a Noetherian monoid. We first prove the theorem under the additional hypothesis that the identity congruence i on S can be represented in the form $i = \rho \cap \rho_1 \cap \cdots \cap \rho_n$, where ρ is cancellative and each ρ_j is primary. For this case, we use induction on n . If $n = 0$, then i is cancellative, S is a cancellative monoid, and Corollary 5.4 shows that S is finitely generated. Assume the result for a given integer n and consider the case where $i = \rho \cap \rho_1 \cap \cdots \cap \rho_{n+1}$. Let $\mu = \rho_1 \cap \cdots \cap \rho_{n+1}$ and let $\{a_j\}_1^m$ be a set of generators for the ideal $[\mu]$ of S . Theorems 5.3 and 5.9 show that one of two cases arises.

Case 1. $S = \langle a_1, \ldots, a_m, [\mu]^c \rangle$.

Case 2. $S = \langle [\mu]^\circ, a_1, \ldots a_m, [\mu]^c \rangle$ and there exist $b, y \in [\mu]$ such that $b = b + y$. For k sufficiently large, kb and ky are in $[\mu]^\circ$ and $kb = kb + ky$, so without loss of generality we assume that $b, y \in [\mu]^\circ$. We prove that $[\mu]^\circ$ is cancellative. Thus, take $s, t, x \in [\mu]^\circ$ so that $s + x = t + x$. By definition of $[\mu]^\circ$, it follows that $s = (s + 0) \mu (s + x)$ and $t \mu (t + x)$. Hence $s \mu t$ Since ρ is cancellative, we also have $s \rho t$. Therefore $s = t$ since $i = \rho \cap \mu$. We have proved that $[\mu]^\circ$ is cancellative. Thus, the equation $b = b + y$ implies that y is an identity element for $[\mu]^\circ$, and part (4) of Theorem

5.1 shows that $[\mu]^{\circ}$ is Noetherian. We conclude from Corollary 5.4 that $[\mu]^{\circ}$ is finitely generated. Thus, in both Case 1 and Case 2, S is generated by $[\mu]^{c}$ and a finite set. Part (3) of Theorem 5.5 shows that $[\mu] = [\cap_1^{n+1} \rho_j] = \cap_1^{n+1} [\rho_j]$ so that $[\mu]^{c} = \cup_1^{n+1} [\rho_j]^{c}$. Therefore, to show that S is finitely generated, it suffices to show that each $[\rho_j]^{c}$ is finitely generated. We examine the identity relation on $[\rho_j]^{c} = A_j$; it is $(\cap_{k=1}^{n+1} \mu_k) \cap \mu$, where $\mu_k = \rho_k|_{A_j}$ and $\mu = \rho|_{A_j}$. By parts (2), (3), and (4) of Theorem 5.6, it follows that μ_k is primary for $k \neq j$, μ_j and μ are cancellative, and hence $\mu_j \cap \mu$ is also cancellative. Thus, the identity congruence on A_j is the intersection of n primary congruences and a cancellative congruence. Since $A_j = [\rho_j]^{c}$ is Noetherian by part (3) of Theorem 5.1 and since $0 \in [\rho_j]^{c}$, the induction hypothesis implies that each $[\rho_j]^{c}$ is finitely generated. Thus, Theorem 5.10 is proved in the case where i admits a primary decomposition in which one of the component congruences is cancellative.

In the general case, let $i = \rho_1 \cap \ldots \cap \rho_n$ be a representative of i as a finite intersection of primary congruences (Theorem 5.7). If $\{a_j\}_1^m$ is a set of generators for the ideal $[i]$, then $S = \langle [i]^{\circ}, a_1, \ldots, a_m, [i]^{c} \rangle = \langle [i]^{\circ}, a_1, \ldots, a_m, \cup_1^n [\rho_j]^{c} \rangle$. Each $[\rho_j]^{c}$ is a Noetherian monoid in which the identity congruence is the intersection of a cancellative congruence and $n - 1$ primary congruences; by the case of Theorem 5.10 proved above, each $[\rho_j]^{c}$ is finitely generated. Finally, $[i]^{\circ}$ contains at most one element, for if $c,d \in [i]^{\circ}$, then $c + 0 = d + 0$. Therefore, the representation $S = \langle [i]^{\circ}, a_1, \ldots, a_m, \cup_1^n [\rho_j]^c \rangle$ shows that S is finitely generated.

The converse of Theorem 5.10 is established in Theorem 7.8, which states that a finitely generated monoid is Noetherian.

Section 5 Remarks

The Primary Decomposition Theorem for congruences on a Noetherian semigroup was proved by Drbohlav [40], who also investigated and proved analogues of classical uniqueness theorems concerning canonical decompositions of ideals of a commutative Noetherian ring. Drbohlav also transferred his results from the setting of primary decomposition of congruences to representation of congruence ideals of Noetherian semigroups (appropriately defined) as a finite intersection of primary congruence ideals (also appropriately defined). The result that a Noetherian monoid is finitely generated was proved by Budach [26].

§6. Factorization in Commutative Monoids

Many of the conditions typically considered in the theory of integral domains, such as unique factorization, are conditions defined in terms of the multiplicative semigroup of nonzero elements of the domain. In this section we define some of these concepts in a cancellative monoid and we derive some of their basic properties. Tradition for multiplicative notation is strong in the area of factorization, and in this section the semigroup operation is usually written multiplicatively. Several proofs in the section generalize familiar proofs in the case of integral domains.

Assume that S is a cancellative monoid, written multiplicatively. An element a of S <u>divides</u> $b \in S$, and we write $a|b$, if there exists $c \in S$ such that $b = ac$; we also say that a is a <u>factor</u> of b and that b is a <u>multiple</u> of a. An <u>irreducible</u> element of S is an element $a \in S$ such that $a = st$ in S implies that either s or t is a <u>unit</u> (that is, an invertible element) of S, and a is <u>prime</u> if $a|st$ implies $a|s$ or $a|t$; as usual, prime elements of S are irreducible. The relations $a|b$ and $b|a$ are satisfied if and only if $a = b\mu$ for some unit μ of S, in which case we say that a and b are <u>associates</u>. The relation on S that identifies associate elements is the congruence of Theorem 4.4, where H (in the notation of (4.4)) is the group of all units of S. If X is a nonempty subset of S, an element a of S is a <u>greatest common divisor</u> of the elements of X (written $a = \gcd\{X\}$) if a divides each element of X and is divisible by each common divisor of the elements of X. A given subset X need not have a greatest common divisor, but if $\gcd\{X\}$ exists, then it is uniquely determined up to associates. The definition of a <u>least common multiple</u> of X, written $\operatorname{lcm}\{X\}$, is as expected, and like the greatest common divisor, $\operatorname{lcm}\{X\}$ is determined up to associates if it exists.

THEOREM 6.1. Assume that $x,y \in S$, a cancellative monoid, and that A and B are nonempty subsets of S.

(1) $\operatorname{lcm}\{xA\}$ exists if and only if $\operatorname{lcm}\{A\}$ exists; if both exist, then $\operatorname{lcm}\{xA\} = x \cdot \operatorname{lcm} A$.

(2) If $\gcd\{xA\}$ exists, then $\gcd\{A\}$ exists and $\gcd\{xA\} = x \cdot \gcd\{A\}$.

(3) If $a = \gcd\{A\}$ and $b = \gcd\{B\}$ exist, then $\gcd\{A \cup B\}$ exists if and only if $\gcd\{a,b\}$ exists; if these two greatest common divisors exists, then $\gcd\{A \cup B\} = \gcd\{a,b\}$. The analogous statement for least common multiples is also valid.

(4) If $c = \operatorname{lcm}\{x,y\}$ exists, then $xy = cs$ in S. Moreover, $s = \gcd\{x,y\}$.

<u>Proof</u>. If $b = \operatorname{lcm}\{A\}$ exists, then it is straightforward to show that $xb = \operatorname{lcm}\{xA\}$. And conversely, if $c = \operatorname{lcm}\{xA\}$ exists, then c is divisible by x, say $c = c_1 x$. It is routine to prove that $c_1 = \operatorname{lcm}\{A\}$.

The proof of (2) is similar to that of (1): if $\gcd\{xA\} = d$, then d is divisible by the common divisor x of xA. If $d = d_1 x$, then d_1 can be shown to be a greatest common divisor of A.

We omit the proof of (3). In (4), xy is a common multiple of x and y, so $xy = cs$ for some $s \in S$. We have $x|c$ and $y|c$, say $xx_1 = c = yy_1$. Since S is cancellative, the equalities $xy = xx_1s = yy_1s$ imply that $x = y_1s$ and $y = x_1s$; hence s is a common divisor of x and y. Let t be any common divisor of x and y and write $x = tx_2$ and $y = ty_2$. Clearly tx_2y_2 is a common multiple of x and y, and hence c divides tx_2y_2, say $tx_2y_2 = cz$. Then $xy = t^2x_2y_2 = tcz = sc$ and $t|s$ since $s = tz$. Therefore $s = \gcd\{x,y\}$.

It is known that the converses of (2) and (4) of Theorem 6.1 may fail, even if S is the multiplicative semigroup of nonzero elements of an integral domain. For example, X^2 and X^3 have greatest common divisor 1 in $Z[X^2,X^3]$, but $\gcd\{X^3X^2,X^3X^3\} = \gcd\{X^5,X^6\}$ and $\operatorname{lcm}\{X^2,X^3\}$ fail to exist in $Z[X^2,X^3]$. On the other hand, if each pair of elements of S has a greatest common divisor in S, we show that each pair of elements of S has a least common multiple in S.

THEOREM 6.2. Let $x,y \in S$, a cancellative monoid. The following conditions are equivalent.

(1) $\operatorname{lcm}\{x,y\} = xy$.

(2) $xS \cap yS = xyS$.

(3) For all z in S, $x|z$ and $y|z$ imply that $xy|z$.

(4) For all $z \in S$, $\gcd\{zx,zy\} = z$.

(5) $\gcd\{x,y\} = 1$ and $\gcd\{xz,yz\}$ exists for each $z \in S$.

(6) For each $z \in S$, $x|yz$ implies that $x|z$.

<u>Proof</u>. The equivalence of conditions (1) – (3) and the implication (4) $\Longrightarrow$ (5) are clear, and part (2) of

Theorem 6.1 shows that (5) implies both (4) and (6) .

(1) $\Longrightarrow$ (5): If $\operatorname{lcm}\{x,y\} = xy$, then $\operatorname{lcm}\{xz,yz\} = xyz$ by part (1) of (6.1) , and hence $\gcd\{xz,yz\} = z$ by part (4) of (6.1) .

(2) $\Longrightarrow$ (6): If $yz = xs$, then $yz \in xS \cap yS = xyS$, so that $z \in xS$ since S is cancellative. Therefore $x | z$.

(6) $\Longrightarrow$ (2): If $xa = yb \in xS \cap yS$, then $x | yb$, and by (6) , $x | b$ so that $b \in xS$ and $yb \in xyS$. Consequently, $xS \cap yS = xyS$.

COROLLARY 6.3. The following conditions are equivalent.

(1) Each pair of elements of S has a greatest common divisor in S .

(2) Each finite subset of S has a greatest common divisor in S .

(3) Each pair of elements of S has a least common multiple in S .

(4) Each finite set of elements of S has a least common multiple in S .

Proof. The implications (2) $\Longrightarrow$ (1) and (4) $\Longrightarrow$ (3) are obvious, and the reverse implications follow by induction from part (3) of Theorem 6.1. Part (4) of (6.1) shows that (3) implies (1) ; we prove that (1) implies (3) . Pick $x,y \in S$, let $d = \gcd\{x,y\}$, and write $x = dx_1$, $y = dy_1$. Then $\gcd\{x_1,y_1\} = 1$ and $\gcd\{zx_1,zy_1\}$ exists for each $z \in S$. Therefore, $\operatorname{lcm}\{x_1,y_1\} = x_1y_1$ by Theorem 6.2, and $\operatorname{lcm}\{x,y\} = \operatorname{lcm}\{dx_1,dy_1\} = dx_1y_1$. This completes the proof of Corollary 6.3.

We call a monoid satisfying the equivalent conditions of Corollary 6.3 a GCD-*monoid*.

THEOREM 6.4. Assume that S is a GCD-monoid.

(1) If $a,b,c \in S$ and $\gcd\{a,b\} = d$, then $\gcd\{a,bc\} = \gcd\{a,dc\}$.

(2) If $a,b_1,\ldots,b_n \in S$ are such that $\gcd\{a,b_i\} = 1$ for each i, then $\gcd\{a,b_1b_2\ldots b_n\} = 1$.

(3) Assume that $a_1,\ldots,a_n \in S$ are such that $\gcd\{a_1,\ldots,a_n\} = 1$ and that $k_1,\ldots,k_n$ are positive integers. Then $\gcd\{a_1^{k_1},\ldots,a_n^{k_n}\} = 1$.

Proof. (1): Let $s = \gcd\{a,bc\}$ and $t = \gcd\{a,dc\}$. It is clear that t divides s. But s divides $\gcd\{ac,bc\} = c \cdot \gcd\{a,b\} = cd$, and hence s divides t. Therefore, $s = \gcd\{a,dc\}$.

(2): By induction, it suffices to consider the case where $n = 2$, and this case follows immediately from (1): $\gcd\{a,b_1b_2\} = \gcd\{a,1 \cdot b_2\} = \gcd\{a,b_2\} = 1$.

(3): It is clear that we need only consider the case where $k_1 = \ldots = k_{n-1} = 1$. Moreover, since $\gcd\{a_1,\ldots,a_n\} = \gcd\{\gcd\{a_1,\ldots,a_{n-1}\},a_n\}$, it suffices to prove the statement in (2) for $n = 2$. This case, however, is a consequence of (2).

Unique factorization into primes can be defined in a cancellative monoid either in terms of elements or in terms of principal prime ideals. We choose the latter approach here because statements of results are a bit cleaner in the language of ideals. Transfer between elements and ideals is easy, being based on three facts: (i) $s \in S$ is prime if and only if sS is a prime ideal of S; (ii) $s \in S$ is irreducible if and only if either s is a unit of S or sS

is maximal in the set of proper principal ideals of S ; and (iii) $sS = tS$ if and only if s and t are associates.

We say that the cancellative monoid S is <u>factorial</u> (or <u>a unique factorization monoid</u>, abbreviated UFM) if each principal ideal of S can be written as a finite product of principal prime ideals of S . The next result shows that such factorization is unique to within order and factors of S .

THEOREM 6.5. Let S be a cancellative monoid that contains a proper principal prime ideal. Let P be the multiplicative semigroup consisting of all principal ideals of S that are expressible as a finite product of proper principal prime ideals of S .

(1) If $xS \in P$, then xS is uniquely expressible, to within order, as a finite product of proper principal prime ideals of S .

(2) If $xS = p_1^{e_1} \cdots p_k^{e_k} S$, where each p_iS is a proper prime ideal of S and $p_iS \neq p_jS$ for $i \neq j$, then $\{p_1^{f_1} \cdots p_k^{f_k} S \mid 0 \leq f_i \leq e_i \text{ for each } i\}$ is the set of principal ideals of S containing xS .

(3) If S is factorial, then S satisfies the ascending chain condition for principal ideals (a.c.c.p.).

<u>Proof</u>. To prove (1) , let $xS = p_1Sp_2S \ldots p_nS = p_1p_2 \ldots p_nS$, where each p_i is a nonunit prime element of S . We obtain uniqueness by induction on n . If $n = 1$, then x is prime; if $xS = x_1S \cdot x_2S = x_1x_2S$ in this case, then $x_1 \in xS$ or $x_2 \in xS$ so that $xS = x_1S$ or $xS = x_2S$.

If $xS = x_1S$, then $x_2S = S$ since S is cancellative. Thus, factorization is unique if $n = 1$. We assume unique factorization for $n = k$ and consider the case where $xS = p_1 \ldots p_k p_{k+1} S$. Assume that $xS = q_1S \cdot \ldots \cdot q_mS$ is another factorization of xS into proper prime ideals. Then $p_1S \supseteq q_jS$ for some j — say $p_1S \supseteq q_1S$. Writing $q_1 = p_1s$, it follows that s is a unit of S and $p_1S = q_1S$. Then canceling p_1 in the equality $p_1 \ldots p_{k+1}S = q_1 \ldots q_mS$ yields $p_2 \ldots p_{k+1}S = q_2 \ldots q_mS$, whence $k = m - 1$ and $p_iS = q_iS$ for $2 \leq i \leq m$ by proper labeling. Thus (1) follows by the principle of mathematical induction.

The proof of (2) follows by the same method as in (1) : if $e = e_1 + \ldots + e_k$, then we use induction on e to prove that each principal ideal of S containing xS is of the required form. For $e = 1$, the proof is essentially given in (1) . If the result holds for $e = n$, then in the case where $e_1 + \ldots + e_k = n + 1$, let yS be a proper principal ideal of S containing xS . We write $x = yz \in p_1S$. Therefore $y \in p_1S$ or $z \in p_1S$. If $y = p_1y_1$, then we have $y_1S \supseteq p_1^{e_1-1} \ldots p_k^{e_k}S$ and the representation of $yS = p_1y_1S$ in the form $p_1^{f_1} \ldots p_k^{f_k}S$, with $0 \leq f_i \leq e_i$ for each i , follows from the induction hypothesis. Similarly, if $z \in p_1S$, then $yS \supseteq p_1^{e_1-1} \ldots p_k^{e_k}S$ and the desired conclusion again follows from the induction hypothesis.

(3): If S is factorial, then (2) shows that only finitely many principal ideals of S contain a fixed principal ideal of S . Thus, a.c.c.p. is satisfied in S .

COROLLARY 6.6. Assume that X is a nonempty subset of the factorial semigroup S .

(1) $\gcd\{X\}$ exists in S .

(2) $\mathrm{lcm}\{X\}$ exists in S iff $\{xS \mid x \in S\}$ is finite iff the elements of X have a common multiple.

Proof. Take $s_1, s_2 \in S$ and choose principal primes $p_1S,\ldots,p_nS$ of S and nonnegative integers a_i, b_i such that $s_1S = p_1^{a_1}\ldots p_n^{a_n}S$ and $s_2S = p_1^{b_1}\ldots p_n^{b_n}S$. Part (2) of Theorem 4.5 implies that s_1 and s_2 have greatest common divisor $p_1^{c_1}\ldots p_n^{c_n}$ in S , where $c_i = \inf\{a_i, b_i\}$ for each i . Thus, S is a GCD-monoid and $\gcd\{X\}$ exists for each finite subset X of S . Consider the case of an arbitrary subset X . If $x \in X$, then only finitely many principal ideals of S contain xS . Hence, among the principal ideals of S that contain yS for each $y \in X$, there exists a minimal one dS . The element d is a common divisor of the elements of X . If c is any common divisor of the elements of X , then $yS \subseteq cS$ for each $y \in X$. Consequently, the ideal $dS \cap cS$ contains each yS . But since S is a GCD-monoid, $dS \cap cS$ is principal, generated by e , where $e = \mathrm{lcm}\{c,d\}$. By choice of dS , it follows that $dS = eS = ds \cap cS \subseteq cS$, and hence c divides d . This proves that $d = \gcd\{X\}$.

(2): Assume that the elements of X have a common multiple m . Then $mS \subseteq xS$ for each $x \in S$, and hence the set $\{xS \mid x \in X\}$ is finite, say $\{xS \mid x \in S\} = \{x_iS\}_{i=1}^n$, where $x_i \in X$ for each i . Then $\cap_{i=1}^n x_iS = \cap\{xS \mid x \in X\}$. Since S is a GCD-monoid, $\cap_{i=1}^n x_iS$ is principal, generated by some $t \in S$. The equality $tS = \cap\{xS \mid x \in X\}$ is equivalent, however, to the assertion that $t = \text{lcm}\{X\}$.

Theorem 6.5 and Corollary 6.6 show that if S is a factorial semigroup, then S satisfies a.c.c.p. and S is a GCD—monoid. It is well known from the theory of integral domains, however, that neither a.c.c.p. nor the GCD—condition implies factoriality. On the other hand, we show that the combination of these two conditions does imply factoriality. On the other hand, we show that the combination of these two conditions does imply factoriality.

THEOREM 6.7. Assume that S is a cancellative monoid.

(1) If S satisfies a.c.c.p., then each proper principal ideal of S can be expressed as a finite product of maximal proper principal ideals of S ; equivalently, each nonunit of S is a finite product of irreducible elements of S. Hence S is factorial if and only if S satisfies a.c.c.p. and irreducible elements of S are prime.

(2) If S is a GCD—monoid, then irreducible elements of S are prime.

(3) S is factorial if and only if S satisfies a.c.c.p. and S is a GCD—monoid.

Proof. If S satisfies a.c.c.p., then the set of proper principal ideals of S admits maximal elements; such an element is what is meant by a maximal proper principal ideal. If (1) fails, then among the proper principal ideals of S not expressible as a finite product of maximal ones, there exists one xS that is maximal within the set. Then xS is not a maxiaml proper principal ideal of S, so there exists $y \in S$ such that $xS < yS < S$. We write $x = yz$, where z is necessarily a nonunit of S. Then each of yS and zS is a finite product of maximal proper principal ideals, hence so is xS. This contradiction establishes (1).

(2): Assume that S is a GCD–monoid and that $s \in S$ is irreducible. If s is a unit, then s is prime. Otherwise, pick $x,y \in S \backslash sS$. Then $\gcd\{x,s\} = \gcd\{y,s\} = 1$, so part (2) of Theorem 6.4 shows that $\gcd\{xy,s\} = 1$. Therefore $xy \notin sS$ and sS is prime as asserted.

We have already observed that a factorial semigroup satisfies a.c.c.p. and is a GCD–monoid; the converse follows from (1) and (2).

Let S be a factorial semigroup and let G be the group of units of S. If $\{p_\alpha\}_{\alpha \in A}$ is a complete set of nonassociate prime elements of S, then the very definition of factoriality implies that S is the internal weak direct sum of G and of the family $\{P_\alpha\}_{\alpha \in A}$ of subsemigroups of S, where $P_\alpha = \{p_\alpha^n\}_{n=0}^{\infty}$ for each α. Thus, S is isomorphic to a monoid of the form $G \dot{\oplus} \Sigma_\alpha^w Z_\alpha$, where G is a group and each Z_α is isomorphic to the additive monoid Z_0 of nonnegative integers. Clearly each monoid of the latter form is factorial, so we have established the following theorem.

THEOREM 6.8. Let G be a group, let $\{Z_\alpha\}_{\alpha \in A}$ be a family of monoids, each isomorphic to the additive monoid of nonnegative integers. The monoid $G \dot{\oplus} \Sigma_\alpha^w Z_\alpha$ is factorial. Conversely, each factorial monoid S is isomorphic to such a monoid $G \dot{\oplus} \Sigma_\alpha^w Z_\alpha$.

We note that G and $|A|$, the cardinality of the set A , determine the monoid $G \dot{\oplus} \Sigma^{w}_{\alpha \in A} Z_{\alpha}$ to within isomorphism, for G is the group of units of the monoid and $|A|$ is the cardinality of a complete set of its nonassociate prime elements. Is there anything special about the case where $|A|$ is finite? Based on what's true for integral domains (a factorial domain with only finitely many nonassociate prime elements is a principal ideal domain), we might conjecture that each ideal of the monoid is principal if $|A|$ is finite, but this statement is false. For example, $S = Z_0 \dot{\oplus} Z_0$ is factorial with exactly two nonassociate primes, but the ideal $Z^+ \dot{\oplus} Z^+$ of S is not principal.

Chapter II

SEMIGROUP RINGS AND THEIR DISTINGUISHED ELEMENTS

This chapter contains five sections. In the first of these, Section 7, semigroup rings are defined and some of their basic properties are established. In particular, Theorem 7.7 is the result that a unitary monoid ring $R[X;S]$ is Noetherian if and only if R is Noetherian and S is finitely generated. Subsequent sections treat zero divisors, nilpotent elements, idempotents, and units of semigroup rings. While each of these classes of elements is of interest in its own right, it is also the case that knowledge of these classes of distinguished elements is frequently an indispensable tool in work on global structure properties of semigroup rings in Chapters III and IV.

§7. Semigroup Rings

Apart from the definition of a semigroup in Section 1, all semigroups considered in the main text up to this point have been assumed to be commutative. We intend to continue to observe that convention, as well as to assume that all rings considered are commutative. But in giving the definition of the semigroup ring of S over R, it seems pointless to place such restrictions on the semigroup S or the ring R. Thus, in this first paragraph of Section 7, we assume that R is an associative ring and that $(S,*)$ is a semigroup. Let T be the set of functions f from S into R that are finitely nonzero, with addition and multiplication defined in T as follows.

$$(f+g)(s) = f(s) + g(s)$$

$$(fg)(s) = \sum_{t*u=s} f(t)g(u) ,$$

where the symbol $\sum_{t*u=s}$ indicates that the sum is taken over all pairs (t,u) of elements of S such that $t*u = s$, and it is understood that $(fg)(s) = 0$ if s is not expressible in the form $t*u$ for any $t,u \in S$. Verification that T is a ring under $+$ and $\cdot$ follows as in the special case $S = Z_0$, with which the reader is no doubt familiar: for $S = Z_0$, T is simply the polynomial ring $R[X]$ in one variable over R. We call T the <u>semigroup ring of</u> S <u>over</u> R; to denote T, we use either $R[X;S]$ or $R[S]$. If the semigroup operation on S is written multiplicatively, we use the notation $R[S]$ for T, and elements of T are written either as $\sum_{s\in S} f(s)s$ or as

$\Sigma_{i=1}^{n} f(s_i)s_i$. Thus, S is a free basis for R[S] as an R-module, and the multiplication in R[S] is determined by distributivity and by setting $r_1 s_1 \cdot r_2 s_2 = r_1 r_2 s_1 s_2$. If the semigroup operation in S is written as + , the elements of T are written either in the form $\Sigma_{s\in S} f(s)X^s$ or in the form $\Sigma_{i=1}^{n} f(s_i)X^{s_i}$, with addition and multiplication defined as for polynomials. Chapter 2 deals primarily with properties of elements of the semigroup ring T , and R[X;S] is the symbol customarily used here to denote T . Subsequent chapters deal, however, with global properties of semigroup rings, there is much less emphasis on the form of elements of T , and in Chapters III–V we use R[S] to denote T . Introduction of the symbol X and the notation X^s has the effect of transforming (S,+) into the multiplicative semigroup $\{X^s \mid s \in S\}$ by means of the isomorphism $s \longrightarrow X^s$. Nevertheless, the notation $\Sigma_{i=1}^{k} f_i X^{s_i}$ is more commonly used in commutative ring theory than $\Sigma_{i=1}^{k} f_i s_i$, no doubt because of customary additive notation in commutative semigroups and because of the suggested analogy between semigroup rings and polynomial rings. Of course, the degree to which R[X] serves as a reasonable model for R[X;S] depends partly upon properties that S has in common with Z_0 . For example, R[X] serves as a good model for generalization to $R[X_1,\ldots,X_n]$ for some ring-theoretic questions, but as a poor model for certain other questions. Yet each polynomial ring $R[\{Y_\lambda\}_{\lambda\in\Lambda}]$ over R in commuting indeterminates Y_λ can easily be seen to be isomorphic to a semigroup ring over R , as follows. Let $F = \Sigma_{\lambda\in\Lambda}^{w} Z_\lambda$, where $Z_\lambda \simeq Z_0$ for each λ , and let $\{e_\lambda\}$ be the standard free basis for F — that is, the λth coordinate of e_λ is 1 and all the other coordinates are 0 .

Each element of F is uniquely expressible in the form $a = \Sigma k_\lambda e_\lambda$ for some $k_\lambda \geq 0$, and identification of $r_a X^a \in R[X;F]$ with $r_a \Pi Y_\lambda^{k_\lambda} \in R[\{Y_\lambda\}]$, with sums $\Sigma_{a \in F} r_a X^a$ in $R[X;F]$ identified with the corresponding sums in $R[\{Y_\lambda\}]$, is easily seen to define an isomorphism of $R[X;F]$ onto $R[\{Y_\lambda\}]$. In the case of a finite set $\{Y_i\}_{i=1}^n$ of indeterminates, isomorphism of $R[Y_1,\ldots,Y_n]$ and $R[X;Z_0^n]$ follows by induction from the next result.

THEOREM 7.1. Assume that R is a commutative ring and that S and T are additive abelian semigroups. Then $R[X;S \dot{\oplus} T]$ is isomorphic to $U = (R[X;S])([X,T])$, the semigroup ring of T over the semigroup ring of S over R .

Proof. Throughout this proof we regard elements of a semigroup ring as finitely nonzero functions from the semigroup into the ring, as defined above. Thus, define $\phi : U \longrightarrow R[X;S \dot{\oplus} T]$ as follows. If $f \in U$ and $(s,t) \in S \dot{\oplus} T$, then $[\phi(f)](s,t) = [f(t)](s)$. The map ϕ is well-defined since each element of U is a well-defined function from T into $R[X;S]$. Moreover, ϕ is surjective, for if $h \in R[X; S \dot{\oplus} T]$, then the element $f \in U$ defined by $[f(t)](s) = h(s,t)$ is such that $\phi(f) = h$. If $f \neq g$, then $f(t) \neq g(t)$ for some $t \in S$. Therefore $[f(t)](s) \neq [g(t)](s)$ for some $s \in S$, and hence $\phi(f) \neq \phi(g)$. To complete the proof, we show that ϕ is a ring homomorphism. Thus, if $f,g \in U$ and $(s,t) \in S \dot{\oplus} T$, then

$$[\phi(f + g)](s,t) = [(f + g)(t)](s) = [f(t) + g(t)](s) =$$

$$[f(t)](s) + [g(t)](s) = [\phi(f)](s,t) + [\phi(g)](s,t) =$$

$$[\phi(f) + \phi(t)](s,t) ,$$

so that $\phi(f + g) = \phi(f) + \phi(g)$. Moreover,

$$[\phi(fg)](s,t) = [fg(t)](s) = [\Sigma_{a+b=t} f(a)g(b)](s)$$

$$= \Sigma_{a+b=t}(\Sigma_{c+d=s}[f(a)](c) \cdot [g(b)](d)) =$$

$$\Sigma_{(c,a)+(d,b)=(s,t)}\{[f(a)](c)\}\{[g(b)](d)\} =$$

$$\Sigma_{(c,a)+(d,b)=(s,t)}\{[\phi(f)](c,a)\}\{[\phi(g)](d,b)\} =$$

$$[\phi(f)\phi(g)](s,t) ,$$

and this shows that $\phi(fg) = \phi(f)\phi(g)$. Therefore ϕ is an isomorphism and the proof is complete.

Henceforth we assume that all semigroups and rings considered are commutative, and we are not hesitant to restrict consideration to monoids and/or commutative unitary rings, for this is the arena with the most interesting action. If S is a monoid, we sometimes refer to $R[X;S]$ as the monoid ring of S over R . Since each semigroup S is imbeddable in a monoid $S^0 = S \cup \{0\}$, restriction to monoids frequently amounts to no loss of generality. Similarly, since each ring R admits unital extension rings $R[e]$ whose structure is generally well known in relation to that of R , the hypothesis that a given coefficient ring is unitary may amount to a matter of convenience, rather than a real restriction. On the other hand, the presence of an identity element in R

plays essentially no role in most of the problems considered in Chapter 2, and in such problems, the unitary hypothesis on R is not automatically imposed. Semigroup rings $A[X;T]$, where A is not unitary or T is not a monoid, may naturally arise, for example, if A is an ideal of a unitary ring R or if T is a subsemigroup or ideal of a monoid S. If S is not a monoid, then $R[X;S]$ can be considered as the ideal of $R[X;S^0]$ consisting of all $f \in R[X;S^0]$ such that $f(0) = 0$. Similarly, if $R^* = R[e]$ is a unital extension of the ring R, then $R[X;S]$ is the ideal of $R^*[X;S]$ consisting of all functions f such that $f(s) \in R$ for each $s \in S$. If S is a monoid, we identify $r \in R$ with $rX^0 \in R[X;S]$, and consider R as a subring of $R[X;S]$.

Each nonzero element f of $R[X;S]$ has a unique representation in the form $\Sigma_{i=1}^{n} f_i X^{s_i}$, where $f_i \neq 0$ and $s_i \neq s_j$ for $i \neq j$; this representation is called the <u>canonical form</u> of f. The subset $\{s_i\}_1^n$ of S is called the <u>support</u> of f, and $<\{s_i\}_1^n>$ is the <u>supporting semigroup</u> of f. The support of f is denoted by Supp(f). The ideal of R generated by $\{f_i\}_1^n$ is called the <u>content</u> of f and is denoted by $c(f)$. It is clear that $f \in R_1[X;S_1]$ for a finitely generated subring R_1 of R and a finitely generated subsemigroup S_1 of S. Since finitely generated rings and finitely generated monoids are Noetherian, certain questions concerning units, idempotents, and nilpotent elements of $R[X;S]$ considered in the rest of this chapter can be reduced to the case where R and S are Noetherian

The next result contains some basic information

concerning homomorphisms of semigroup rings.

THEOREM 7.2. Let μ be a homomorphism of R into the ring R_0, let A be the kernel of μ, and let ϕ be a homomorphism of S into the semigroup S_0. Define the mappings μ^*, ϕ^*, and τ on $R[X;S]$ as follows.

$$\mu^* : R[X;S] \longrightarrow R_0[X;S] \; ; \; \mu^*(\Sigma_1^n r_i X^{s_i}) = \Sigma_1^n \mu(r_i) X^{s_i} ,$$

$$\phi^* : R[X;S] \longrightarrow R[X;S_0] \; ; \; \phi^*(\Sigma r_i X^{s_i}) = \Sigma r_i X^{\phi(s_i)} ,$$

$$\tau \; : R[X;S] \longrightarrow R_0[X;S_0] \; ; \; \tau(\Sigma r_i X^{s_i}) = \Sigma \mu(r_i) X^{\phi(s_i)} .$$

Then the following statements hold.

(1) μ^* is a homomorphism with kernel $A[X;S]$; μ^* is surjective if μ is surjective.

(2) ϕ^* is a homomorphism and its kernel is the ideal I of $R[X;S]$ generated by $\{rX^a - rX^b \mid \phi(a) = \phi(b)\}$; ϕ^* is surjective if ϕ is surjective.

(3) τ is a homomorphism with kernel $A[X;S] + I$; τ is surjective if μ and τ are surjective.

Proof. It is routine to verify that μ^*, ϕ^*, and τ are homomorphisms, as well as the statements concerning surjectivity of these three maps. That $A[X;S]$ is the kernel of μ^* is clear. We verify the equality $I = \ker \phi^*$. Since $\phi^*(rX^a - rX^b) = rX^{\phi(a)} - rX^{\phi(b)} = 0$ if $\phi(a) = \phi(b)$, it follows that $I \subseteq \ker \phi^*$. To prove the reverse inclusion, take a nonzero element $f = \Sigma_{i=1}^m f_i X^{s_i} \in \ker \phi^*$. It's easy to see that $m > 1$. If $m = 2$, then clearly $\phi(s_1) = \phi(s_2)$ and $f_1 = -f_2$, so $f = f_1 X^{s_1} - f_1 X^{s_2} \in I$. For $m > 2$, there exist distinct elements s_i, s_j of $\mathrm{Supp}(f)$ such that

$\phi(s_i) = \phi(s_j)$. Thus, $g = f - (r_i X^{s_i} - r_i X^{s_j})$ is an element of ker ϕ^* with fewer than m elements in its support. By induction, g is in I , and hence so is f . Therefore $I = \ker \phi^*$. The assertion concerning $\ker \tau$ follows from (1) and (2) .

COROLLARY 7.3. Assume that A is an ideal of the ring R and that $\sim$ is a congruence on the semigroup S . Let $I = (\{rX^a - rX^b \mid r \in R \text{ and } a \sim b\})$. Then

$$R[X;S]/A[X;S] \simeq (R/A)[X;S] ,$$

$$R[X;S]/I \simeq R[X;S/\sim] , \text{ and}$$

$$R[X;S]/(A[X;S]+I) \simeq (R/A)[X;S/\sim] .$$

The ideal I of Corollary 7.3 is called the kernel ideal of the congruence $\sim$. The proof of part (2) of Theorem 7.2 shows that I consists of all finite sums of elements of the form $rX^a - rX^b$, where $r \in R$ and $a \sim b$. If the ring R is unitary, then $\{X^a - X^b \mid a \sim b\}$ is, of course, a set of generators for I . Ideals of the semigroup ring that are kernel ideals have some interesting properties, as we proceed to show.

THEOREM 7.4. Assume that $A, B, \{A_\lambda\}_{\lambda \in \Lambda}$ are ideals of the ring R and I is the kernel ideal of the congruence $\sim$ on the semigroup S .

(1) $\cap_{\lambda \in \Lambda}\{A_\lambda[X;S]+I\} = (\cap_{\lambda \in \Lambda} A_\lambda)[X;S] + I$.

Moreover, if R is unitary, then (2) and (3) are satisfied.

(2) $A[X;S] \cap I = (A[X;S]) \cdot I = AI$.

(3) $A[X;S] \cap (B[X;S]+I) = (A \cap B)[X;S] + AI$.

Proof. (1): Since I is the kernel of the canonical homomorphism of $R[X;S]$ onto $[R X;S/\sim]$, (1) is equivalent to the assertion that $\cap(A_\lambda[X;S/\sim]) = (\cap A_\lambda)[X;S/\sim]$, and this equality is clearly satisfied.

(2): Since R is unitary, $A[X;S] = A \cdot R[X;S]$. Thus $(A[X;S]) \cdot I = AI$. The inclusion $AI \subseteq (A[X;S]) \cap I$ is patent. The proof of the reverse inclusion is similar to the proof of (2) in Theorem 7.2. Thus, take a nonzero element $f = \Sigma_{i=1}^{m} f_i X^{s_i} \in I \cap A[X;S]$. Then $f \in I$ implies $m > 1$, and $m = 2$ implies $f = f_1(X^{s_1} - X^{s_2})$, where $f_1 \in A$ and $s_1 \sim s_2$, from which it follows that $f \in AI$. For $m > 2$, there exists distinct $s_i, s_j \in \mathrm{Supp}(f)$ such that $s_i \sim s_j$; moreover, $f \in A[X;S]$ implies that $f_i \in A$. The induction hypothesis implies that $f - f_i(X^{s_i} - X^{s_j}) \in AI$, and since $f_i(X^{s_i} - X^{s_j}) \in AI$, we have $f \in AI$.

(3): In view of (1) and (2) , assertion (3) amounts to saying that intersection distributes over sum in (3) . This is known to hold if $B \subseteq A$. But we can reduce to that case since

$$A[X;S] \cap (B[X;S]+I) =$$

$$A[X;S] \cap (A[X;S]+I) \cap (B[X;S]+I) =$$

$$A[X;S] \cap \{(A \cap B)[X;S]+I\} .$$

This completes the proof of Theorem 7.4.

Theorem 7.5 establishes a close relationship between the lattice of congruences on S and the set of kernel ideals of $R[X;S]$. In the statement of Theorem 7.5, we use $\vee$ and $\wedge$, respectively, to denote the least upper bound and greatest lower bound of two congruences on S .

THEOREM 7.5. Let R be a ring and let S be a semigroup. With each congruence $\sim$ on S , associate the kernel ideal $I(\sim)$ of I . The mapping $\theta : \sim \longrightarrow I(\sim)$ from the lattice of congruences on S into the lattice of ideals of $R[X;S]$ has the following properties.

(1) θ is injective.

(2) θ preserves order — that is, $\rho_1 \leq \rho_2$ if and only if $I(\rho_1) \subseteq I(\rho_2)$.

(3) If ρ_1 and ρ_2 are congruences on S , then $I(\rho_1 \wedge \rho_2) \subseteq I(\rho_1) \cap I(\rho_2)$; the inclusion may be proper. Moreover, if R is unitary and S is a monoid, then (4) and (5) are satisfied.

(4) $I(\rho_1 \vee \rho_2) = I(\rho_1) + I(\rho_2)$ for congruences ρ_1, ρ_2 on S .

(5) The congruence $\sim$ is finitely generated if and only if the ideal $I(\sim)$ is finitely generated. In fact, $\{(s_i, t_i)\}_{i=1}^n$ generates $\sim$ if and only if $\{X^{s_i} - X^{t_i}\}_{i=1}^n$ generates $I(\sim)$.

Proof. If ρ_1 and ρ_2 are distinct congruences on S , then we may assume that there exist $s,t \in S$ such that $s\rho_1 t$ while $(s,t) \notin \rho_2$. Then $X^s - X^t$ is in $I(\rho_1)$, but not in $I(\rho_2)$. This proves (1) . That $I(\rho_1) \subseteq I(\rho_2)$ implies $\rho_1 \leq \rho_2$ is clear, and the other part of (2) follows from the fact that $\{rX^a - rX^b \mid a\rho_1 b\}$ generates $I(\rho_1)$. The inclusion $I(\rho_1 \wedge \rho_2) \subseteq I(\rho_1) \cap I(\rho_2)$ in (3) follows from (2). Part (3) of Theorem 7.6 implies that the inclusion may be proper. The inclusion $I(\rho_1) + I(\rho_2) \subseteq I(\rho_1 \vee \rho_2)$ in (4) also follows from (2). For the reverse inclusion, it suffices to show that $X^a - X^b \in I(\rho_1) + I(\rho_2)$ for each $(a,b) \in \rho_1 \vee \rho_2$. From the description of $\rho_1 \vee \rho_2$ given in Section 4, it follows that there exist elements $s_1, s_2, \ldots, s_n \in S$ such that $a = s_1, b = s_n, s_1\rho_1 s_2, s_2\rho_2 s_3, s_3\rho_1 s_4, \ldots, s_{n-1}\rho_2 s_n$. Then $X^a - X^b = \Sigma_{i=1}^{n-1}(X^{s_i} - X^{s_{i+1}})$, where each $X^{s_i} - X^{s_{i+1}}$ is in $I(\rho_1) + I(\rho_2)$. Therefore $X^a - X^b \in I(\rho_1) + I(\rho_2)$, and the equality $I(\rho_1 \vee \rho_2) = I(\rho_1) + I(\rho_2)$ is established.

Assume for the moment that the second statement in (5) has been established. We observe that the first statement then follows. It is clear from the second statement that $\sim$ finitely generated implies that $I(\sim)$ is finitely generated. For the converse, note that if $I(\sim)$ is finitely generated, then the generating set $\{X^a - X^b \mid a \sim b\}$ for $I(\sim)$ contains a finite set of generators for $I(\sim)$, whence $\sim$ is finitely generated by the second statement in (5). We proceed to prove the second statement. Thus, assume that $\{(s_i,t_i)\}_1^n$ generates $\sim$. Let J be the ideal of $R[X;S]$ generated by $\{X^{s_i} - X^{t_i}\}_{i=1}^n$. The inclusion $J \subseteq I(\sim)$ is clear. We define a congruence μ on S by setting $s \mu t$ if $X^s - X^t \in J$. Clearly $J = I(\mu)$. Since $s_i \mu t_i$ for each i, $\sim \le \mu$, and hence $I(\sim) \subseteq I(\mu) \subseteq I(\sim)$. It follows that $I(\sim) = I(\mu)$ is generated by $\{X^{s_i} - X^{t_i}\}_1^n$. Conversely, assume that $\{X^{s_i} - X^{t_i}\}_1^n$ generates $I(\sim)$ and let μ be the congruence on S generated by $\{(s_i,t_i)\}_1^n$. We have just proved that $I(\mu)$ is generated by

$\{X^{s_i} - X^{t_i}\}_1^n$, and hence $I(\mu) = I(\sim)$, implying that $\mu = \sim$ and that $\sim$ is generated by $\{(s_i,t_i)\}_1^n$.

Theorem 7.5 has a number of important consequences. Before looking at some of these, we restate (7.5) in the case where $S = G$ is a group. In that case, the congruences on G correspond to the subgroups, and $G/\sim$, for $\sim$ the congruence corresponding to the subgroup H , is G/H . If R is unitary, the corresponding ideal $I(\sim)$ of $R[X;G]$ is generated by $\{1 - X^h \mid h \in H\}$ in this case, and $1 - X^g \in I(\sim)$ if and only if $g \in H$. For a subset A of G , it follows easily that A generates H if and only if $\{1 - X^a \mid a \in A\}$ generates $I(\sim)$. These observations, together with Theorem 7.5, yield the following result.

THEOREM 7.6. Let R be a ring and let G be a group. Define a mapping ϕ from the lattice of subgroups of G into the lattice of ideals of $R[X;G]$ as follows. If H is a subgroup of G , then $\phi(H)$ is the kernel of the canonical map from $R[X;G]$ onto $R[X;G/H]$. The mapping ϕ has the following properties.

(1) ϕ is injective.

(2) For subgroups H_1, H_2 of G , $H_1 \subseteq H_2$ if and only if $\phi(H_1) \subseteq \phi(H_2)$.

(3) $\phi(H_1 \cap H_2) \subseteq \phi(H_1) \cap \phi(H_2)$, with equality holding if and only if $H_1 \subseteq H_2$ or $H_2 \subseteq H_1$.

Moreover, if R is unitary, then (4) and (5) are satisfied.

(4) $\phi(H_1 + H_2) = \phi(H_1) + \phi(H_2)$.

(5) The subgroup H of G is finitely generated if

and only if $\phi(H)$ is finitely generated. In fact, a subset $\{h_i\}_1^n$ of G generates H if and only if $\{1 - X^{h_i}\}_1^n$ generates $\phi(H)$.

Proof. Only the assertion in (3) that $\phi(H_1 \cap H_2) = \phi(H_1) \cap \phi(H_2)$ implies $H_1 \subseteq H_2$ or $H_2 \subseteq H_1$ requires proof. Assume that $H_2 \nsubseteq H_1$ and choose $g \in H_2 \backslash H_1$. Let $h \in H_1$ and let $r \in R$. Then $f = rX^0 - rX^h + rX^{h+g} - rX^g$ is in $\phi(H_1) \cap \phi(H_2)$, hence in $\phi(H_1 \cap H_2)$. Since $h + g$ and g are not in $H_1 \cap H_2$, it follows that $h \in H_1 \cap H_2$, and hence $H_1 \subseteq H_2$.

We remark that the lattice of subgroups of the nontrivial abelian group G is linearly ordered if and only if G is cyclic of prime-power order or a quasicyclic group (Theorem 19.3), and hence examples where the inclusion in (3) is proper abound. If $\sim$ is the universal congruence $S \times S$ on S , then $S/\sim$ is the semigroup with only one element and the semigroup ring $R[X;S/\sim]$ is isomorphic to R . We call the corresponding homomorphism $\Sigma_1^n r_i X^{s_i} \longrightarrow \Sigma_1^n r_i$ of $R[X;S]$ onto R the augmentation map on $R[X;S]$; its kernel I is called the augmentation ideal of $R[X;S]$, and is the ideal generated by $\{rX^a - rX^b \mid r \in R$ and $a,b \in S\}$.

THEOREM 7.7. Let R be a unitary ring and let S be a monoid. The following conditions are equivalent.

(1) The monoid ring $R[X;S]$ is Noetherian.

(2) R is Noetherian and S is finitely generated.

Proof. Assume that $R[X;S]$ is Noetherian. Since R is a homomorphic image of $R[X;S]$, it is also Noetherian.

Moreover, $R[X;S]$ satisfies a.c.c. on congruence ideals, so Theorem 7.5 implies that S satisfies a.c.c. on congruences — that is, S is Noetherian. Thus, S is finitely generated by Theorem 5.10.

Conversely, if R is Noetherian and $S = \langle s_1, \ldots, s_n \rangle^0$ is finitely generated, then $R[X;S] = R[X^{s_1}, \ldots, X^{s_n}]$ is a finitely generated ring extension of R, and hence is a Noetherian ring.

We remark that the case of Theorem 7.7 where S is cancellative is much easier than Theorem 7.7 itself. This is so because the proof that a cancellative Noetherian monoid is finitely generated (Corollary 5.4) is less profound than the proof of the general case (Theorem 5.10). The next result is the converse of Theorem 5.10.

THEOREM 7.8. A finitely generated monoid is Noetherian.

<u>Proof</u>. Let S be a finitely generated monoid. The monoid ring $Z[X;S]$ is Noetherian by Theorem 7.7, and this implies, by Theorem 7.5, that S satisfies a.c.c. on congruences.

A monoid S is said to be <u>finitely presented</u> (or <u>finitely definable</u>) if $S \simeq F/\rho$ for some finitely generated free monoid F and some finitely generated congruence ρ on F. The next result is a theorem of Redei [122, Theorem 72].

THEOREM 7.9. A finitely generated monoid is finitely presented.

Proof. If $S = \langle s_1, \ldots, s_n \rangle^0$ is finitely generated, then S is the homomorphic image of the free monoid $F = Z_0^n$; hence $S \simeq F/\rho$ for an appropriate ρ , and Theorem 7.8 shows that ρ is finitely generated.

Note that since a free monoid is cancellative, the proof of Theorem 7.9 can be obtained without appeal to any of the material in Section 5 beyond Corollary 5.4.

For semigroups that need not be monoids, Theorem 7.10 gives partial results concerning conditions under which a semigroup ring is Noetherian.

THEOREM 7.10. Let R be a ring and let S be a semigroup.

(1) If S is finite, then $R[X;S]$ is Noetherian if and only if R is Noetherian.

(2) If S is infinite and $R[X;S]$ is Noetherian, then R is a Noetherian unitary ring; the convere fails, even for S finitely generated.

Proof. Since R is a homomomorphic image of $R[X;S]$, the Noetherian property in $R[X;S]$ is inherited by R . Moreover, if R is Noetherian and S is finite, then $R[X;S]$ is a finitely generated (hence Noetherian) R-module, so $R[X;S]$ is also Noetherian as a ring.

In (2) , we first prove that $R[X;S]$ Noetherian implies that R is unitary. Consider the Noetherian ring $(R/R^2)[X;S]$. Since multiplication in this ring is trivial, the additive group of $(R/R^2)[X;S]$ is Noetherian, hence finitely generated. It is also true, however, that the additive group of $(R/R^2)[X;S]$ is the weak direct sum of

$|S|$ copies of the additive group of R/R^2. This implies that $R/R^2 = \{0\}$, so $R = R^2$. Since R is Noetherian, it then follows that R has an identity element. For the converse, the proof of part (2) of Theorem 20.7 shows that $R[X;Z^+]$ is Noetherian if and only if the additive group of R is finitely generated.

The final result of this section is in the same vein as Theorem 7.9 in that it describes all monoid rings over a fixed unitary ring R as certain homomorphic images of the free monoid rings (that is, polynomial rings) over R. For the statement of Theorem 7.11, we introduce the _ad hoc_ terminology _pure difference binomial_ for a polynomial $g \in R[\{X_\lambda\}]$ of the form $g = X_{\lambda_1}^{e_1} \dots X_{\lambda_n}^{e_n} - X_{\lambda_1}^{f_1} \dots X_{\lambda_n}^{f_n}$ for some nonnegative integers e_i, f_i.

THEOREM 7.11. Let R be a unitary ring. To within isomorphism, the class of monoid rings over R can be characterized as the class of all residue class rings $R[\{X_\lambda\}_{\lambda \in \Lambda}]/A$, where A is an ideal of $R[\{X_\lambda\}]$ generated by pure difference binomials. If S is a finitely generated monoid, then $R[X;S]$ is isomorphic to such a residue class ring $R[\{X_\lambda\}_{\lambda \in \Lambda}]/A$, where Λ is finite and A is generated by a finite set of pure difference binomials.

Proof. Theorem 7.11 is primarily a translation of results already proved into slightly different language. Thus, let S be a monoid and let $\{s_\lambda\}_{\lambda \in \Lambda}$ be a generating set for S. If F is the free monoid $\Sigma^w_{\lambda \in \Lambda} Z_\lambda$, where $Z_\lambda \simeq Z_0$ for each λ, then $S \simeq F/\rho$ for an appropriate congruence ρ on F. But Theorem 7.2 shows that the kernel I of the canonical homomorphism of $R[X;F]$ onto $R[X;F/\sim]$ is generated by $\{X^g - X^h | g \sim h\}$; moreover, under the canonical isomorphism of $R[X;F]$ onto $R[\{X_\lambda\}_{\lambda\in\Lambda}]$, the elements $X^g - X^h$ are precisely the elements of $R[X;F]$ that map to pure difference binomials in $R[\{X_\lambda\}]$. This proves that each monoid ring over R is of the form $R[\{X_\lambda\}]/A$, where A is generated by pure difference binomials. If S is finitely generated, then Λ can be taken to be finite, in which case the congruence $\sim$ is finitely generated by Theorem 7.8. Part (5) of Theorem 7.5 then implies that I is generated by a finite set $\{X^{g_i} - X^{h_i}\}_{i=1}^m$, whence A is also generated by a finite set of pure difference binomials.

Conversely, any such residue class ring $R[\{X_\lambda\}]/A$ is isomorphic to $R[X;F]/(\{X^{g_\alpha} - X^{h_\alpha} \mid g_\alpha, h_\alpha \in F\})$ for suitable subsets $\{g_\alpha\}$, $\{h_\alpha\}$ of F . Let ρ be the congruence on F generated by $\{(g_\alpha, h_\alpha)\}$. Part (5) of Theorem 7.5 shows that $R[X;F]/(\{X^{g_\alpha} - X^{h_\alpha}\}) \simeq R[X;F/\rho]$, and hence $R[\{X_\lambda\}]/A$ is, to within isomorphism, a monoid ring over R . This completes the proof of Theorem 7.11.

Section 7 Remarks

We have mentioned in this section that a general semigroup ring $R[X;S]$ can be defined as an appropriate ideal of a monoid ring $R'[X;S']$ over a unitary ring R' . Two advantages of working with R' and S' are that X can be regarded as an element of $R'[X;S']$ in this case and, in the definition of multiplication in $R'[X;S']$, each $s \in S'$ can always be expressed in the form $t*u$ for some $t,u \in S'$. (In the proof in Theorem 7.1 that ϕ preserves multiplication, for example, the possibility that $s \notin S + S$ or that $t \notin T + T$ has been ignored. Not to fear, however; the proof still works in those cases.)

The term unital extension of R has been used in the paragraph following the proof of Theorem 7.1 without an explicit definition. The term means a unitary extension ring S of R such that $S = R + Ze$, where e is the identity element of S. The papers [25] and [11] contain information concerning the possible unital extensions of a given ring R, even in the case where R is already unitary.

Under what conditions is $R[X;S]$ unitary? A reasonable conjecture would seem to be that $R[X;S]$ is unitary if and only if R is unitary and S is a monoid. Indeed, the conditions on R and S imply that $R[X;S]$ is unitary, and since R is a homomorphic image of $R[X;S]$, it is also true that $R[X;S]$ unitary implies R is unitary. If S contains a cancellative element, it is also true that $R[X;S]$ unitary implies that S is a monoid, but not in general. For example, let $S = \{a,b,c\}$ be the semigroup with the following Cayley table.

+	a	b	c
a	a	c	c
b	c	b	c
c	c	c	c

For any unitary ring R, the element $X^a + X^b - X^c$ is an identity element for $R[X;S]$, but S is not a monoid. For more on this topic, see [74].

Theorems 5.10 and 7.8 show that a monoid S is

Noetherian if and only if S is finitely generated. For a semigroup, I do not know if either of these conditions implies the other. Similarly, I know of no results beyond Theorems 7.10 and 20.7 concerning the problem of determining conditions under which a semigroup ring is Noetherian.

§8. Zero Divisors

Two problems are considered in this section. The first is that of determining conditions under which $R[X;S]$ is free of nontrivial zero divisors —— that is, $R[X;S]$ is an integral domain. This problem is easily settled in Theorem 8.1. In the second problem we take a fixed element $f \in R[X;S]$ and we seek necessary and sufficient conditions, usually in terms of the coefficients of f, in order that f should be a zero divisor in $R[X;S]$. The more definitive results on the second problem require some type of restriction on S or on R.

Theorem 8.1. Assume that $R \neq \{0\}$. The semigroup ring $R[X;S]$ is an integral domain if and only if R is an integral domain and S is torsion-free and cancellative.

<u>Proof</u>. Assume first that R is an integral domain and S is torsion-free and cancellative. By Corollary 3.4, S admits a total order $\leq$ compatible with the semigroup operation. Let f,g be nonzero elements of $R[X;S]$ and write $f = \Sigma_{i=1}^{m} f_i X^{s_i}$, $g = \Sigma_{i=1}^{n} g_i X^{t_i}$, where $s_1 < \ldots < s_m$, $t_1 < \ldots < t_n$, and f_1 and g_1 are nonzero. Then $s_1 + t_1 \in \text{Supp}(fg)$ and $f_1 g_1 X^{s_1+t_1}$ is the corresponding term in fg. In particular, $fg \neq 0$ so $R[X;S]$ is an integral domain.

If R is not an integral domain, then choose nonzero elements $a,b \in R$ so that $ab = 0$. If $s \in S$, then $aX^s \cdot bX^s = 0$ in $R[X;S]$, so $R[X;S]$ is not an integral domain. Similarly, if S is not cancellative and if $s,t,u \in S$ are such that $s + t = s + u$ but $t \neq u$, then for $r \in R\backslash\{0\}$ we have $rX^s(rX^t - rX^u) = 0$, where rX^s and $rX^t - rX^u$ are nonzero. Finally, assume that R is an integral domain and that S is cancellative, but not torsion–free. Let $s,t \in S$ be such that $s \neq t$ while $ns = nt$ for some $n \in Z^+$, and choose $k \in Z^+$ minimal so that $ks = kt$. If $r \in R$, $r \neq 0$, then

$$0 = r^2X^{ks} - r^2X^{kt} = (rX^s - rX^t)(\Sigma_{i=0}^{k-1} rX^{(k-i-1)s + it}) .$$

Since S is cancellative, the choice of k implies that $(k - i_1 - 1)s + i_1t \neq (k - i_2 - 1)s + i_2t$ for $0 \leq i_1 < i_2 \leq k - 1$.

Thus, $\Sigma_{i=0}^{k-1} rX^{(k-i-1)s + it} \neq 0$, so again $R[X;S]$ is not an integral domain in this case.

COROLLARY 8.2. Assume that A is a proper ideal of R . Then $A[X;S]$ is prime in $R[X;S]$ if and only if (1) A is prime in R , (2) S is cancellative, and (3) S is torsion–free.

Proof. Corollary 8.2 follows from Theorem 8.1 and from the isomorphism $R[X;S]/A[X;S] \simeq (R/A)[X;S]$.

COROLLARY 8.3. Assume that S is torsion–free and cancellative and that $f = \Sigma f_iX^{s_i}$ and $g = \Sigma g_jX^{t_j}$ are such that $fg = 0$. Then f_ig_j is nilpotent for all i and j .

Proof. Let P be a prime ideal of R . Then $fg \in P[X;S]$, which is a prime ideal. Therefore $f \in P[X;S]$ or $g \in P[X;S]$, and in either case $f_ig_j \in P$. Since P is arbitrary, it follows that f_ig_j is nilpotent.

If the semigroup S is ordered under a fixed total order, then the usual notions of degree and order of elements of $R[X;S]$ can be defined. Thus, if $f = \Sigma_{i=1}^{n} f_i X^{s_i}$ is the canonical form of the nonzero element $f \in R[X;S]$, where $s_1 < s_2 < \ldots < s_n$, then s_n is called the degree of f and we write $\deg f = s_n$. Similarly, s_1 is the order of f and we write $\operatorname{ord} f = s_1$; the element f is monic if $f_n = 1$. If R is unitary, the set of monic elements of $R[X;S]$ forms a multiplicative system in $R[X;S]$. If R is an integral domain, then we have, as usual,

$$\deg(fg) = \deg(f) + \deg(g) \quad \text{and}$$

$$\operatorname{ord}(fg) = \operatorname{ord}(f) + \operatorname{ord}(g) \quad \text{for} \quad f,g \in R[X;S]\backslash\{0\} .$$

It must be understood that $\deg f$ and $\operatorname{ord} f$ are not, in general, intrinsically determined by $f \in R[X;S]$; they depend upon the relation under which S is totally ordered.

What can be said about zero divisors of $R[X;S]$ if one of the three conditions

(1) R is an integral domain,

(2) S is cancellative,

(3) S is torsion-free,

is not satisfied? It turns out that the zero divisors of $R[X;S]$ can still be characterized — at least in terms of the structure of R — if hypothesis (1) is dropped; condition (2) is the most crucial of the three conditions in a determination of the zero divisors of $R[X;S]$. In the case of polynomial rings, McCoy [104] proved that $f \in R[X]$ is a zero divisor if and only if $cf = 0$ for some nonzero element $c \in R$. This result extends to the case of semigroup rings $R[X;S]$, where S is torsion-free and cancellative, but in order to prove a bit more, we introduce the notion of an S-graded ring.

Let R be a ring and let S be a semigroup. We say that R is S-<u>graded</u> if for each $s \in S$, there exists a subgroup R_s of the additive group of R such that

(1) $R = \Sigma^w_{s \in S} R_s$ is the weak direct sum, as an abelian group, of the family $\{R_s\}$, and

(2) $R_s R_t \subseteq R_{s+t}$ for $s,t \in S$.

The family $\{R_s\}_{s \in S}$ is called an S-<u>grading</u> of R . A given ring R may admit more than one S-grading for a given S , so the phrase "assume that $\{R_s\}_{s \in S}$ is an S-grading of R" is more descriptive than "assume that R is S-graded" . If $\{R_s\}_{s \in S}$ is an S-grading of R , then elements of $\cup_{s \in S} R_s$ are said to be <u>homogeneous</u>, while elements of $R \setminus (\cup_{s \in S} R_s)$ are said to be <u>nonhomogeneous</u>. There is a natural S-grading of the semigroup ring $A = R[X;S]$ —— namely, that where we take $A_s = \{rX^s \mid r \in R\}$ for each $s \in S$.

THEOREM 8.4. Assume that $\{R_s\}_{s \in S}$ is an S-grading of R , where S is torsion-free and cancellative. If $f \in R$ is a zero divisor, then f is annihilated by a nonzero homogeneous element of R .

<u>Proof</u>. A nonzero element g of R has a unique decomposition $g = \Sigma^n_{i=1} g_i$, where each g_i is a nonzero homogeneous element and where g_i and g_j do not belong to the same R_s for $i \neq j$. To prove the result, we choose

a nonzero element $g \in \mathrm{Ann}(f)$ whose number n of homogeneous components is as small as possible. The conclusion of the theorem amounts to saying that $n = 1$. Let $\leq$ be a total order on S compatible with the semigroup operation. Let $f = \Sigma_{i=1}^{m} f_i$ and $g = \Sigma_{j=1}^{n} g_j$ be the homogeneous decompositions of f and g, respectively, where $f_i \in R_{s_i}$ with $s_1 < \ldots < s_m$ and $g_j \in R_{t_j}$ with $t_1 < t_2 < \ldots < t_n$. Clearly $f_m g_n = 0$. We consider $f_m g$. If $f_m g \neq 0$, we obtain a contradiction to the choice of g, for $f_m g$ annihilates f and it has at most $n - 1$ components in its homogeneous decomposition. Therefore $f_m g = 0$ and $fg = (f_1 + \ldots + f_{m-1})g = 0$. Consequently, $f_{m-1} g_n = 0$, and the argument just given shows that $f_{m-1} g = 0$. By induction it follows that $f_i g = 0$ for each i, so $fg_n = 0$. By choice of g, we have $g = g_n$, and hence $n = 1$. This completes the proof.

Before stating some consequences of Theorem 8.4, we note that if $\{R_s\}_{s \in S}$ is an S–grading of R, then each surjective homomorphism $\phi : S \longrightarrow T$ of ϕ onto a semigroup T induces a T–grading of R, as follows. For $t \in T$, let $A_t = \Sigma\{R_s \mid \phi(s) = t\}$. That $R = \Sigma_{t \in T}^{w} A_t$ is clear. Moreover, if $t, v \in T$, then $A_t A_v$ is generated as an abelian group by $\{R_s R_u \mid \phi(s) = t$ and $\phi(u) = v\}$. Since $R_s R_u \subseteq R_{s+u}$ and since $\phi(s + u) = \phi(s) + \phi(u) = t + v$, it follows that $A_t A_v \subseteq A_{t+v}$. Hence $\{A_t\}_{t \in T}$ is a T–grading of R.

COROLLARY 8.5. Let Δ be the smallest congruence on the semigroup S such that S/Δ is torsion–free and cancellative. If $f \in R[X;S]$ is a zero divisor, then there exists a nonzero element $g \in R[X;S]$ such that $\mathrm{Supp}(g)$ is contained in an equivalence class of Δ and $fg = 0$.

Proof. The canonical homomorphism ϕ of S onto $T = S/\Delta$ induces a T-grading of $R[X;S]$. Since T is torsion-free and cancellative, Theorem 8.4 implies that f is annihilated by a nonzero T-homogeneous element g. But T-homogeneity of g is equivalent to the statement that $\mathrm{Supp}(g)$ is contained in an equivalence class of Δ. Incidentally, if $f = \Sigma_{i=1}^{r} f_i$ is the T-homogeneous decomposition of f, then $fg = 0$ implies $f_i g = 0$ for each i.

COROLLARY 8.6. Assume that S is torsion-free and cancellative and that $f = \Sigma_{i=1}^{n} f_i X^{s_i}$ is the canonical form of the nonzero element $f \in R[X;S]$. The following conditions are equivalent.

(1) f is a zero divisor in $R[X;S]$.

(2) There exists a nonzero element $r \in R$ such that $rf_i = 0$ for each i.

Proof. That (2) implies (1) is clear. The converse follows from Corollary 8.5, for that result implies that there exists $r \in R\backslash\{0\}$ and $s \in S$ such that $0 = rX^{s} \cdot f = \Sigma_{i=1}^{n} rf_i X^{s+s_i}$. Since S is cancellative, it follows that $rf_i = 0$ for each i.

COROLLARY 8.7. Assume that A is a proper ideal of the unitary ring R.

(1) If $A[X;S]$ is a primary ideal of $R[X;S]$, then A is primary in R and S is cancellative.

(2) If A is primary in R and if S is torsion-free

and cancellative, then $A[X;S]$ is primary in $R[X;S]$. Moreover, if $P = rad(A)$, then $P[X;S] = rad(A[X;S])$.

Proof. If $A[X;S]$ is primary in $R[X;S]$, then A is primary in R since $A = A[X;S] \cap R$. If $s \in S$ is not cancellative, then $X^s(X^t - X^u) = 0$ for some $t,u \in S$ with $t \neq u$. Thus $X^s(X^t - X^u) \in A[X;S]$, $X^t - X^u \notin A[X;S]$, and no power of X^s is in $A[X;S]$. This contradiction to the assumption that $A[X;S]$ is primary shows that S is cancellative.

By passage to $R[X;S]/A[X;S]$ in (2), it suffices to consider the case where $A = (0)$. Thus, if $f = \Sigma_{i=1}^n f_i X^{s_i}$ is a zero divisor in $R[X;S]$, then Corollary 8.6 implies that there exists a nonzero element $r \in R$ such that $rf_i = 0$ for each i. Since (0) is primary in R, each f_i is nilpotent, and hence f is also nilpotent. This proves (2). It is clear that $P[X;S] \subseteq rad(A[X;S])$, and the reverse inclusion follows since $P[X;S]$ is prime in $R[X;S]$ by Corollary 8.2.

In (1) of (8.7), the semigroup S need not be torsion-free. For example, if K is a field of characteristic $p \neq 0$ and G is the cyclic group of order p^n, then the group ring $K[G]$ is isomorphic to $K[X]/(X^{p^n} - 1) \simeq K[X]/(X - 1)^{p^n}$, a special principal ideal ring. Hence each ideal of $K[G]$ is primary, but G is a torsion group. The same example shows that $rad(A[X;S])$ may properly contain $rad(A)[X;S]$ if S is not torsion-free.

In order to continue our investigation of zero divisors of $R[X;S]$, we return to Corollary 8.5 and the notation used there. Thus, assume that $fg = 0$, where f and g are nonzero and g is T-homogeneneous, with $T = S/\Delta$ and Δ the congruence on S defined by $a \Delta b$ if $na + c = nb + c$ for some $n \in Z^+$ and some $c \in S$. If $f = \Sigma_{i=1}^n f_i$ is the decomposition of f into T-homogeneous components, then $f_i g = 0$ for each i. Let $\phi^*: R[X;S] \longrightarrow R[X;T]$ be the homomorphism induced by the canonical mapping $\phi: S \longrightarrow T$. Then ϕ^* maps each T-homogeneous element $\Sigma h_i X^{a_i}$ of $R[X;S]$ onto a monomial $(\Sigma h_i) X^{[a_i]}$ of $R[X;T]$. It is easy to show, however, that a monomial rX^s is a zero divisor if and only if either r is a zero divisor in R or s is not cancellative in S. Since T is a cancellative semigroup, we are able to conclude: if $f_i = \Sigma f_{ij} X^{s_{ij}}$ and $g = \Sigma g_j X^{a_j}$, then $(\Sigma f_{ij})(\Sigma g_j) = 0$ for each i. We note that Σf_{ij} and Σg_j are $\mu(f_i)$ and $\mu(g)$, respectively, where μ is the augmentation map on $R[X;S]$. If R is an integral domain, it follows that either $\mu(g) = 0$ or $\mu(f_i) = 0$ for each i. In the case where S is cancellative, we prove a form of the converse in Theorem 8.9. The proof of (8.9) uses an auxillary result.

THEOREM 8.8. Let N be the nilradical of the ring R. If $x \in R$ is such that x is a zero divisor modulo N, then x is a zero divisor.

Proof. Choose $y \in R \backslash N$ such that $xy \in N$, say $x^k y^k = 0$. If $xy^k = 0$, then x is a zero divisor. And if $xy^k \neq 0$, we choose i maximal so that $x^i y^k \neq 0$. Since $x \cdot x^i y^k = 0$, it follows that x is a zero divisor.

THEOREM 8.9. Assume that D is an integral domain and

S is cancellative. Let Δ be the smallest congruence on S such that S/Δ is torsion-free and cancellative and let I be the kernel ideal of the congruence Δ. Then I consists of zero divisors in $D[X;S]$.

Proof. We prove the result first in the case where D is unitary. Denote by ϕ the canonical homomorphism of S onto S/Δ and by ϕ^* the induced homomorphism on $D[X;S]$. Since $D[X;S/\Delta]$ is an integral domain, I is prime in $D[X;S]$. Let $f \in I$. The proof of part (2) of Theorem 7.2 shows that f is expresible in the form $\Sigma_{i=1}^{m} f_i$, where each f_i is of the form $r_i(X^{s_i} - X^{t_i})$ with $s_i \,\Delta\, t_i$. Let n_i be the smallest positive integer such that $n_i s_i = n_i t_i$; we assume that $n_i > 1$ for each i. Before proving that f is a zero divisor, we consider certain multiples in $D[X;S]$ of $X^s - X^t = h$, where $s \,\Delta\, t$. If k is any positive integer, then $X^{ks} - X^{kt} = hg$, where $g = \Sigma_{j=0}^{k-1} X^{(k-j-1)s+jt}$. We note that $(k-j-1)s + jt \,\Delta\, (k-1)s$ for each j since $s \,\Delta\, t$. Hence $\phi^*(g) = kX^{(k-1)s}$ and $\phi^*(g) \neq 0$ if $\operatorname{char} D = 0$ or if $\operatorname{char} D = p \neq 0$ and $p \nmid k$; thus, $g \notin I$ in either of these two cases. We return to the proof that f is a zero divisor, considering separately the cases where the characteristic of D is, or is not, zero.

If $\operatorname{char} D = 0$, then the proof above shows that for $1 \leq i \leq m$, there exists $g_i \in D[X;S] \setminus I$ such that $f_i g_i = r_i(X^{n_i s_i} - X^{n_i t_i}) = 0$. Therefore $fg_1 \ldots g_m = 0$ and $g_1 \ldots g_m \neq 0$ since this product is not in I.

If $\operatorname{char} D = p \neq 0$, then there are two cases to consider, depending upon n_i. If n_i is divisible by p, we write $n_i = m_i p^{e_i}$, where $e_i > 0$ and m_i is relatively prime to p . There exists $g_i \notin I$ such that $f_i g_i = r_i(X^{m_i s_i} - X^{m_i t_i})$, a nilpotent element since $(f_i g_i)^{p^{e_i}} = r_i^{p^{e_i}}(X^{n_i s_i} - X^{n_i t_i}) = 0$. If n_i is not divisible by p , then there exists $g_i \notin I$ such that $f_i g_i = 0$. Since I is prime, it follows that there exists $g \notin I$ such that $fg = \Sigma_1^m f_i g$ is nilpotent. Hence f is a zero divisor modulo the nilradical of $D[X;S]$, and Theorem 8.8 shows that f is a zero divisor. This completes the proof in the case where D is unitary.

If D is not unitary, let $D^* = D[e]$ be the subring of the quotient field of D generated by D and its identity element e . If I^* is the kernel ideal of Δ on $D^*[X;S]$, then $I = I^* \cap D[X;S]$. Thus, if $f \in I$, there exists $g \in D^*[X;S]\backslash\{0\}$ such that $fg = 0$. If $d \in D\backslash\{0\}$, then dg is a nonzero element of $D[X;S]$ that annihilates f .

The proof of Theorem 8.9 abounds with examples of zero divisors in $D[X;S]$ that need not be in I . In fact, if D has characteristic 0 and S is not torsion-free, then $D[X;S]$ contains a zero divisor of the form $\Sigma_{j=0}^{k-1} X^{(k-j-1)s+jt}$ that is not in I . Theorem 8.9 enables us to determine what subsets of S can be realized in the form $\mathrm{Supp}(f)$ for some zero divisor f of $D[X;S]$, at least in the case where $|D| > 2$.

THEOREM 8.10. Assume that D is an integral domain with $|D| > 2$, that S is a cancellative semigroup, and that $A = \{a_i\}_{i=1}^n$ is an n-element subset of S . Then A is the support of a zero divisor of $D[X;S]$ if and only if each a_i is Δ-related to some a_j with $j \neq i$.

Proof. The condition given is necessary, for suppose $A = \mathrm{Supp}(f)$, where f is a zero divisor. If $f = \Sigma_{i=1}^{m} f_i$ is the T-homogeneous decomposition of f, where $T = S/\Delta$, then Corollary 8.5 implies that each f_i is a zero divisor in $D[X;S]$. Since a nonzero monomial is not a zero divisor, it follows that each $a_i \in A$ is Δ-related to some a_j with $j \neq i$.

To prove the converse, Theorem 8.10 shows that it is sufficient to prove the following statement: if $B = \{b_1,\ldots,b_k\}$ is a k-element subset of S, where $k > 1$ and any two elements of B are Δ-related, then B is the support of an element g of I. If $k = 2r$ is even, take $g = \Sigma_{i=1}^{r}(dX^{b_{2i}} - dX^{b_{2i-1}})$, where $d \in D\backslash\{0\}$. If $k = 2r + 1$ is odd, this is where the assumption that $|D| > 2$ comes in. Thus, choose nonzero elements c,d of D such that $c + d \neq 0$ and let $g = \Sigma_{i=1}^{r}(dX^{b_{2i}} - dX^{b_{2i-1}}) + (cX^{b_{2r}} - cX^{b_{2r+1}})$.

The assumption that $|D| > 2$ is necessary in the statement of Theorem 8.11, for if D is the field of two elements and $G = \{0,a,b,c\}$ is the Klein four-group, then $A = \{0,a,b\}$ is a set of Δ-related elements of G, but $f = X^0 + X^a + X^b$, the only element of $D[X;G]$ with support A, is a unit of $D[X;G]$ of order 2.

For a cancellative semigroup S, questions concerning zero divisors of $R[X;S]$ can sometimes be reduced to the case where S is a group. Theorem 8.11 provides two techniques for accomplishing such reductions.

THEOREM 8.11. Assume that S is a cancellative

semigroup with quotient group G, and let $f \in R[X;S]$.

(1) f is a zero divisor in $R[X;S]$ if and only if f is a zero divisor in $R[X;G]$.

(2) Assume that S is a monoid and H is a subgroup of S containing 0. If $f \in R[X;H]$, then f is a zero divisor of $R[X;S]$ if and only if f is a zero divisor of $R[X;H]$.

Proof. If R^* is a unital extension of R, then $R[X;G]$ is a subring of $R^*[X;G]$ and X^s is a unit of $R^*[X;G]$ for each $s \in S$. Hence $X^s b = 0$ implies $b = 0$ for $b \in R^*[X;G]$. We use this fact in proving both (1) and (2). Thus, assume that $fh = 0$, where h is a nonzero element of $R[X;G]$. There exists $s \in S$ such that $X^s h \in R[X;S]$, and $X^s h \neq 0$. Since $fX^s h = 0$, it follows that f is a zero divisor in $R[X;S]$. The other assertion in (1) is patent.

Similarly, f a zero divisor in $R[X;H]$ implies f is a zero divisor in $R[X;S]$. For the converse, let b be a nonzero element of $R[X;S]$ such that $fb = 0$; without loss of generality, we assume that $S = G$. The group $T = G/H$ induces a grading of $R[X;G]$; let $b = \Sigma_{i=1}^{n} b_i$ be the T–homogeneous decomposition of b. Since $f \in R[X;H]$, the T–homogeneous decomposition of $0 = fb$ is $\Sigma_{i=1}^{n} fb_i$. Hence $fb_i = 0$ for each i. If $s \in \operatorname{Supp}(b_1)$, then $b_1 = X^s b^*$ for some $b^* \in R[X;H]$. Therefore $0 = fb_1 = fX^s b^*$, and as in (1), $fb^* = 0$.

The proof of Theorem 8.11 establishes the following corollary.

COROLLARY 8.12. Assume that S is a cancellative monoid and H is a subgroup of S containing 0. If $f \in R[X;H]$, then the annihilator of f in $R[X;S]$ is generated as an ideal by the annihilator of f in $R[X;H]$.

As an application of Corollary 8.12, we determine conditions under which $1 - X^s$ is a zero divisor, as well as the annihilator of this element.

THEOREM 8.13. Assume that R is unitary and S is a cancellative monoid. The element $f = 1 - X^s$ is a zero divisor in $R[X;S]$ if and only if $ns = 0$ for some $n \in Z^+$. If n is minimal so that $ns = 0$, then $g = 1 + X^s + \ldots + X^{(n-1)s}$ generates the annihilator of f.

Proof. If f is a zero divisor, then Corollary 8.5 shows that each Δ-homogeneous component of f is a zero divisor. Since 1 is not a zero divisor, it follows that $0 \,\Delta\, s$ — that is, $ns = 0$ for some $n \in Z^+$. On the other hand, the proof of Theorem 8.1 (or the next paragraph) shows that f is a zero divisor if s and 0 are Δ-related.

If n is minimal so that $ns = 0$, then $\langle s \rangle = H$ is a subgroup of S of order n, and H contains 0. By Corollary 8.12, it therefore suffices to prove that g generates the annihilator of f in the case where $S = H$. That $fg = 0$ is clear. And if $(1 - X^s)(a_0 + a_1X^s + \ldots + a_{n-1}X^{(n-1)s}) = 0$, then a calculation of the coefficients in the product leads to the system of equations $0 = a_0 - a_{n-1} = a_1 - a_0 = \ldots = a_{n-1} - a_{n-2}$. Therefore $a_0 = a_1 = \ldots = a_{n-1}$ and $\Sigma a_iX^{is} = a_0g$ is in the ideal generated by g.

COROLLARY 8.14. If $G = \{g_i\}_{i=1}^{k}$ is a finite group and R is a unitary ring, then $h = \Sigma_1^k X^{g_i}$ annihilates $1 - X^g$ for each $g \in G$.

Proof. Let $f = 1 - X^g$ and assume that g has order n. If $\{a_i\}_{i=1}^{m}$ is a complete set of coset representatives of (g) in G, then $h = (\Sigma_{i=1}^{k} X^{a_i})(\Sigma_{i=1}^{n-1} X^{ig})$. Thus, Theorem 8.13 shows that h annihilates f.

Corollary 8.14 yields an extension of Theorem 8.9 to the case of a coefficient ring with zero divisors.

THEOREM 8.15. Assume that R is a ring and S is a cancellative semigroup. If $\{s_i\}_1^n \cup \{t_i\}_1^n$ is a subset of S such that $s_i \Delta t_i$ for each i, then $f = \Sigma_1^n r_i(X^{s_i} - X^{t_i})$ is a zero divisor of $R[X;S]$ for each subset $\{r_i\}_1^n$ of R.

Proof. Let R^* be a unital extension of R and let G be the quotient group of S. If $g_i = t_i - s_i \in G$, then $\{g_i\}_1^n$ generates a finite subgroup H of G. Let $H = \{h_i\}_{i=1}^{k}$ and let $h = \Sigma_1^k X^{h_i}$. Corollary 8.14 shows that h annihilates each $1 - X^{g_i}$, and hence $hf = 0$ as well. If $r \in R\backslash(0)$ and if $s = \Sigma_1^n s_i$, it follows that $rX^s h$ is a nonzero annihilator of f in $R[X;S]$.

Section 8 Remarks

The reader has probably noted the crucial role played by the assumption that S is cancellative in this section. In fact, Corollary 8.5 is essentially the only result about zero divisors in the section that does not use the cancellative hypothesis on S. The topic of zero divisors of

$R[X;S]$, in the case where S is not cancellative, is largely unexplored territory.

Theorems 8.9 and 8.10 stem from the doctoral dissertation of Janeway [81]. Janeway's thesis contains additional results on zero divisors that have not been included here.

Another result that we have not included is a form of the Dedekind-Mertens Lemma for semigroup rings. For polynomial rings, the Dedekind-Mertens Lemma states that if $f,g \in R[\{X_\lambda\}]$, then there exists a positive integer k such that $c(f)^{k+1}c(g) = c(f)^k c(fg)$. This lemma can be used to give an alternate proof of McCoy's Theorem on zero divisors in $R[\{X_\lambda\}]$. For a torsion-free cancellative semigroup S , it's true that if $f,g \in R[X;S]\setminus\{0\}$ and if $k = |\mathrm{Supp}(g)|$, then $c(f)^k c(g) = c(f)^{k-1} c(fg)$. The proof hasn't been included in Section 8 because of limited applicability of the result. For additional information on the Dedekind-Mertens Lemma and for its proof, the interested reader may consult [111], [51, Section 28], or [54].

§9. Nilpotent Elements

In considering nilpotent elements of $R[X;S]$, there are again two basic problems. One is the global problem of determining conditions under which $R[X;S]$ is reduced. This is closely related to the problem of determining conditions under which $N[X;S]$ is the nilradical of $R[X;S]$, for since $N[X;S]$ is clearly contained in the nilradical of $R[X;S]$, it follows that $N[X;S]$ is the nilradical if and only if $(R/N)[X;S]$ is reduced. The local problem is that of determining, for a given element $f = \Sigma r_i X^{s_i}$, conditions under which f is nilpotent. One of our main approaches to the local problem is to attempt to determine a generating set for the nilradical of $R[X;S]$. Our first result indicates three ways in which nilpotent elements of $R[X;S]$ may arise. Recall from Section 4 that elements $a,b \in S$ are said to be asymptotically equivalent if there exists $K \in Z^+$ such that $na = nb$ for each $n \geq K$, while a and b are p—equivalent for a prime p if $p^k a = p^k b$ for some $k \geq 0$.

THEOREM 9.1. Assume that R is a ring with nilradical N and that S is a semigroup.

(1) Each element of $N[X;S]$ is nilpotent.

(2) If $a,b \in S$ are p—equivalent and if $r \in R$ is such that pr is nilpotent, then $rX^a - rX^b$ is nilpotent.

(3) If $a,b \in S$ are asymptotically equivalent, then $rX^a - rX^b$ is nilpotent for each $r \in R$.

Proof. Statement (1) is clear. In proving (2) and (3) , there is no loss of generality in assuming that R is unitary since $R[X;S]$ is a subring of $R^*[X;S]$, where R^*

is a unital extension of R. In (2), if $p^k a = p^k b$, then $(X^a - X^b)^{p^k} \in pR[X;S]$, and hence $[r(X^a - X^b)]^{p^k} \in pr^{p^k}R[X;S]$, a nilpotent ideal. Therefore $r(X^a - X^b)$ is nilpotent. To prove (3), we show that $X^a - X^b$ is nilpotent. Let $K \in Z^+$ be such that $na = nb$ for each $n \geq K$. If $m = 2K + 1$, we show that $(X^a - X^b)^m = 0$. If u and v are nonnegative integers such that $u + v = m$, then either $u \geq K$ or $v \geq K$, and hence $ua = ub$ or $vb = va$. In either case, $ua + vb = (u + v)a = ma$. Therefore

$$(X^a - X^b)^m = \Sigma_{i=0}^m (-1)^i \binom{m}{i} X^{(m-i)a+ib} =$$

$$X^{ma}\Sigma_{i=0}^m (-1)^i \binom{m}{i} = X^{ma}(1 - 1)^m = 0 .$$

Several necessary conditions in order that $R[X;S]$ should be reduced follow from Theorem 9.1, but rather than state them here, we allow them to appear in various theorems as conditions that are both necessary and sufficient in order that $R[X;S]$ should be reduced. There is a natural dichotomy into the cases of nonzero characteristic and characteristic 0 in considering the nilradical of $R[X;S]$. We begin with the case where char $R \neq 0$.

THEOREM 9.2. Assume that R is a ring of prime characteristic p and with nilradical N. Then $N[X;S]$ is the nilradical of $R[X;S]$ if and only if either (1) $R = N$ — that is, R is a nil ring, or (2) $R \neq N$ and S is p—torsion free.

<u>Proof</u>. In view of part (2) of Theorem 9.1, it follows that (1) or (2) is satisfied if $N[X;S]$ is the nilradical of $R[X;S]$. For the converse, it suffices, by

passage to $(R/N)[X;S]$, to prove that $R[X;S]$ is reduced if R is reduced and S is p–torsion–free. Thus, let $f = \Sigma f_i X^{s_i}$ be the canonical form of the nonzero element f. For any positive integer k, the canonical form of f^{p^k} is $\Sigma f_i^{p^k} X^{p^k s_i}$; this is true since $f_i^{p^k} \neq 0$ for each i and since s_i and s_j are not p–equivalent for $i \neq j$. Therefore $f^{p^k} \neq 0$ for each k, and this implies that $R[X;S]$ is reduced.

COROLLARY 9.3. If D is a nonzero integral domain of characteristic $p \neq 0$, then $D[X;S]$ is reduced if and only if S is p–torsion–free.

As in Section 4, we use $\underset{p}{\sim}$ in the statement of Theorem 9.4 to denote the congruence of p–equivalence on the semigroup S.

THEOREM 9.4. Let R be a ring of prime-power characteristic p^n. The nilradical of $R[X;S]$ is the ideal $N[X;S] + I$, where N is the nilradical of R and I is the kernel ideal of the congruence $\underset{p}{\sim}$.

Proof. Theorem 9.1 shows that $N[X;S] + I$ is contained in the nilradical of $R[X;S]$. The reverse inclusion is clear if $R = N$, and if $R \neq N$, we establish the reverse inclusion by showing that $R[X;S]/(N[X;S] + I)$ is reduced. By Corollary 7.3, this residue class ring is isomorphic to $(R/N)[X;S/\underset{p}{\sim}]$. Since R has characteristic p^n, $pR \subseteq N$, and hence R/N has characteristic p. Because $S/\underset{p}{\sim}$ is p–torsion–free by Theorem 4.5, it follows from Theorem 9.2 that $(R/N)[X;S/\underset{p}{\sim}]$ is reduced. This completes the proof.

If G is an abelian group of finite exponent n and if $n = p_1^{e_1} \dots p_k^{e_k}$ is the prime decomposition of n, then it is well known that G decomposes as a direct sum $G = G_1 \oplus \dots \oplus G_k$, where $G_i = \{g \in G \mid p_i^{e_i} g = 0\}$. The group G_i has exponent $p_i^{e_i}$ and is called the p_i-<u>primary component of</u> G. If G is, in fact, a ring considered as a group under addition, then each G_i is an ideal of G. We use these facts in the statement of Theorem 9.5.

THEOREM 9.5. Assume that R is a ring of nonzero characteristic $n = p_1^{e_1} \dots p_k^{e_k}$, and let $R = R_1 \oplus \dots \oplus R_k$ be the decomposition of R into primary components, where $\operatorname{char} R_i = p_i^{e_i}$. If N is the nilradical of R, then $N[X;S] + \Sigma_{i=1}^k(\{rX^a - rX^b \mid r \in R_i \text{ and } a \sim_{p_i} b\})$ is the nilradical of $R[X;S]$.

<u>Proof</u>. The decomposition $R = \Sigma_1^k \oplus R_i$ induces the decomposition $R[X;S] = \Sigma_1^k \oplus R_i[X;S]$ of $R[X;S]$, and the nilradical of $R[X;S]$ is the direct sum of the nilradicals of the components $R_i[X;S]$. If N_i is the nilradical of R_i, then Theorem 9.4 shows that $N_i[X;S] + (\{rX^a - rX^b \mid r \in R_i, a \sim_{p_i} b\})$ is the nilradical of $R_i[X;S]$; here ( ) is used for "ideal of $R_i[X;S]$ generated by the given set", but this is the same as the ideal of $R[X;S]$ generated by the set. Thus, the nilradical of $R[X;S]$ is

$$\Sigma_{i=1}^k(N_i[X;S] + (\{rX^a - rX^b \mid r \in R_i \text{ and } a \sim_{p_i} b\})) =$$
$$N[X;S] + \Sigma_{i=1}^k(\{rX^a - rX^b \mid r \in R_i \text{ and } a \sim_{p_i} b\}) .$$

COROLLARY 9.6. Let the notation and hypothesis be as in Theorem 9.5. Then $N[X;S]$ is the nilradical of $R[X;S]$ if and only if S is p_i-torsion free for each i such that R_i is not a nil ring; $R[X;S]$ is reduced if and only if R is reduced and S is p_i-torsion-free for each i.

The results leading up to Theorem 9.5 provide an effective device for determining whether a given $f \in R[X;S]$ is nilpotent, in the case where $\operatorname{char} R = n \neq 0$. The procedure is as follows. The decomposition $f = \Sigma_{i=1}^{k} f_i$ of f with respect to the decomposition $R[X;S] = \Sigma_{i=1}^{k} \oplus R_i[X;S]$ can be effectively determined; f is nilpotent if and only if each f_i is nilpotent. Thus, it suffices to consider the case where R has prime power characteristic p^e. First, we can subtract from $f = \Sigma_{j=1}^{m} a_j X^{s_j}$ the sum g of all monomials $a_j X^{s_j}$ such that a_j is nilpotent. If $g = f$, then f is nilpotent. Otherwise, f is nilpotent if and only if $f - g$ is nilpotent. Assume that the notation is such that $f - g = \Sigma_{j=1}^{n} a_j X^{s_j}$. We examine s_1. If it is p-equivalent to no s_j for $j > 1$, then f is not nilpotent. If $s_1 \underset{p}{\sim} s_j$ for $j > 1$, then f is nilpotent if and only if $f - g - (a_1 X^{s_1} - a_1 X^{s_j})$ is nilpotent, and the latter element has its support of smaller cardinality than $f - g$ does. Continuing this process, we either settle the question of nilpotency of f before we reach the $(n - 1)$st step or, at this step we reach a monomial bX^s to test for nilpotency. Alternately, everything can be handled at the $f - g$ stage, as follows. The congruence $\underset{p}{\sim}$ induces a composition $f - g = \Sigma_{i=1}^{u} g_i$ of $f - g$ into components such that all elements of $\operatorname{Supp}(g_i)$ are p-equivalent for each i, while elements of $\operatorname{Supp}(g_i)$ are not p-equivalent to elements of $\operatorname{Supp}(g_j)$ for $j \neq i$. (This is the same as the T-homogeneous decomposition of f considered in Section 8, where $T = S/\tilde{p}$.) If μ denotes the augmentation map on $R[X;S]$, then an examination of the proofs of Theorem 9.2 and 9.4 shows that f is nilpotent if and only if $\mu(g_i) \in N$ for each i .

We turn to a consideration of nilpotent elements of $R[X;S]$ in the case where $\operatorname{char} R = 0$. Most questions can be reduced to the unital case, for if R^* is a unital extension of R , then the nilradical of $R[X;S]$ is the contraction to $R[X;S]$ of the nilradical of $R^*[X;S]$. If $f = \Sigma_{i=1}^{n} a_i X^{s_i}$, then $f \in R_0[X;S]$, where $R_0 = Z[a_1,\ldots,a_n]$ is a finitely generated ring extension of the ring Z of integers. In what follows, we need to use some of the properties of such rings $Z[a_1,\ldots,a_n]$. The required properties of $Z[a_1,\ldots,a_n]$ follow from the fact that it is a Hilbert FMR-ring. We therefore interrupt our treatment of nilpotent elements in order to present a basic theory of such rings. We stick with the statement made in the preface to the effect that the reader is expected to be familiar with the results of [83]. Thus, Hilbert rings are treated in Section 1–3 and elsewhere in [83] and we do not repeat proofs of results that appear there.

To recall the definition, a <u>Hilbert ring</u> is a unitary ring R such that each proper prime ideal of R is an intersection of maximal ideals of R . It is clear from this definition that the class of Hilbert rings is closed under taking homomorphic images. One of the basic results in the theory is that the class is also closed under (finite) polynomial ring extension; hence, each finitely generated ring extension of a Hilbert ring is again a Hilbert ring, and this applies to the rings $Z[a_1,\ldots,a_n]$ since Z is a Hilbert ring. The other properties of the class of Hilbert rings we need are contained in following theorem.

THEOREM 9.7. The following conditions are equivalent in a unitary ring R.

(1) R is a Hilbert ring.

(2) $R[X]$ is a Hilbert ring.

(3) For each maximal ideal M of $R[X]$, the contraction $M \cap R$ of M to R is maximal in R.

The unitary ring R is said to be an FMR-<u>ring</u> (FMR for finite maximal residue class rings) if R/M is finite for each maximal ideal M of R. The rings $Z[a_1,\ldots,a_n]$ have this property by the next result.

THEOREM 9.8. Assume that R is a Hilbert FMR-ring. Then each finitely generated extension $R[y_1,\ldots y_n]$ of R is also a Hilbert FMR-ring.

<u>Proof</u>. By induction, it suffices to resolve the case $n = 1$. Moreover, since $R[y_1]$ is a homomorphic image of $R[Y]$ and since homomorphic images of Hilbert FMR-rings have these two properties, it suffices to show that $R[Y]$ is an FMR-ring. Let M be a maximal ideal of $R[Y]$ and let $M_0 = M \cap R$. Then M_0 is maximal in R, $F = R/M_0$ is finite, and $M/M_0[Y]$ is maximal in $R[Y]/M_0[Y] \simeq F[Y]$. Since each maximal ideal of $F[Y]$ is generated by an irreducible polynomial, the associated residue field is finite.

Hence $R[Y]/M$ is finite, as we wished to show.

With Theorems 9.7 and 9.8 in tow, we return to the question of nilpotents of $R[X;S]$.

THEOREM 9.9. If S is torsion-free, then $N[X;S]$ is the nilradical of $R[X;S]$, where N is the nilradical of R .

Proof. If R has nonzero characteristic, the result follows from Corollary 9.6. Assume that $\operatorname{char} R = 0$. If R^* is the ring obtained by canonically adjoining an identity element to R , then N is the nilradical of R^* , so it suffices to prove (9.9) under the assumption that R is unitary. Assume that $f = \Sigma_{i=1}^{n} f_i X^{s_i} \in R[X;S]$ is nilpotent. Then $f \in R_0[X;S]$, where $R_0 = Z[f_1,\ldots,f_n]$ is a Hilbert FMR-ring. The nilradical N_0 of R_0 is an intersection of maximal ideals since R_0 is a Hilbert ring — say $N_0 = \cap M_\lambda$. For each λ , R_0/M_λ has nonzero characteristic, and hence $(R_0/M_\lambda)[X;S] \simeq R_0[X;S]/M_\lambda[X;S]$ is a reduced ring. Therefore $f \in M_\lambda[X;S]$ for each λ , and consequently, $f \in \cap(M_\lambda[X;S]) = N_0[X;S] \subseteq N[X;S]$.

The following ancillary result is used in the proof of Theorem 9.11.

THEOREM 9.10. Assume that D is an integral domain of characteristic 0 and that $\{P_\lambda\}_{\lambda\in\Lambda}$ is a family of prime ideals of D such that $\cap P_\lambda = (0)$. Let $\{p_i\}_{i=1}^{n}$ be a finite set of prime integers and let Λ_0 be the subset of Λ consisting of all λ such that the characteristic of D/P_λ is distinct from each p_i . Then $\cap_{\lambda\in\Lambda_0} P_\lambda = (0)$.

Proof. Let $a \in \cap_{\lambda \in \Lambda_0} P_\lambda$. Then $p_1 p_2 \cdots p_n a \in P_\lambda$ for each $\lambda \in \Lambda$, and hence $p_1 \cdots p_n a = 0$. This implies that $a = 0$, for if not, the characteristic of D would be a divisor of $p_1 p_2 \cdots p_n$.

THEOREM 9.11. Assume that D is an integral domain of characteristic 0 . Then $D[X;S]$ is reduced if and only if S is free of asymptotic torsion.

Proof. Theorem 9.1 shows that S is free of asymptotic torsion if $D[X;S]$ is reduced. Conversely, assume that S is free of asymptotic torsion. If $f = \Sigma_{i=1}^{n} f_i X^{s_i}$, then as in the proof of Theorem 9.9, $f \in D_0[X;S]$ for a Hilbert FMR-ring D_0 of characteristic 0 . Thus, to prove that $f = 0$, there is no less of generality in assuming that D is a Hilbert FMR-ring. We observe that if s and t are distinct elements of S , then $s \underset{p}{\sim} t$ for at most one prime p ; this follows from Theorem 4.6 since s and t are not asymptotically equivalent. Thus, there are only finitely many primes p such that $s_i \underset{p}{\sim} s_j$ for some i,j with $1 \le i < j \le n$; we label this set as $\{p_t\}_{t=1}^{m}$. Since D is a Hilbert ring, $(0) = \cap_{\lambda \in \Lambda} M_\lambda$, where $\{M_\lambda\}_{\lambda \in \Lambda}$ is the family of maximal ideals of D . Theorem 9.10 shows that if Λ_0 is the subset of Λ consisting of all λ such that the characteristic of D/M_λ is distinct from each p_t , then $(0) = \cap_{\lambda \in \Lambda_0} M_\lambda$. To complete the proof, we show that $f \in M_\lambda[X;S]$ for each $\lambda \in \Lambda_0$; this is sufficient since $\cap_{\lambda \in \Lambda_0} M_\lambda [X;S] = (0)$. Thus, assume that M is maximal in D and that D/M has characteristic $q \notin \{p_t\}_1^m$. Let $\mu : D \longrightarrow D/M$ be the canonical homomorphism, and let μ^* be the canonical extension of μ to $D[X;S]$. Since $\mu^*(f)$ is nilpotent, $[\mu^*(f)]^{q^k} = \Sigma_1^n \mu(f_i)^{q^k} X^{q^k s_i} = 0$ for some $k \in Z^+$ By choice of q and the set $\{p_j\}_1^m$, the elements $q^k s_i$ are distinct for each i . Therefore $\mu(f_i) = 0$ for each i and $f \in M[X;S]$ as we wished to show. This completes the proof of Theorem 9.11.

COROLLARY 9.12. Assume that D is a domain of characteristic 0 . The nilradical of $D[X;S]$ is the kernel ideal of the congruence of asymptotic equivalence on S .

A semigroup S is said to be <u>separative</u> if $2x = x + y = 2y$ implies $x = y$ for $x,y \in S$. The class of separative semigroups is well known in semigroup theory. The next result shows that separative semigroups are the same as the semigroups that are free of asymptotic torsion.

THEOREM 9.13. Let S be a semigroup. Then S is separative if and only if S is free of asymptotic torsion. In particular, a cancellative semigroup is free of asymptotic torsion.

Proof. If S is not separative, then there exist $x,y \in S$ such that $2x = x + y = 2y$ and $x \neq y$. Then $3x = 2x + y = 2y + y = 3y$, and hence x and y are asymptotically equivalent. This shows that S is separative if S is free of asymptotic torsion.

Conversely, if there exist $x,y \in S$ with $x \sim y$ but $x \neq y$, then choose $k > 1$ minimal such that $nx = ny$ for all $n \geq k$. Then $2(k - 1)x = 2(k - 1)y$. Let $z = (k - 1)x + (k - 1)y$. If $z = 2(k - 1)x$, then S is not separative. On the other hand, if $z \neq 2(k-1)x$, then S still fails to be separative, for $2z = 2(k-1)x + 2(k-1)y = 4(k-1)x$ and $z + 2(k-1)x = 3(k-1)x + (k-1)y = 4(k-1)y = 4(k-1)x$.

It is clear that a cancellative semigroup is separative.

COROLLARY 9.14. If S is cancellative and D is a domain of characteristic 0, then $D[X;S]$ is a reduced ring.

In view of Theorem 9.5 and Corollary 9.12, a description of the nilradical of $R[X;S]$ is missing only in the case where R has characteristic 0 and is not an integral domain. Theorem 9.16 fills this gap without explicit mention either of the characteristic of R or of any hypothesis concerning absence of zero divisors in R. The strategy of the proof of Theorem 9.16 is easy to describe. Let $\{P_\lambda\}_{\lambda\in\Lambda}$ be the set of proper prime ideals of R and let $p_\lambda = \text{char}\,(R/P_\lambda)$ for each λ. Let I_{p_λ} be the kernel ideal of the congruence $\tilde{p}_\lambda$ on S, where $\tilde{0}$ denotes asymptotic equivalence on S. It follows from Theorems 9.4 and 9.11 that $R[X;S]/(P_\lambda[X;S] + I_{p_\lambda})$ is a reduced ring for each λ. Hence, the nilradical of $R[X;S]$ is contained in $\bigcap_\lambda(P_\lambda[X;S] + I_{p_\lambda})$. Theorem 9.16 shows that, conversely, each element f of this intersection is nilpotent. What one needs to show, of course, is that $f \in \bigcap_\lambda(P_\lambda[X;S] + I_{p_\lambda})$ implies that some fixed power f^n of f belongs each $P_\lambda[X;S]$. Theorem 9.15 treats two special cases of this problem.

THEOREM 9.15. Assume that P is a prime ideal of R

such that R/P has characteristic $p > 0$ and let I be the kernel ideal of the congruence $\underset{p}{\sim}$ on S. Let $f = \Sigma_{i=1}^{m} f_i X^{s_i} \in R[X;S]$.

(1) If $f \in P[X;S] + I$ and if s_i is not p-equivalent to s_j for $i \neq j$, then $f \in P[X;S]$.

(2) If the positive integer k is such that $p^k s_i = p^k s_j$ for any $s_i, s_j \in \text{Supp}(f)$ that are p-equivalent, then the relation $f^{p^k} \in P[X;S] + I$ implies that $f^{p^k} \in P[X;S]$.

Proof. (1): Under the canonical map from $R[X;S]$ onto $R[X;S/\underset{p}{\sim}]$, f maps to $\Sigma_1^m f_i X^{[s_i]}$, which is in $P[X;S/\underset{p}{\sim}]$. Since $[s_i] \neq [s_j]$ for $i \neq j$, it follows that $f_i \in P$ for each i.

(2) Let μ be the canonical homomorphism of $R[X;S]$ onto $(R/P)[X;S]$. Then $\mu(f^{p^k}) = [\mu(f)]^{p^k}$ belongs to $\mu(I) = (\{r^* X^a - r^* X^b \mid r^* \in R/P \text{ and } a \underset{p}{\sim} b\})$, the kernel ideal of $\underset{p}{\sim}$ in $(R/P)[X;S]$. By choice of p^k, $[\mu(f)]^{p^k}$ satisfies the hypothesis of (1), where the ideal P in (1) is replaced by the zero ideal of R/P. Therefore (1) implies that $\mu(f^{p^k}) = 0$, which means that $f^{p^k} \in P[X;S]$.

THEOREM 9.16. Let the notation be as in the paragraph preceding the statement of Theorem 9.15. The nilradical of $R[X:S]$ is

$$\cap_{\lambda \in \Lambda} \{P_\lambda[X;S] + I_{p_\lambda}\}.$$

Proof. We have previously observed that the nilradical is contained in the intersection described. For the converse, observe that since $N[X;S] + I_0$ is contained both in the nilradical of $R[X;S]$ and in each $P_\lambda[X;S] + I_{p_\lambda}$, we can pass to the residue class ring $(R/N)[X;S/\underset{0}{\sim}]$. Thus, without loss of generality we assume that R is reduced and that S is free of asymptotic torsion. We take an element $f = \Sigma_{i=1}^{m} f_i X^{s_i}$ in $\cap_\lambda \{P_\lambda[X;S] + I_{p_\lambda}\}$. As in the proof of Theorem 9.11, it follows that the set A of primes p such that $s_i \underset{p}{\sim} s_j$ for some i,j with $1 \le i < j \le m$ is finite, say $A = \{p_i\}_1^n$ (we presently address the case where A is empty). Part (1) of Theorem 9.15 shows that $f \in P_\lambda[X;S]$ for each λ such that $p_\lambda \notin A$. Thus, if $A = \phi$, then $f \in \cap_\lambda P_\lambda[X;S] = (0)$ and f is nilpotent. If $A \neq \phi$, then we choose k large enough so that $s_i \underset{p}{\sim} s_j$, with $p \in A$, implies $p^k s_i = p^k s_j$. Take a prime ideal P_λ such that $p_\lambda \in A$. Part (2) of Theorem 9.15 implies that $f^{p_\lambda^k} \in P_\lambda[X;S]$. If $t = \sup\{p_i^k\}_{i=1}^n$, it then follows that $f^t \in \bigcap_\lambda P_\lambda[X;S] = (0)$. This completes the proof.

Theorem 9.16 generalizes the three previous results — (9.4), (9.5), and (9.12) — of this section characterizing the nilradical of $R[X;S]$ in special cases. The way to obtain a given special case from (9.16) may not be clear, however. It is of some help to have alternate descriptions of the intersection $\cap_\lambda (P_\lambda[X;S] + I_{p_\lambda})$. One such description follows from part (1) of Theorem 7.4. Thus, let $p_0 = 0$ and for $i > 0$, let p_i be the <u>ith</u> prime. Define $\Lambda_i = \{\lambda \in \Lambda \mid p_\lambda = p_i\}$. If $C_{p_i} = \cap_{\lambda\in\Lambda_i} P_\lambda$, then part (1) of (7.4) shows that

$$\cap_{\lambda\in\Lambda_i}(P_\lambda[X;S] + I_{p_\lambda}) = C_{p_i}[X;S] + I_{p_i} ,$$

and hence

$$\cap_{\lambda \in \Lambda}(P_\lambda[X;S] + I_{p_\lambda}) =$$

$$\cap_{i=0}^{\infty} \cap_{\lambda \in \Lambda_i}(P_\lambda[X;S] + I_{p_\lambda}) =$$

$$\cap_{i=0}^{\infty}(C_{p_i}[X;S] + I_{p_i}) .$$

A third description of the nilradical — the one that yields (9.12) as a consequence of (9.16) — follows from an examination the proof of (9.16). To wit, if f and A are as defined in (9.16) and if $B_A = \cap\{P_\lambda \mid p_\lambda \notin A\}$, then the proof of (9.16), together with (1) of Theorem 7.4, show that $f \in (B_A[X;S] + I_0) \cap \{\cap_{p \in A}(C_p[X;S] + I_p)\}$. It is clear that each such set is contained in the nilradical, and hence a third description of the nilradical is

$$\Sigma\{(B_A[X;S] + I_0) \cap [\cap_{p \in A}(C_p[X;S] + I_p)]\} ,$$

where the sum is taken over all finite subsets A of the set of positive primes.

The description of the nilradical in (9.16) is strong enough to enable us to determine conditions under which R[X;S] is reduced, and under which N[X;S] is the nilradical of R[X;S] . To state these results, we introduce one item of new terminology. The integer n is said to be <u>regular</u> on the ring R if $nx = 0$ implies $x = 0$ for $x \in R$; if R is unitary, this amounts to saying that n is a regular element of R , and in general it means that n is not divisible by the additive order of a nonzero element of R .

THEOREM 9.17. The semigroup ring R[X;S] is reduced if and only if (1) R is reduced, (2) S is free of asymptotic torsion, and (3) p is regular on R for each prime

p such that S is not p-torsion-free.

Proof. Theorem 9.1 shows that conditions (1)–(3) are satisfied if $R[X;S]$ is reduced. Conversely, assume that conditions (1)–(3) are satisfied. Let W be the set of primes p such that S is not p-torsion-free. If W is empty, then S is torsion-free and Theorem 9.9 shows that $R[X;S]$ is reduced. Assume that $W \neq \phi$. Since R is reduced, the characteristic c of R is either 0 or a square-free integer. Let $R^* = R \oplus Ze$ be the unital extension of R obtained by canonically adjoining an identity element e of characteristic c. It is straightforward to show that R^* is reduced and that each element of W is regular in R^*. Thus, we assume without loss of generality that R is unitary. Let V be the regular multiplicative system in R generated by W. To show that $R[X;S]$ is reduced, it suffices to show that $R_V[X;S]$ is reduced. Thus, we further assume that p is a unit of R for each p such that S is not p-torsion-free. In this case we examine the representation $\cap(P_\lambda[X;S] + I_{p_\lambda})$ for the nilradical of $R[X;S]$ in Theorem 9.16. We observe that $I_{p_\lambda} = (0)$ for each λ. If $p_\lambda = 0$, then $I_{p_\lambda} = (0)$ since S is free of asymptotic torsion. If $p_\lambda \neq 0$, then p_λ is a nonunit of R, and hence S is p_λ-torsion-free — that is, $I_{p_\lambda} = (0)$. Therefore, the nilradical of $R[X;S]$ is $\cap_\lambda P_\lambda[X;S] = N[X;S] = (0)$. This completes the proof.

COROLLARY 9.18. Let N be the nilradical of the ring R. In order that $N[X;S]$ should be the nilradical of $R[X;S]$, the following two conditions are necessary and sufficient.

(1) S is free of asymptotic torsion.

(2) p is regular on R/N for each prime p such that S is not p–torsion–free.

Section 9 Remarks

Section 31 of [51] is essentially self-contained, and it represents another source for the basic theory of Hilbert rings. For additional information on separative semigroups and some reasons they're studied in semigroup theory, see Section 4.3 of [32] or Section 4 of [74].

The proofs of both Theorem 9.9 and Theorem 9.17 refer to the canonical adjunction of an identity element to a ring R. Besides the sources mentioned in Section 7 remarks concerning unital extensions of a ring R, [51, Section 1] also contains information on the adjunction of an identity element to a ring.

For noncommutative rings R, the term <u>semiprime ring</u> is used instead of reduced ring. Connell in [36, Theorem 5] proves the following result. If R is a commutative unitary ring and G is a (possibly nonabelian) group, then the group ring R[G] is semiprime if and only if R is semiprime and n is regular in R for each $n \in Z^+$ that is the order of an element of G.

Alternate versions of Theorem 9.17, using some of the properties of separative semigroups, may be found in [132], [129], and [30].

Much of the material in Section 9 first appeared in [117].

§10. Idempotents

Given a unitary ring R, the two extremes in terms of idempotent elements of R are that R consists of idempotents — that is, R is a Boolean ring — or 0 and 1 are the only idempotents of R. In the second case, R is said to be indecomposable, for in this case R is not expressible as a nontrivial internal direct sum of ideals. In Section 17 we examine the problem of determining conditions under which R[X;S] is Boolean or, more generally, regular in the sense of von Neumann. We concentrate here on determining conditions under which R[X;S] is indecomposable. This is a special case of the problem of determining conditions under which the idempotents of R[X;S] are in R, and this more general problem is essentially no more difficult to resolve. If $f = \Sigma_{i=1}^{n} f_i X^{s_i} \in R[X;S]$, then we determine several necessary conditions on Supp(f) and c(f) in order that f should be idempotent. On the other hand, Theorem 10.1 is one of the few results that give sufficient conditions for f to be idempotent.

THEOREM 10.1. Assume that R is a unitary ring and S is an additive semigroup.

(1) A monomial rX^s of R[X;S] is idempotent if and only if r and s are idempotent.

(2) If $H = \{h_i\}_{i=1}^{n}$ is a finite subgroup of S and if $n = |H|$ is a unit of R, then $n^{-1}\Sigma_1^n X^{h_i}$ is idempotent.

Proof. (1) is obvious. We note, however, that r is considered in (1) as an element of the multiplicative semigroup of R, rather than the additive group of R. In

(2) , let $f = \Sigma_1^n X^{h_i}$. Since $X^{h_j} f = f$ for each j , it follows that $f^2 = nf$ and $(n^{-1}f)^2 = n^{-1}f$.

We remark that each idempotent of $R[X;S]$ is in R if and only if each idempotent of $R[X;S]$ is a monomial. That the first condition implies the second is clear. And if $R[X;S]$ contains an idempotent $f \notin R$, then either f or $1 - f$ is an idempotent that is not a monomial.

If $f = \Sigma_{i=1}^n f_i X^{s_i}$ is idempotent, what conditions must the coefficients f_i and the elements s_i of Supp(f) satisfy? Some conditions on $\{f_i\}_1^n$ are easy to establish (Theorem 10.12). The main result concerning Supp(f) is Theorem 10.6, which shows that $\langle s_i \rangle$ is a subgroup for each i . In this connection, we recall from Section 2 that $\langle s \rangle$ is a subgroup of S if and only if $s = s + ks$ for some $k > 0$. The set S^* of elements $s \in S$ such that $\langle s \rangle$ is a group forms a subsemigroup of S if S^* is nonempty; to see this, note that if $s = s + k_1 s$ and $t = t + k_2 t$, then $s + t = s + t + k_1 k_2 (s + t)$. This fact will be used in the statement of Theorem 10.6. The proof of (10.6) uses two preliminary results, Theorems 10.2 and 10.4.

THEOREM 10.2. Let R be a ring and let S be an additive semigroup.

(1) Assume that $\{s_i\}_1^n$ is a finite subset of S and that $k > 1$ is an integer such that $\{s_i\}_1^n = \{ks_i\}_1^n$. Then for $1 \le i \le n$, there exists a positive integer $m_i \le n$ such that $s_i = k^{m_i} s_i$.

(2) Assume that R has prime characteristic p and that $e = \Sigma_{i=1}^n r_i X^{s_i}$ is the canonical form of the idempotent

e of $R[X;S]$. For each i between 1 and n , there exist positive integers $m_i, k_i \le n$ such that $s_i = p^{m_i} s_i$ and $r_i = r_i^{p^{k_i}}$.

<u>Proof</u>. (1): The equality $\{s_i\}_1^n = \{ks_i\}_1^n$ implies that each s_i is uniquely expressible as ks_j for some j . Fix i between 1 and n , define $b_1 = i$, and define b_{r+1} inductively by $s_{b_r} = ks_{b_{r+1}}$. Since the set $\{s_j\}_1^n$ has n elements, there exist positive integers h, q with $h < q \le n + 1$ such that $b_h = b_q$. Then

$$s_{b_h} = ks_{b_{h+1}} = \ldots = k^{q-h} s_{b_q} = k^{q-h} s_{b_h} .$$

We conclude that

$$k^{q-h} s_i = k^{q-h} k^{h-1} s_{b_h} = k^{h-1} s_{b_h} = s_i .$$

Moreover, $q - h \le n$ since $q \le n + 1$.

To prove (2) , we note that $e = e^2 = e^p = \Sigma_{i=1}^n r_i^p X^{ps_i}$. Therefore $\{s_i\}_1^n = \{ps_i\}_1^n$ and $\{r_i\}_1^n = \{r_i^p\}_1^n$. The assertions of (2) then follow from (1) , considering $\{r_i\}_1^n$ as a subset of the multiplicative semigroup of R .

COROLLARY 10.3. If R has prime characteristic p and if S is a p–group, then idempotents of $R[X;S]$ are in R .

<u>Proof</u>. Assume that $e = \Sigma_1^n r_i X^{s_i}$ is idempotent. Choose m_i as in Theorem 10.2. Then $s_i = p^{km_i} s_i$ for each positive integer k , and for k sufficiently large, $p^{km_i} s_i = 0$. Therefore each $s_i = 0$ and $e \in R$.

An alternate elementary proof of (10.3) avoids the use of (10.2). To wit, if $m \in Z^+$ is such that $p^m s_i = 0$ for each i . then

$$e = e^{p^m} = \Sigma_1^n r_i^{p^m} x^{p^m s_i} = \Sigma_1^n r_i^{p^m} \in R .$$

THEOREM 10.4. Assume that T is a unitary ring, that $u \in T$, and that U is the subring of T generated by u.

(1) If A is a nil ideal of T such that $u + A$ is idempotent in T/A, then there exists $q \in A \cap U$ such that $u + q$ is idempotent.

(2) Conversely, if $q \in T$ is such that $u + q$ is idempotent and q is nilpotent, then $q \in U$.

Proof. (1): Choose k such that $(u - u^2)^k = u^k(1 - u)^k = 0$. Then $T = (u^k) \oplus (1 - u)^k$. We first determine the decomposition of 1 with respect to this direct sum. We have $(1 - u)^k = 1 - b$, where $b \in (u)$. Moreover, $1 - b$ divides $1 - b^k$, where $b^k \in (u^k)$. Thus, the decomposition of 1 is $1 = b^k + (1 - b^k)$, where b^k and $1 - b^k$ are idempotent. We have $u = b^k + (u - b^k)$, and to complete the proof, we show that $u - b^k$ is in $A \cap U$. We have $u - b^k = (u - ub^k) + (ub^k - b^k)$. Moreover,

$$u - ub^k = u\{1 - [1 - (1 - u)^k]^k\} =$$

$$u\{1 - [1 - k(1 - u)^k + \dots + (-1)^k (1 - u)^{k^2}]\} ,$$

which is clearly in $U \cap (u - u^2) \subseteq U \cap A$; also,

$$ub^k - b^k = (u - 1)[1 - (1 - u)^k]^k =$$

$$(u - 1)\{1 - [1 - ku + \dots + (-u)^k]\}^k ,$$

which again is in $U \cap (u - u^2) \subseteq U \cap A$. This completes the proof of (1).

(2): We have $0 = (u + q) - (u + q)^2$, and hence $u - u^2 = 2uq + q^2 - q$ is nilpotent; say $(u - u^2)^k = 0$. Expanding $(u - u^2)^k$, it follows that

$$u^k = -\Sigma_{i=1}^{k}\binom{k}{i}u^{k-i}(-u^2)^i$$

is an element of the ideal of U generated by u^{k+1}. Consequently, the ideal I of U generated by u^k is idempotent. Let e be an idempotent generator for I. Consider the decomposition $T = Te \oplus T(1 - e)$ of T. Since $u + q$ is idempotent, it follows that $(u + q)e$ and $(u + q)(1 - e)$ are orthogonal idempotents. We observe that ue is a unit of the ring Te since $Tu^ke^k = Te^{k+1} = Te$. As qe is nilpotent, it follows that $ue + qe$ is an idempotent unit of the ring Te, and hence $ue + qe = e$. The element $u(1 - e)$ is nilpotent since u^k annihilates $1 - e$. Therefore $u(1 - e) + q(1 - e)$ is an idempotent element of $T(1 - e)$ that is also nilpotent; whence $u(1 - e) + q(1 - e) = 0$. Therefore $u + q = (u + q)e = e$, and $q = e - u$ is in U, as we wished to prove.

COROLLARY 10.5. Assume that A is a nil ideal of the unitary ring R. Then idempotents of $R[X;S]$ are in R if and only if idempotents of $(R/A)[X;S]$ are in R/A.

Proof. Let f be the canonical homomorphism of $R[X;S]$ onto $(R/A)[X;S]$. If idempotents of $R[X;S]$ are in R, then we take $h \in R[X;S]$ such that $f(h)$ is idempotent. Since $A[X;S]$ is a nil ideal of $R[X;S]$, part (1) of Theorem 10.4 implies that there exists $q \in A[X;S]$ such that $h + q$ is idempotent. Thus $h + q \in R$ and

$f(h) = f(h + q) \in R/A$.

For the converse, assume that idempotents of $(R/A)[X;S]$ are in R/A . If e is an idempotent of $R[X;S]$, we conclude that $e = r + a$ for some $r \in R$, $a \in A[X;S]$. By part (2) of (10.4), it then follows that a , and hence e , belongs to the subring of $R[X;S]$ generated by r . Since $r \in R$, this implies that $e \in R$, which completes the proof of (2) .

THEOREM 10.6. Assume that R is unitary and that S is a monoid. Let S^* be the submonoid of S consisting of all elements $\langle s \rangle$ such that $\langle s \rangle$ is a group. Then each idempotent of $R[X;S]$ belongs to $R[X;S^*]$.

Proof. Let $e = \Sigma_1^n r_i X^{s_i}$ be a nonzero idempotent of $R[X;S]$. Consider e as an element of $T[X;S]$, where $T = \Pi[r_1,\ldots,r_n]$ and Π is the prime subring of R . The ring T is a Hilbert FMR-ring. Let $\{M_\lambda\}_{\lambda \in \Lambda}$ be the family of maximal ideals of T , and for each λ , let f_λ be the canonical homomorphism of $T[X;S]$ onto $(T/M_\lambda)[X;S]$.

Fix i between 1 and n . If $r_i \notin M_\lambda$ for some λ , then using the fact that $f_\lambda(e)$ is an idempotent of $(T/M_\lambda)[X;S]$, where T/M_λ has prime characteristic, we conclude from part (2) of Theorem 10.2 that $s_i \in S^*$. On the other hand, if $r_i \in M_\lambda$ for each i , then r_i is nilpotent because T is a Hilbert ring. We can therefore write e as $q + h$, where h is nilpotent and $g \in R[X;S^*]$. Part (2) of Theorem 10.4 then shows that e belongs to the subring of $R[X;S]$ generated by g , and consequently, $e \in R[X;S^*]$ as asserted.

COROLLARY 10.7. Assume that S is a cancellative semigroup. If $R[X;S]$ contains a nonzero idempotent f, then the supporting semigroup of f is a finite group. In particular, S is a monoid.

Proof. If G is the quotient group of S and if R^* is a unital extension of R, then considering f as a nonzero element of $R^*[X;G]$, it follows from Theorem 10.6 that $\text{Supp}(f) \subseteq H$, the torsion subgroup of G. Thus, the supporting semigroup T of f is a finite group. Since $T \subseteq S$ by definition, we conclude that $0 \in S$, and hence S is a monoid.

COROLLARY 10.8. If R is a ring and S is an aperiodic semigroup, then idempotents of $R[X;S]$ are in R. Moreover, if S is not a monoid, then 0 is the only idempotent of $R[X;S]$.

Proof. Let T be a unital extension of R and let S^0 be the monoid obtained by adjoining an identity element to S. Since S^0 is also aperiodic, Theorem 10.6 shows that idempotents of $T[X;S^0]$ are in T. Therefore idempotents of $R[X;S]$ are in $R[X;S] \cap T$. If S is a monoid, then $R[X;S] \cap T = R$; in the contrary case, $R[X;S] \cap T = (0)$. This completes the proof.

Under what conditions are idempotents of $R[X;S]$ in R? Corollary 10.8 shows that a sufficient condition for this to occur is that S is aperiodic. If S is not aperiodic, Theorem 10.9 shows, in the case where R is unitary and S is a monoid, that the question can be reduced to that in which S is a group.

THEOREM 10.9. Assume that R is unitary and that S is a monoid. Let G be the group of invertible elements of S.

(1) If the only idempotents of $R[X;S]$ are those of R, then each periodic element of S is in G.

(2) Conversely, if G contains each periodic element of S, then idempotents of $R[X;S]$ are in R if and only if idempotents of $R[X;G]$ are in R.

Proof. To prove (1), let s be a periodic element of S. The semigroup $\langle s \rangle$ contains a unique idempotent t. Thus $1 - X^t$ is an idempotent in $R[X;S]$. The hypothesis in (1) implies that $t = 0$; hence s is invertible.

Assume that G contains each periodic element of S. We show that each idempotent of $R[X;S]$ is in $R[X;G]$; this is sufficient to prove (2). If $G = S$, the assertion is clear. If $G \neq S$, then $I = S \backslash G$ is a prime ideal of S and we have $R[X;S] = R[X;G] + R[X;I]$. Considered as abelian groups, this sum is direct, and $R[X;I]$ is ideal of $R[X;S]$. Let e be an idempotent of $R[X;S]$ and write $e = e_1 + e_2$, where $e_1 \in R[X;G]$ and $e_2 \in R[X;I]$. Then $e^2 = e_1^2 + (2e_1e_2 + e_2^2) = e$, implying that $e_1 = e_1^2$ and $e_2 = 2e_1e_2 + e_2^2$. Therefore $(1 - e_1)e_2 = (1 - e_1)2e_1e_2 + (1 - e_1)e_2^2 = (1 - e_1)e_2^2 = [(1 - e_1)e_2]^2$ is an idempotent of $R[X;I]$. Since I is aperiodic and contains no identity element, Corollary 10.8 shows that $(1 - e_1)e_2 = 0$, and hence $e_2 = 2e_1e_2 + e_2^2 = 2e_2 + e_2^2$. It follows that $-e_2 = e_2^2 = (-e_2)^2$ is an idempotent of $R[X;I]$, and again we conclude that $-e_2 = 0$. Therefore $e = e_1 \in R[X;G]$.

The result just established in the proof of (2) of (10.9) is of sufficient interest to deserve a separate statement.

COROLLARY 10.10. Assume that R is unitary and S is a monoid. Let $I \neq S$ be a prime ideal of S that contains no periodic element. Then each idempotent of $R[X;S]$ is in $R[X;S\backslash I]$.

If G is a group, Corollary 10.7 implies that the idempotents of $R[X;G]$ are in $R[X;H]$, where H is the torsion subgroup of G . Moreover, a fixed idempotent $e = \Sigma_1^n r_i X^{h_i}$ of $R[X;H]$ belongs to $R[X;K]$, where K is the (finite) subgroup of H generated by Supp(e) . In Theorem 10.14 we determine necessary and sufficient conditions in order that each idempotent of $R[X;G]$ should be in R , but first we establish two results that are used in the proof of (10.14). The first of these, Theorem 10.12, is of interest in its own right.

THEOREM 10.12. Assume that f is an idempotent element of $R[X;S]$.

(1) The content ideal $c(f)$ is idempotent.

(2) Assume that R is indecomposable. If R is unitary, then either $f = 0$ or $c(f) = R$. If R is not unitary, then $f = 0$.

Proof. It is clear that $c(f^2) \subseteq [c(f)]^2$. Since f is idempotent, we conclude that $c(f) = [c(f)]^2$. Because $c(f)$ is idempotent and finitely generated, it is principal and is generated by an idempotent element e. Assume that R is indecomposable. If R is unitary, then $e = 0$ or 1 and $f = 0$ or $c(f) = R$; if R is not unitary, then $e = 0$ and $f = 0$.

THEOREM 10.13. Assume that A is a proper ideal of the Noetherian ring R. Define $B_1 = \cap_{n=1}^{\infty} A^n$, $B_2 = \cap_1^{\infty} B_1^n, \ldots$. There exists a positive integer k such that $B_k = B_k^2$. If R is indecomposable, then $B_k = (0)$ for some k.

Proof. If R^* is a unital extension of R, then R^* is Noetherian and A is an ideal of R^*. Hence we assume without loss of generality that R is unitary. Let $(0) = Q_1 \cap \ldots \cap Q_t$ be a shortest representation of (0) in R, where Q_i is P_i-primary. Then B_1 is the intersection of the subfamily C_1 of $C = \{Q_i\}_1^t$ consisting of those primary components Q_i such that A is not relatively prime to P_i; assume that $|C_1| = k_1$, where $1 \leq k_1 \leq t$. If $k_1 = t$, then $B_1 = (0)$ and we're finished. If $k_1 < t$, then either $B_1 = B_1^2$ and the proof is complete, or else $B_2 < B_1$ and B_2 is the intersection of a subfamily C_2 of C, where $C_2 > C_1$. Since the set C is finite, it follows that $B_k = B_k^2$ for some k.

Since R is Noetherian, the equality $B_k = B_k^2$ implies that B_k is generated by an idempotent element. Hence $B_k = (0)$ if R is indecomposable.

Theorem 10.12 shows that if $R[X;S]$ has a nonzero idempotent f, then R contains a nonzero idempotent e, and hence R has a nonzero unitary subring. This observation shows that the assumption in the statement of Theorem 10.14

that the set C is nonempty is necessary.

THEOREM 10.14. Assume that G is an abelian group. Let Σ be the set of primes p such that G contains an element of order p, and let C be the family of nonzero unitary subrings of R. Assume that C is nonempty. The following conditions are equivalent.

(1) The idempotents of $R[X;G]$ are those of R.

(2) For each $p \in \Sigma$ and each $T \in C$, p is a nonunit of T.

Proof. If (2) fails, then there exists $g \in G$ of order p and a nonzero subring T of R with identity element e such that pe is a unit of T. Part (2) of Theorem 10.1 then shows that $(pe)^{-1}\Sigma_{i=0}^{p-1} eX^{ig}$ is an idempotent of $R[X;G]$ that is not in R.

(2) $\Longrightarrow$ (1): We consider a nonzero idempotent $f = \Sigma_{i=1}^{n} f_i X^{g_i} \in R[X;G]$ and seek to prove that $f \in R$. If e is an idempotent generator for $c(f)$, then $ef_i = f_i$ for each i. Hence if Π is the subring of R generated by e and if $T = \Pi[f_1,\ldots,f_n]$, then $T \in C$ and T is Noetherian.

Moreover, the subgroup H of G generated by $\{g_i\}_1^n$ is finite, and $f \in T[X;H]$. Each Noetherian ring is a finite direct sum of indecomposable rings. Thus, write $T = T_1 \oplus \ldots \oplus T_r$, where each T_i is nonzero and indecomposable. With respect to the decomposition $T[X;H] = \Sigma_1^r \oplus T_i[X;H]$, the idempotent f has a decomposition $f = \Sigma_1^n f_i$, where each f_i is idempotent. We note that if $f_i \in T_i$ for each i, then $f \in T \subseteq R$. Also, since each

T_i is unitary, (2) implies that each prime divisior of $|H|$ is a nonunit in each T_i. To establish (1), it is therefore sufficient to prove the following statement (*).

(*) Assume that R is a nonzero indecomposable Noetherian unitary ring and that H is a finite group such that each prime divisor of $|H|$ is a nonunit of R. Then the group ring $R[X;H]$ is indecomposable.

To prove (*), we express H as a direct sum $H_1 \oplus \ldots \oplus H_k$ of cyclic groups of prime-power order, and we use induction on k. For $k = 1$, H is cyclic of order p^v, where p is a nonunit of R. Let M be a maximal ideal of R containing p and let ϕ be the canonical homomorphism of $R[X;H]$ onto $(R/M)[X;H]$. If f is a nonzero idempotent of $R[X;H]$, then $\phi(f)$ is idempotent in $(R/M)[X;H]$. Moreover, part (2) of Theorem 10.12 shows that $c(f) = R$, and hence $\phi(f) \neq 0$. The field R/M has characteristic p and H is a p–group. Therefore $\phi(f) = 1 + M$ by Corollary 10.3. Whence $f = 1 - m$, where $m \in M[X;H]$. Since $m = 1 - f$ is idempotent, $m \in M^t[X;H]$ for each $t \in Z^+$. Thus $m \in M_1[X;H]$, where $M_1 = \cap_1^\infty M^t$. Similarly, $m \in M_2[X;H]$, where $M_2 = \cap_1^\infty M_1^t$. By Theorem 10.13, we conclude that $m = 0$. Hence $f = 1$, and $R[X;H]$ is indecomposable if $k = 1$.

Assume that $R[X;H]$ is indecomposable for $k = r$. If H is the direct sum of $r + 1$ cyclic groups $H_1, \ldots H_{r+1}$ of prime-power order, where $|H_{r+1}| = p^c$, then we consider $R[X;H]$ as the group ring of H_{r+1} over $R[X;K]$, where $K = H_1 \oplus \ldots \oplus H_r$. By assumption, $R[X;K]$ is indecomposable.

Since K is finite, $R[X;K]$ is Noetherian and is an integral extension of R. Consequently, p a nonunit of R implies p is a nonunit of $R[X;K]$. The case $k = 1$ then yields the desired conclusion that $R[X;H]$ is indecomposable. This completes the proof of Theorem 10.14.

The analogue of Theorem 10.14 for monoid rings is the following result.

THEOREM 10.15. Assume that S is a monoid. Let C be the family of nonzero unitary subrings of R, and assume that C is nonempty. The following conditions are equivalent.

(1) The idempotents of $R[X;S]$ are those of R.

(2) The idempotents of $T[X;S]$ are those of T for each $T \in C$.

(3) The set G of periodic elements of S is a subgroup of S, and if the prime p is the order of an element of G, then p is a nonunit of T for each $T \in C$.

Proof. The equivalence of (2) and (3) follows from Theorems 10.6, 10.9 and 10.14. It is clear that (1) implies (2), and the reverse implication follows from Theorem 10.12: if f is a nonzero idempotent in $R[X;S]$, then $f \in T[X;S]$ for some $T \in C$.

COROLLARY 10.16. Assume that S is a monoid and that R is a unitary ring of nonzero characteristic n. The following conditions are equivalent.

(1) The idempotents of $R[X;S]$ are those of R.

(2) Either (i) S is aperiodic or (ii) $n = p^k$ is a

prime power and the set of periodic elements of S is a p-group.

Proof. That (2) implies (1) follows from Theorems 10.6 and 10.15, for the characteristic of each subring of R is a divisor of the characteristic of R. For the converse, assume (1) and assume that the set G of periodic elements of S is nonzero — that is, S is not aperiodic. Theorem 10.6 shows that G is a group. Let p be a prime that is the order of an element of G. Theorem 10.15 implies that p is a nonunit of each nonzero unitary subring of R, and this implies that n is a power of p, for R has a nonzero unitary subring of characteristic d for each prime-power divisor $d > 1$ of n. Since $n = p^k$ for some k, (10.15) then implies that G is a p-group.

For a unitary ring R and a monoid S, it is clear that $R[X;S]$ is indecomposable if and only if R is indecomposable and the idempotents of $R[X;S]$ are those of R. Since the identity element of each nonzero unitary subring of an indecomposable unitary ring is the same as the identity element of the ring itself, the next result follows from Theorem 10.15.

COROLLARY 10.17. For a unitary ring R and a monoid S, the ring $R[X;S]$ is indecomposable if and only if R is indecomposable, the set of periodic elements of S is a subgroup of S, and the order of each nonzero periodic element of S is a nonunit of R.

The characteristic of an idecomposable ring is either 0

or a prime power. In the case of a ring of prime-power characteristic, Corollary 10.17 translates to the following statement.

COROLLARY 10.18. Assume that R is an indecomposable unitary ring of prime-power characteristic p^k. Let S be a monoid, and denote by S^* the set of periodic elements of S. Then $R[X;S]$ is indecomposable if and only if S^* is a p-subgroup of S.

We conclude the section with a result for group rings that has the familiar ring of (10.14)–(10.18).

THEOREM 10.19. Assume that R is an indecomposable unitary ring and that G is a group. Let B be the set of primes p that are units of R and let $H = \Sigma_{p \in B} G_p$, where G_p denotes the p-primary component of G. Then each idempotent of $R[X;G]$ is in $R[X;H]$.

Proof. If $f \in R[X;G]$ is idempotent, then $f \in R[X;G_1]$ for some finitely generated subgroup G_1 of G. Moreover, $G_1 = H_1 \oplus H_2$, where H_1 is a subgroup of H and where $(H_2)_p = \{0\}$ for each $p \in B$. Considering $R[X;G_1]$ as the group ring of H_2 over $R[X;H_1]$, Theorem 10.14 implies that $f \in R[X;H_1]$ if each prime integer q that is a nonunit of R is also a nonunit of each nonzero unitary subring T of $R[X;H_1]$. Let e be the identity element of T and assume that qe is a unit of T — say $e = qe \cdot t$. Computing contents of e, qe, and t as elements of $R[G]$, part (2) of Theorem 10.12 shows that $c(e) = R$. But $c(qe \cdot t) = qc(e \cdot t) \subseteq qR$, contrary to the fact that q is a nonunit of

R . This completes the proof.

Section 10 Remarks

The literature is rich in material concerning idempotents of group rings R[G] , where R is a commutative unitary ring and G is nonabelian. Two general references for some of this material are Chapter 1 of [121] and Chapters 2 and 4 of [125]. It is unknown in the nonabelian case, for example, whether R[G] has only trivial idempotents if R is indecomposable and p is a nonunit of R for each prime p that is the order of an element of G (cf. Conjecture 2.18 of [125, p. 23]); Formanek in [43] has proved some results related to this question.

There is an analogue of (10.15) for semigroups S that are not monoids. It involves use of Theorem 10.6 and introduction of the family of subsemigroups of S that are monoids.

In general, presentation of the material in this section has followed that of Section 2 of [61]. The description of $\cap_{n=1}^{\infty} A^n$ used in the proof of Theorem 10.13 can be found in Section 7, Chapter IV of [136].

The condition in (10.14) and (10.15) that p is a nonunit of each nonzero unitary subring of R is equivalent to the condition that the ideal pR of R contains no nonzero idempotent.

§11. Units

All rings considered in this section are assumed to be unitary, and all semigroups are assumed to be monoids. The area of units of monoid rings is vast — large enough to be the topic of a monograph in itself — and hence some restrictions on the scope of treatment here are necessary. The main restriction we impose is in considering primarily torsion-free monoids. In this case, we are interested in characterizing the units $u = \Sigma_1^n r_i X^{s_i}$ of $R[X;S]$ in terms of the coefficients r_i and the elements s_i of $\mathrm{Supp}(u)$. We also seek a description of the Jacobson radical of $R[X;S]$ in this case. Another problem of significant interest is that of giving conditions under which each unit of $R[X;S]$ is trivial, where by a <u>trivial unit</u>, we mean one of the form rX^s where, necessarily, r is a unit of R and s is invertible in S. To begin, we consider the case where S is both torsion-free and cancellative.

THEOREM 11.1. Assume that D is an integral domain with identity and S is a nonzero torsion-free cancellative monoid.

(1) If $f,g \in D[X;S]\backslash(0)$ and if fg is a monomial, then f and g are monomials.

(2) $D[X;S]$ admits only trivial units.

(3) $D[X;S]$ is semisimple.

<u>Proof</u>. (1): Let $\leq$ be a total order on S compatible with the monoid operation. Since $\mathrm{ord}(f) + \mathrm{ord}(g) = \mathrm{ord}(fg) = \deg(fg) = \deg(f) + \deg(g)$, it follows that $\mathrm{ord}(f) = \deg(f)$ and $\mathrm{ord}(g) = \deg(g)$ — that is, f and g are monomials.

Statement (2) follows from (1). To prove (3), take a nonzero element $h = \Sigma_1^n r_i X^{s_i}$ in $D[X;S]$. Since S is infinite and cancellative, there exists $s \in S$ such that $0 \notin \mathrm{Supp}(X^s h)$. Consequently, $1 + X^s h$ is not a unit of $D[X;S]$, and h is not in the Jacobson radical of $D[X;S]$.

The statement of Theorem 11.3 uses one equivalence which is purely-ring theoretic, and which we choose to state as a separate result.

THEOREM 11.2. Let $\{a_i\}_{i=1}^n$ be a finite subset of the unitary ring R. The following conditions are equivalent.

(1) $R = (a_1,\ldots,a_n)$ and $a_i a_j$ is nilpotent for $i \neq j$.

(2) There exists a positive integer k such that $R = (a_1^k)\oplus\ldots\oplus(a_n^k)$.

(3) There exists a finite subset $\{b_i\}_{i=1}^n$ of R such that $\Sigma_1^n a_i b_i = 1$ and $a_i b_j$ is nilpotent for $i \neq j$.

<u>Proof</u>. (1) $\Longrightarrow$ (2): Choose k large enough so that $(a_i a_j)^k = 0$ for all $i \neq j$. Since $R = (a_1,\ldots,a_n)$, then $R = (a_1^k,\ldots,a_n^k)$. Moreover, for $1 \leq i \leq n$,

$$(a_i^k) \cap \Sigma_{j\neq i}(a_j^k) = (a_i^k)\Sigma_{j\neq i}(a_j^k) = (0) .$$

Therefore $R = (a_1^k)\oplus\ldots\oplus(a_n^k)$.

(2) $\Longrightarrow$ (3): For $1 \leq i \leq n$, let $a_i = \Sigma_{j=1}^n c_{ij}$ be the decomposition of a_i with respect to the decomposition $R = \Sigma_{j=1}^n \oplus (a_j^k)$ of R. Then $a_i^k = \Sigma_{j=1}^n c_{ij}^k$ implies that $c_{ii}^k = a_i^k$ and $c_{ij}^k = 0$ for each $j \neq i$. Therefore c_{ij} is nilpotent for $i \neq j$ and c_{ii} is a unit of the ring (a_i^k). Let b_i be the inverse of c_{ii} in (a_i^k)

for each i. Then $b_i a_j = b_i c_{ji}$ is nilpotent for $j \neq i$. Moreover, $b_i a_i = b_i c_{ii}$, so $\Sigma_1^n b_i a_i = \Sigma b_i c_{ii} = 1$.

(3) $\Longrightarrow$ (1): The equality $\Sigma_1^n a_i b_i = 1$ implies that $R = (a_1, \ldots, a_n)$, and multiplication of the equation by $a_u a_v$ for $u \neq v$ yields $a_u a_v = \Sigma_{i=1}^n a_i b_i a_u a_v$. For each i either $b_i a_u$ or $b_i a_v$ is nilpotent. Therefore $a_u a_v$ is also nilpotent.

THEOREM 11.3. Let R be a unitary ring with nilradical N and with set $\{P_\lambda\}_{\lambda \in \Lambda}$ of proper prime ideals. Let S be a torsion-free cancellative monoid, let G be the group of invertible elements of S, and for each λ, let ϕ_λ denote the canonical homomorphism of $R[X;S]$ onto $(R/P_\lambda)[X;S]$. Assume that $f \in R[X;S]$, and write f in the form $g + h$, where the canonical form $g = \Sigma_{i=1}^n g_i X^{s_i}$ of g is such that $g_i \notin N$ for each i, $h = \Sigma_{i=1}^m h_i X^{t_i} \in N[X;S]$, and $\mathrm{Supp}(g) \cap \mathrm{Supp}(h) = \phi$. The following conditions are equivalent.

(1) f is a unit of $R[X;S]$.

(2) g is a unit of $R[X;S]$.

(3) $\phi_\lambda(g)$ is a unit of $(R/P_\lambda)[X;S]$ for each $\lambda \in \Lambda$.

(4) $c(g) = R$, $g_i g_j \in N$ for $i \neq j$, and $\mathrm{Supp}(g) \subseteq G$.

(5) There exists a positive integer k such that $R = (g_1^k) \oplus \ldots \oplus (g_n^k)$ and $\mathrm{Supp}(g) \subseteq G$.

(6) $\mathrm{Supp}(g) \subseteq G$ and there exists a subset $\{b_i\}_{i=1}^n$ of R such that $1 = \Sigma_1^n b_i g_i$, where $b_i g_j$ is nilpotent for $i \neq j$.

<u>Proof</u>. The equivalence of (1) and (2) is well known, and the equivalence of (4), (5), and (6) is the content

of Theorem 11.2. The implication (2) $\Longrightarrow$ (3) is clear.

(3) $\Longrightarrow$ (4): Fix $\lambda \in \Lambda$. Theorem 11.1 shows that $\phi_\lambda(g) = \Sigma_1^n \phi_\lambda(g_i)X^{s_i}$ is a trivial unit of $(R/P_\lambda)[X;S]$. It follows that there exists a unique i between 1 and n such that $g_i \notin P_\lambda$, and for this i , $s_i \in G$. Therefore $c(g) = (g_1,\ldots,g_n) = R$, and $g_i g_j$ is nilpotent for $i \neq j$. To see that each $s_k \in G$, note that $g_k \notin N$ implies there exists a prime P_α of R such that $g_k \notin P_\alpha$. Then $s_k \in G$ since $\phi_\alpha(g)$ is a unit of $(R/P_\alpha)[X;S]$.

(6) $\Longrightarrow$ (2): Under the hypothesis of (6) , we let $w = \Sigma_{i=1}^n b_i X^{-s_i}$. The hypothesis that $b_i g_j$ is nilpotent for $i \neq j$ implies that

$$gw = (\Sigma g_i X^{s_i})(\Sigma b_i X^{-s_i}) = (\Sigma_1^n g_i b_i) + v = 1 + v ,$$

where $v \in N[X;S]$. Therefore gw is a unit, and g is also a unit.

The statement of Theorem 11.3 can be simplified in certain cases where it is unnecessary to pass from f to the decomposition $g + h$.

COROLLARY 11.4. Assume that R is a unitary ring and that S is a torsion-free cancellative monoid. Let $f = \Sigma_1^n a_i X^{s_i}$ be the canonical form of $f \in R[X;S]$, and let G be the group of invertible elements of S .

(1) If R is reduced, then f is a unit if and only if $R = (a_1)\oplus\ldots\oplus(a_n)$ and $\text{Supp}(f) \subseteq G$.

(2) If S is a group, then f is a unit if and only if $c(f) = R$ and $a_i a_j$ is nilpotent for $i \neq j$.

(3) If R is indecomposable, then f is a unit if and

only if there exists i between 1 and n such that $a_iX^{s_i}$ is a trivial unit of $R[X;S]$ and $f - a_iX^{s_i}$ is nilpotent.

COROLLARY 11.5. Assume that R is a unitary ring with nilradical N and that S is a nonzero torsion-free cancellative monoid. The Jacobson radical of $R[X;S]$ is $N[X;S]$, the nilradical of $R[X;S]$.

Proof. Theorem 9.9 implies that $N[X;S]$ is the nilradical of $R[X;S]$. By passage to $(R/N)[X;S]$, we can prove Corollary 11.5 by showing that $R[X;S]$ is semisimple if R is reduced. Thus, take a nonzero element $f \in R[X;S]$ and choose $s \in S$ such that $0 \notin \mathrm{Supp}(X^s f)$; we note that $X^s f \neq 0$ since X^s is a regular element of $R[X;S]$. It then follows from part (1) of Corollary 11.4 that $1 + X^s f$ is not a unit of $R[X;S]$. Consequently, $f \notin J(R[X;S])$ and $R[X;S]$ is semisimple, as we wished to show.

THEOREM 11.6. Let R, S, and G be as described in the statement of Theorem 11.3. The monoid ring $R[X;S]$ has only trivial units if and only if R is reduced and either (1) or (2) is satisfied:

(1) R is indecomposable.

(2) $G = \{0\}$.

Proof. It is immediate from part (1) of Corollary 11.4 that $R[X;S]$ has only trivial units if R is reduced and (1) or (2) is satisfied. If R is not reduced and if r is a nonzero nilpotent of R, then $1 + rX^s$ is a nontrivial unit of $R[X;S]$ for each nonzero element s of S. On the other hand, if R is decomposable and $G \neq \{0\}$, then take a decomposition $1 = e_1 + e_2$ into nonzero orthogonal idempotents of R and consider a nonzero element g of G. Then $e_1 + e_2X^g$ is a nontrivial unit of $R[X;S]$ with inverse $e_1 + e_2X^{-g}$. This completes the proof of Theorem 11.6.

In seeking to extend Theorem 11.4 and its consequences, we have a choice of weakening the hypothesis that S is torsion-free or the assumption that S is cancellative. For our purposes, it is more reasonable to weaken the cancellative hypothesis; as soon as one drops the assumption that a cancellative monoid is torsion-free, then the quotient group of S admits a nontrivial finite subgroup H, and the problem of determining the units of $R[X;H]$ arises. This is an interesting problem that is not at all intractable, but it's also a very large problem area that has limited applicability to topics subsequently covered in this text. Hence we retain the assumption that S is torsion-free. To give an indication of the generalization of Theorem 11.4 that we seek, we reinterpret that result as follows. If $I = S\backslash G$, then I is a prime ideal of S and the monoid ring decomposes as $R[X;S] = R[X;G] + R[X;I]$, where $R[X;I]$ is an ideal of $R[X;S]$. **Each** $f \in R[X;S]$ has a decomposition $f_1 + f_2$, where $f_1 \in R[X;G]$ and $f_2 \in R[X;I]$; we call f_1 the G-<u>component</u> of f and f_2 the I-<u>component</u> of f. (More generally, we speak of the I- and $(S\backslash I)$-components of f for any proper prime ideal I of S.) If $G = S$, then we interpret $R[X;I]$ as (0). In this terminology, (11.4) states that for S torsion-free and cancellative, f is a unit of $R[X;S]$ if and only if f_1 is a unit of $R[X;G]$ and $f_2 \in N[X;I]$, where N is the nilradical of R. In

Theorem 11.13, we extend this result to the case where S is torsion-free and aperiodic. Note that a torsion-free cancellative monoid is aperiodic, for its quotient group is torsion-free. The converse fails. For example, if $S = Z_0 \dot{\oplus} Z_0 \dot{\oplus} Z_0$ and if $\sim$ is the congruence on S defined by $(a,b,c) \sim (d,e,f)$ if $a = d > 0$ and $b + c = e + f$, then $S/\sim$ is torison-free and aperiodic, but not cancellative; to within isomorphism, $S/\sim$ can be thought of as the multiplicative monoid of terms $x^i y^j z^k$ in $R[X,Y,Z]/(XY - XZ)$. The next result gives an equivalent description of a torsion-free aperiodic monoid.

THEOREM 11.7. The torsion-free monoid S is aperiodic if and only if 0 is the only idempotent of S.

Proof. Clearly 0 is the only idempotent of S if S is aperiodic. For the converse, assume that 0 is the only idempotent of S and let s be a periodic element of S. The semigroup $\langle s \rangle$ contains an idempotent ks, and $ks = 0 = k0$ by hypothesis. Since S is torsion-free, it follows that $s = 0$.

If S contains a nonzero idempotent s, then Theorem 11.4 need not extend to the monoid ring $R[X;S]$. For example, $1 - 2X^s \in Z[X;S]$ is a unit of order 2 and $s \notin G$, but -2 is not nilpotent in Z.

Each of Theorems 11.9 through 11.12 is a special case of Theorem 11.13. We assume throughout these results that R is a unitary ring, that S is a torsion-free aperiodic monoid, and that G is the set of invertible elements of S. For the sake of brevity, we do not repeat these hypotheses in the statements of (11.9)–(11.12). The proof of (11.9) uses an auxillary result that has already been proved in (2) of Theorem 1.1; we repeat the statement of the result in the form in which it is used here.

THEOREM 11.8. Assume that T is a subsemigroup of the monoid S and that I is an ideal of S disjoint from T. Then I is contained in an ideal J of S maximal with respect to failure to meet T, and each such J is a prime in S.

THEOREM 11.9. Assume that R is a field of characteristic $p \neq 0$ and assume that $S^* = S - \{0\}$ is the only proper prime ideal of S. If $1 + f$ is a unit of $R[X;S]$, where $f \in R[X;S^*]$, then $f = 0$.

Proof. We first claim that the set $A = \{s \in S \mid$ there exists $t \in S^*$ such that $s = s + t\}$ is empty. If not, then there exists $t \in S^*$ such that $I = \{s \in S \mid t + s = s\}$ is also nonempty. The set I is a proper ideal of S, and since S^* is the unique proper prime ideal of S, it follows that I meets the semigroup $\langle t \rangle$ of S. Therefore $t + kt = kt$ for some $k \in Z^+$, contrary to the fact that S is aperiodic. We conclude that A is empty.

Let $\sim$ be the smallest congruence on S such that the factor semigroup $T = S/\sim$ is cancellative. We prove that T is torsion-free. Thus, if $[s], [t] \in T$ are such that $n[s] = n[t]$, then $ns + c = nt + c$ for some $c \in S$. Therefore $n(s + c) = n(t + c)$ and $s + c = t + c$ since S is torsion-free. Thus $[s] = [d]$ and T is torsion-free. Let $[x]$ be an invertible element of T, say

$[0] = [x] + [y] = [x + y]$. Then $x + y \sim 0$, and since A is empty, we conclude that $x + y = 0$. Because S^* is the only proper prime ideal of S , it follows that 0 is the only invertible element of S . Therefore $x = y = 0$, and $[0]$ is the only invertible element of T . Summarizing, T is a torsion-free cancellative monoid with no invertible element other than 0 . It then follows from Theorem 11.6 that the only units of $R[X;T]$ are those of R . Thus, if ϕ is the canonical homomorphism of $R[X;S]$ onto $R[X;T]$, then $\phi(1 + f) = 1 + \phi(f) \in R$. But $[0] \notin \mathrm{Supp}(\phi(f))$ since A is empty. Therefore $\phi(f) = 0$. Let $f = \Sigma_1^n r_i X^{s_i}$. If $s_i, s_j \in \mathrm{Supp}(f)$ and if $s_i \sim s_j$, then we choose $t_{ij} \in T$ such that $s_i + t_{ij} = s_j + t_{ij}$. Let $t = \Sigma t_{ij}$, where the sum is taken over all $s_i , s_j \in \mathrm{Supp}(f)$ such that $s_i \sim s_j$. It then follows that $s_i \sim s_j$ if and only if $s_i + t = s_j + t$. We consider $X^t f = \Sigma_1^n r_i X^{s_i + t} = \Sigma_1^m a_j X^{b_j}$, where $b_1, \ldots, b_m$ are distinct and where $b_i \not\sim b_j$ for $i \neq j$. Then $0 = \phi(X^t)\phi(f) = \phi(X^t f) = \Sigma_1^m a_j X^{[b_j]}$ and since $b_i \not\sim b_j$ for $i \neq j$, it follows that $a_1 = \ldots = a_m = 0$. Therefore $X^t f = 0$. In particular, $f = 0$ if $t = 0$.

Assume that $t \neq 0$. It is easy to see that the inverse of $1 + f$ is of the form $1 + g$ for some $g \in R[X;S^*]$. The contrapositve of Theorem 11.8 implies that each ideal of S meets $\langle s \rangle$ for $s \in S^*$. In particular, $t + S$ meets $\langle s \rangle$ for each $s \in \mathrm{Supp}(g)$. Choose k large enough so that $p^k s \in t + S$ for each $s \in \mathrm{Supp}(g)$. Then $1 = (1 + f)(1 + g)$ so $0 = f + g + fg = (f + g + fg)^{p^k} = f^{p^k} + g^{p^k} + f^{p^k} g^{p^k}$. By choice of k , X^t is a factor of g^{p^k} , and hence $f^{p^k} g^{p^k} = 0$. Consequently, $f^{p^k} = - g^{p^k}$ and

$f^{2p^k} = -f^{p^k}g^{p^k} = 0$. Hence $f = 0$, for Corollary 9.3 shows that the ring $R[X;S]$ is reduced.

THEOREM 11.10. Assume that R is a field of characteristic $p \neq 0$ and that $G = \{0\}$. If $1 + f$ is a unit of $R[X;S]$, where $f \in R[X;S - G]$, then $f = 0$.

<u>Proof</u>. Assume that Theorem 11.10 is false. We write $(1 + f)^{-1} = 1 + g$, where $g \in R[X;S - G]$. We may assume that $|\text{Supp}(f)| + |\text{Supp}(g)|$ is minimal among all counterexamples to the statement of the theorem. Replacing S by the submonoid of S generated by $U = \text{Supp}(f) \cup \text{Supp}(g)$, we also assume without loss of generality that the subsemigroup of S generated by U is $S^* = S \setminus \{0\}$. Since $S^* = S \setminus G$, it follows that S^* is a prime ideal of S . From Theorem 11.9, we conclude that there exists a prime ideal P of S properly contained in S^* .

As $S^* = (S^* \setminus P) \cup P$ and S^* is generated as a semigroup by U , it is clear that $U \not\subseteq P$ and $U \not\subseteq (S^* \setminus P)$. We write $1 + f = 1 + f_1 + f_2$, where $1 + f_1$ is the $(S \setminus P)$–component of $1 + f$ and f_2 is the P–component of $1 + f$. Similarly, $1 + g = 1 + g_1 + g_2$, where $1 + g_1$ is the $(S \setminus P)$–component of $1 + g$. It follows that $1 = (1 + f)(1 + g) = (1 + f_1)(1 + g_1)$ and $(1 + f)g_2 + f_2(1 + g_1) = 0$. Therefore $1 + f_1$ is a unit of the monoid ring $R[X;S \setminus P]$ with inverse $1 + g_1$. Since $U \not\subseteq S^* \setminus P$, either f_2 or g_2 is nonzero, and $|\text{Supp } f_1| + |\text{Supp } g_1| < |\text{Supp}(f)| + |\text{Supp}(g)|$. Moreover, as $U \not\subseteq P$, either f_1 or g_1 is nonzero. Thus $1 + f_1$ and $1 + g_1$ provide a counterexample to Theorem 11.10 that contradicts the assumption that $|\mathrm{Supp}(f)| + |\mathrm{Supp}(g)|$ is minimal. This completes the proof of Theorem 11.10.

THEOREM 11.11. If R is a field of characteristic $p \neq 0$, then each unit of $R[X;S]$ is in $R[X;G]$.

Proof. Let u be a unit of $R[X;S]$ with inverse v and write $u = u_1 + u_2$, $v = v_1 + v_2$, where u_1 and v_1 are the G—components of u and v, respectively. Then u_1 is a unit of $R[X;G]$ with inverse v_1, and hence $v_1u = 1 + v_1u_2$ is a unit of $R[X;S]$ with inverse $u_1v = 1 + u_1v_2$. As v_1u_2 and u_1v_2 are in $R[X;S\backslash G]$, it follows that $1 + v_1u_2$ is, in fact, a unit of the monoid ring $R[X;(S\backslash G) \cup \{0\}]$. Applying Theorem 11.10, we conclude that $0 = v_1u_2$, and hence $u_2 = 0$. Therefore $u = u_1 \in R[X;G]$ as was to be proved.

THEOREM 11.12. If R is a reduced ring, then each unit of $R[X;S]$ is in $R[X;G]$.

Proof. Let $f = \Sigma_1^n f_i X^{s_i}$ be a unit of $R[X;S]$ with inverse $g = \Sigma_1^m g_i X^{t_i}$. Let Π be the prime subring of R and let $T = \Pi[\{f_i\},\{g_i\}]$. Then T is a Hilbert FMR—ring. Let $\{M_\lambda\}_{\lambda\in\Lambda}$ be the family of maximal ideals of T and for each $\lambda \in \Lambda$, let ϕ_λ be the canonical homomorphism of $T[X;S]$ onto $(T/M_\lambda)[X;S]$. By Theorem 11.11, $\phi_\lambda(f) \in (T/M_\lambda)[X;G]$ for each λ. Thus, if h is the $(S\backslash G)$—component of f, then $h \in \cap_\lambda (M_\lambda[X;S\backslash G]) = N[X;S\backslash G]$, where N is the nilradical of T. Because R is reduced, $N = (0)$ and $f \in R[X;G]$ as asserted.

Theorem 11.12 is the last preliminary result needed for

the proof of Theorem 11.13. In stating (11.13), we repeat hypotheses on R and S that have sometimes been assumed implicitly in the statements of (11.9)–(11.12).

THEOREM 11.13. Assume that R is a unitary ring and that S is a torsion-free aperiodic monoid. Let N be the nilradical of R and let G be the group of invertible elements of S. An element $f \in R[X;S]$ is a unit if and only if the G-component of f is a unit of $R[X;G]$ and the $S\backslash G$-component of f is in $N[X;S\backslash G]$.

Proof. The only part of the statement of Theorem 11.13 that requires proof is the assertion that the $(S\backslash G)$-component of a unit f is in $N[X;S\backslash G]$. The image $\phi(f)$ of f under the canonical map ϕ onto $(R/N)[X;S]$ is in $(R/N)[X;G]$ by Theorem 11.12. Let $h \in R[X;G]$ be a preimage of $\phi(f)$. Then $f - h = g \in N[X;S]$, so the $(S\backslash G)$-component of g is in $N[X;S\backslash G]$. Since $f = h + g$, it follows that f and g have the same $(S\backslash G)$-component. Therefore the $(S\backslash G)$-component of f is in $N[X;S\backslash G]$, as asserted.

Corollary 11.5 and Theorem 11.6 extend easily to the case where S is torsion-free and aperiodic. We state these extensions without proof.

THEOREM 11.14. If R is unitary and $S \neq \{0\}$ is a torsion-free aperiodic monoid, then the Jacobson radical of $R[X;S]$ is $N[X;S]$, where N is the nilradical of R. Thus, the nilradical and the Jacobson radical of $R[X;S]$ coincide.

THEOREM 11.15. Let the hypothesis on R and S be as

in Theorem 11.14. The monoid ring $R[X;S]$ has only trivial units if and only if R is reduced and either (1) or (2) is satisfied.

(1) R is reduced.

(2) $G = \{0\}$.

We return briefly to the considerations in Theorem 11.13 to develop a notion that will be used in Section 22 in examining the R–automorphisms of $R[X;S]$, and in Section 24 in considering the relationship between rings R_1 and R_2 if the monoid rings $R_1[X;S]$ and $R_2[X;S]$, for a fixed S , are isomorphic. Let R, S, and G be as in the statement of Theorem 11.13. If $1 = \Sigma_1^n e_i$ is a decomposition of 1 into orthogonal idempotents e_i and if u is a unit of $R[X;S]$, then we say that the set $E = \{e_i\}_1^n$ <u>splits</u> u if there exists a unit $v \in R$, elements $s_1,\ldots,s_n \in G$ (not necessarily distinct), and a nilpotent element w such that $u = v(\Sigma_1^n e_i X^{s_i}) + w$. The value of this concept is the following: if E splits u , then in the decomposition $u = \Sigma_1^n u_i$ of u with respect to the decomposition $R[X;S] = \Sigma_1^n \oplus Re_i[X;S]$ of $R[X;S]$, each u_i has the form

$$(\text{trivial unit of } Re_i[X;S]) + (\text{nilpotent element of } Re_i[X;S]) .$$

A unit of this form is frequently easier to work with than a general unit. In Theorems 11.16 and 11.17 and in Corollary 11.18, we continue to use the hypothesis of Theorem 11.13 on R, S, and G ; thus R is unitary, S is a torsion–free aperiodic monoid, and G is the group of invertible elements of S .

THEOREM 11.16. Assume that $\{u_i\}_{i=1}^k$ is a finite set of units of $R[X;S]$. There exists a decomposition E of the identity of R that splits each u_i.

Proof. We use induction on k. For $k = 1$, take a unit $u = \Sigma_{i=1}^n g_i X^{s_i} + h$ in $R[X;S]$, where each s_i is in G, h is nilpotent, and no g_i is nilpotent. By Theorem 11.3, there exists a decomposition $1 = e_1 + \ldots + e_n$ of 1 into orthogonal idempotents e_i such that $Re_i = Rg_i^k$ for each i, where $k \in Z^+$. Moreover, if $g_i = \Sigma_{j=1}^n g_{ij}$ is the decomposition of g_i with respect to the decomposition $R = \Sigma_1^n \oplus Re_i$ of R, then the proof of Theorem 11.2 shows that g_{ii} is a unit of Re_i, while g_{ij} is nilpotent for $i \neq j$. Therefore $u = \Sigma_{i=1}^n g_{ii} X^{s_i} + h^*$, where h^* is nilpotent. Let $v = \Sigma_1^n g_{ii}$. Then v is a unit of R and $v(\Sigma_1^n e_i X^{s_i}) + h^* = \Sigma_1^n v e_i X^{s_i} + h^* = \Sigma_1^n g_{ii} X^{s_i} + h^* = u$. Consequently, $E = \{e_i\}_1^n$ splits u.

For $k = 2$, it is easy to show that if $E = \{e_i\}_1^n$ splits u_1 and $F = \{f_j\}_1^m$ splits u_2, then $B = \{e_i f_j \mid 1 \leq i \leq n, 1 \leq j \leq m\}$ splits both u_1 and u_2. It follows by the same reasoning that if E splits $u_1, \ldots, u_r$ and F splits u_{r+1}, then the set of all pairwise products ef, for $e \in E$ and $f \in F$, splits each of $u_1, \ldots, u_{r+1}$.

THEOREM 11.17. Let $E = \{e_i\}_1^n$ be a decomposition of the identity of R. The set of units $u \in R[X;S]$ that are split by E is a subgroup of the multiplicative group of $R[X;S]$.

Proof. First, $1 = 1(\Sigma e_i X^0) + 0$ is split by E. If

$u = v(\Sigma e_i X^{s_i}) + w$ and $u^* = v^*(\Sigma e_i X^{t_i}) + w^*$ are split by E, then $uu^* = vv^*(\Sigma e_i X^{s_i+t_i}) + wu^* + w^*v(\Sigma e_i X^{s_i})$, and hence uu^* is split by E. Let u be as in the preceding sentence. Let N be the nilradical of R and let ϕ be the canonical homomorphism of $R[X;S]$ onto $(R/N)[X;S]$. Then

$$\phi(u^{-1}) = [\phi(u)]^{-1} = [\phi(v)\Sigma_1^n \phi(e_i)X^{s_i}]^{-1} = \phi(v^{-1})\Sigma\phi(e_i)X^{-s_i} =$$

$$\phi(v^{-1}\Sigma e_i X^{-s_i}) .$$

Therefore $u^{-1} - v^{-1}(\Sigma e_i X^{-s_i}) \in N[X;S]$ and E splits u^{-1}.

COROLLARY 11.18. If H is a finitely generated subgroup of the multiplicative group of units of $R[X;S]$, then there exists a decomposition of 1 that splits each element of H.

Section 11 Remarks

Most of the material in this section on units of $R[X;S]$, in the case where S is torsion-free and aperiodic, comes from [55]. Units of $R[X;S]$ of finite order are also determined in [55].

Since our treatment of units is modest in terms of what is known, it seems appropriate that we give a few references to additional material in the literature. Again the books of Passman [121] and Sehgal [125] are general references on units of (not necessarily commutative) group rings. Papers of Higman [76], Cohn and Livingstone [34], Ayoub and Ayoub [16], and May [102] deal with the group of units of commutative group rings, with particular emphasis in the first two papers on determining conditions under which a group ring $R[G]$ has only trivial units. Higman's classical result for an abelian torsion group G is that the integral group ring $Z[G]$ has only trivial units if and only if the exponent of G divides either 4 or 6. Ayoub and Ayoub determine the structure of the group of units of $Z[G]$ for a finite group G, as well as the structure of the group ring $Q[G]$ over the field Q of rational numbers.

Our restricted treatment of units of $R[X;S]$ in this section has some consequences in Section 25, where the question of whether isomorphism of $R[X;S_1]$ and $R[X;S_2]$ implies isomorphism of S_1 and S_2 is considered. Partly because some needed results concerning units are not available in the text, Section 25 is merely a survey of some of the known results on the isomorphism problem; no proofs are presented in Section 25.

For a periodic monoid S and a unitary ring R, Gilmer and Teply in [60] determine (1) those monoid rings $R[S]$ for which each unit of $R[S]$ is in $R[G]$, where G is the group of invertible elements of S, and (2) the monoid rings $R[S]$ with only trivial units.

CHAPTER III

RING-THEORETIC PROPERTIES OF MONOID DOMAINS

For a number of ring-theoretic properties E, it is natural to investigate conditions on R and S under which the semigroup ring $R[X;S]$ has property E. For example, Theorem 7.7 shows that if R is unitary and S is a monoid, then $R[X;S]$ is Noetherian if and only if R is Noetherian and S is finitely generated; Theorem 9.17 gives equivalent conditions under which an arbitrary semigroup ring is reduced. In Chapters III and IV we consider several other ring-theoretic properties E. The conditions considered in Chapter III are of interest primarily for integral domains — for example, integral closure, the factorial property, and the conditions that a domain should be a Prüfer domain or a Krull domain. In Chapter IV we consider properties that are also of interest in the case of rings with zero divisors.

In contrast with Chapter II, restriction to the case of unitary rings R and monoids S is more common in the rest of the text. In particular, in Chapter III R is usually assumed to be a unitary integral domain and S a torsion-free cancellative monoid, so that the monoid ring $R[X;S]$ is a unitary integral domain. There is also an abbreviation of the Chapter II notation $R[X;S]$ in the remainder of the text; we shorten it to $R[S]$, for the primary stress will shift from elements and their canonical representations to concepts more global and ring-theoretic in nature. If T is a unitary extension ring of the ring R and if E is a subset of T (not necessarily closed under $+$ or $\cdot$), then the notation $R[E]$ is sometimes used in commutative algebra to denote the subring of T generated by $R \cup E$. There is a consistency between this notation and the notation $R[S]$ for the monoid ring of S over R, for $R[X;S]$ is generated as a ring by $R \cup \{X^s | s \in S\}$.

§12. Integral Dependence for Monoid Rings

In the remaining sections of this chapter, we determine equivalent conditions in order that a monoid domain should be factorial, a Prüfer domain, or a Krull domain. A common feature of the domains of each of these classes is that they are integrally closed. Hence, this section is devoted to determining the integral closure or complete integral closure of one monoid ring in another, and to determining conditions under which a monoid domain is integrally closed or completely integrally closed. The nature of the results is such that we need not restrict to a consideration of integral domains only, although that condition is imposed throughout most of the other sections of this chapter. We do assume, however, that all rings and semigroups considered in the section have identity elements. And to repeat, the customary notation $R[X;S]$ of Chapter II is abbreviated to $R[S]$ here; we continue to write elements of $R[S]$ in the form $\Sigma_{i=1}^{n} r_i X^{s_i}$.

The first two results of the section are related to questions of integral dependence, but they have much wider application than that. In fact, various parts of Theorem 12.2 are cited throughout Chapters III–V. These results are used primarily to transfer various properties from one monoid ring to a subring or to an extension ring. If R is a subring of the ring T , we say that R is an <u>algebraic retract</u> of T if there exists a homomorphism $T \longrightarrow R$ that is the identity mapping on the ring R . It is straightforward to show that R is an algebraic retract of T if and only if there exists an ideal B of T such that $T = R + B$ and $R \cap B = (0)$.

THEOREM 12.1. Assume that R is a unitary ring and S is a monoid.

(1) R is an algebraic retract of $R[S]$.

(2) $R[S\backslash P]$ is an algebraic retract of $R[S]$ for each proper prime ideal P of S .

(3) Assume that H is a subgroup of S containing 0 and let $\{s_\alpha\}_{\alpha\in A}$ be a complete set of representatives for the cosets of H in S . As a module over $R[H]$, $R[S]$ is the direct sum of the modules $R[H]X^{s_\alpha}$; if S is cancellative, then $\{X^{s_\alpha}\}_{\alpha\in A}$ is free. In any case, $R[H]$ is a direct summand of $R[S]$.

<u>Proof</u>. The augmentation mapping on $R[S]$ defines R as an algebraic retract of $R[S]$. Moreover, the decomposition $R[S] = R[S\backslash P] + R[P]$ of $R[S]$ shows that $R[S\backslash P]$ is an algebraic retract of $R[S]$.

To prove (3) , take $s \in S$. Then $s = s_\alpha + h$ for some $\alpha \in A$, $h \in H$. Hence $X^s = X^h X^{s_\alpha} \in R[H]X^{s_\alpha}$ and $R[S] = \Sigma_\alpha R[H]X^{s_\alpha}$. To show that the sum is direct, assume that $\Sigma_1^n f_i X^{s_{\alpha_i}} = 0$, where $f_i \in R[H]$ for each i . By choice of $\{s_\alpha\}$, $\mathrm{Supp}(f_i X^{s_{\alpha_i}}) \cap \mathrm{Supp}(f_j X^{s_{\alpha_j}}) = \phi$ for $i \neq j$. Hence $f_i X^{s_{\alpha_i}} = 0$ for each i and the sum $\Sigma_\alpha R[H]X^{s_\alpha}$ is direct. If S is cancellative, then X^{s_α} is regular in $R[S]$, so $f_i = 0$ for each i . Thus $\{X^{s_\alpha}\}$ is a free basis for $R[S]$ over $R[H]$ if S is cancellative. If s_{α_0} is the representative of the coset H in S , then $R[H]X^{s_{\alpha_0}} = R[H]$, so $R[H]$ is a direct summand of $R[S]$.

THEOREM 12.2. Assume that R is a unitary ring and

that T is a unitary extension ring of R . Consider the following conditions.

(1) R is an algebraic retract of T .

(2) R is a direct summand of T as an R–module.

(3) Each ideal of R is contracted from T .

(4) T meets the total quotient ring of R in R .

(5) If T satisfies the ascending chain condition on principal ideals, then so does R .

(6) If T is a GCD–domain, then so is R .

(7) If T is a factorial domain, then so is R .

The following diagram indicates implications that exist among these seven conditions.

$$\begin{array}{ccccccc} (6) & & & & & & \\ \uparrow & & & & & & \\ (1) & \longrightarrow & (2) & \longrightarrow & (3) & \longrightarrow & (4) \\ \downarrow & & & & \downarrow & & \\ (7) & & & & (5) & & \end{array}$$

<u>Proof</u>. If R is an algebraic retract of T , then there exists an ideal B of T such that $T = R + B$, where $B \cap R = (0)$. In particular, B is an R–submodule of T and (1) implies (2) . To see that (2) implies (3) , let $T = R \oplus W$, where W is an R–submodule of T . If B is an ideal of R , then $BT = BR \oplus BW = B \oplus BW$, and hence $BT \cap R = B$.

(3) $\Longrightarrow$ (4) and (5) : Assume that each ideal of R is contracted from T . Let U be the total quotient ring of R and let $t = a/b \in T \cap U$, where $a,b \in R$ and b is regular in R . Then $a \in bT \cap R = bR$, and since b is regular in U , it follows that $t = a/b \in R$. Thus (4) is

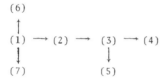

satisfied. Moreover, if $\{a_i R\}_1^\infty$ is an ascending sequence of principal ideals of R, then $\{a_i T\}$ is an ascending sequence in T. If T satisfies a.c.c.p., then $a_k T = a_{k+1} T = \ldots$ for for some k, and upon contracting to R we have $a_k R = a_{k+1} R = \ldots$.

(1) $\Longrightarrow$ (6): Assume that $\phi : T \longrightarrow R$ defines R as an algebraic retract of T and assume that T is a GCD–domain. Take $a, b \in R$ and let $cT = at \cap bT$. We show that $\phi(c)R = aR \cap bR$. We have $\phi(c)R = \phi(cT) = \phi(aT \cap bT) \subseteq \phi(aT) \cap \phi(bT) = aR \cap bR = \phi(aR \cap bR) \subseteq \phi(aT \cap bT)$. Therefore $\phi(c)R = aR \cap bR$, and R is also a GCD–domain.

(1) $\Longrightarrow$ (7): Factoriality of an integral domain J is defined solely in terms of the multiplicative semigroup of nonzero elements of J, and hence Theorem 6.7 implies that J is factorial if and only if J is a GCD–domain and a.c.c.p. is satisfied in J. Thus (1) implies (7) since it implies both (5) and (6).

Theorems 12.1 and 12.2 cover the applications we need, save one. That application concerns the case where S is a cancellative monoid and H is a subgroup of S containing 0. Theorem 12.3 places the result needed in a general context; we note that the extension $R[H] \subseteq R[S]$ satisfies the hypothesis of (12.3).

THEOREM 12.3. Assume that T is a unitary extension of the unitary ring R and that as an R–module, T has a free basis $\{1\} \cup \{t_\alpha\}_{\alpha \in A}$, where each t_α is regular in T. If $\{B_\lambda\}_{\lambda \in \Lambda}$ is a family of ideals of R, then $(\cap B_\lambda)T = \cap(B_\lambda T)$.

<u>Proof</u>. We have $B_\lambda T = B_\lambda \oplus (\Sigma_\alpha \oplus B_\lambda t_\alpha)$ for each λ, and hence

$$\cap_\lambda(B_\lambda T) = (\cap_\lambda B_\lambda) \oplus \Sigma_\alpha(\cap_\lambda B_\lambda t_\alpha) .$$

Because each t_α is regular in T, $(\cap_\lambda B_\lambda t_\alpha) = (\cap_\lambda B_\lambda)t_\alpha$ for each α. Consequently,

$$\cap_\lambda(B_\lambda T) = (\cap_\lambda B_\lambda) \oplus \Sigma_\alpha(\cap B_\lambda)t_\alpha = (\cap_\lambda B_\lambda)T .$$

Let T be a monoid and let S be a submonoid of T containing 0. An element $t \in T$ is said to be <u>integral</u> over S if $nt \in S$ for some $n \in Z^+$. The set S_0 of elements $t \in T$ that are integral over S is a submonoid of T containing S; S_0 is called the <u>integral closure of</u> S <u>in</u> T. If $S = S_0$, we say S is <u>integrally closed</u> in T. It is easy to see that S_0 is integrally closed in T. In case S is cancellative and T is the quotient group of S, S_0 is called the <u>integral closure of</u> S and S is said to be <u>integrally closed</u> if $S = S_0$. In defining the concepts corresponding to integrality for almost integrality, we restrict to the case where S is cancellative and T is a submonoid of the quotient group of S. In this case, an element $t \in T$ is <u>almost integral</u> over S if there exists $s \in S$ such that $s + nt \in S$ for each $n \in Z^+$. The set S^* of all $t \in T$ that are almost integral over S is called the <u>complete integral closure of</u> S <u>in</u> T; it is a submonoid of T containing S, and S is <u>completely integrally closed in</u> T if $S = S^*$. In general, S^* need not be completely integrally closed in T. If T is the quotient group of S, then S^* is called the <u>complete integral closure</u> of S, and S is <u>completely integrally closed</u> if $S = S^*$. If $t \in T$ is integral over S, then t is almost integral over S. The next result establishes the converse under appropriate hypotheses on S and T .

THEOREM 12.4. Let T be a cancellative monoid, let S be a submonoid of T , and let R be a unitary ring.

(1) $X^t \in R[T]$ is integral over $R[S]$ if and only if t is integral over S .

(2) Assume that T is a submonoid of the quotient group of S . If $t \in T$ is almost integral over S and if S satisfies a.c.c. for ideals, then t is integral over S .

(3) If R is an integral domain and T is a submonoid of the quotient group of S , then considered as an element of q.f.$(R[S])$, an element X^t is almost integral over $R[S]$ if and only if t is almost integral over S .

<u>Proof</u>. (1): We show that t is integral over S if X^t is integral over $R[S]$; the converse is clear. Thus, let $X^{nt} + h_{n-1}X^{(n-1)t}+\dots+h_0 = 0$, where each $h_i \in R[S]$. It follows that $nt \in \mathrm{Supp}(h_i X^{it})$ for some $i < n$. Therefore $nt = s + it$ for some $s \in S$ and $(n - i)t = s$ since T is cancellative. This proves that t is integral over S .

(2): Let $s \in S$ be such that $s + nt \in S$ for each $n \in Z^+$. The ideal $\cup_{n=1}^{\infty}[(s + nt) + S]$ of S is generated by $\{s + nt\}_{n=1}^{m}$ for some m . Hence $s + (m + 1)t \in s + nt + S$ for some $n < m + 1$. It follows that $(m + 1 - n)t \in S$, and therefore t is integral over S .

(3): The assertion that X^t is almost integral over $R[S]$ if t is almost integral over R is clear. We prove the converse. Thus, let $f \in R[S]\backslash(0)$ be such that

$fX^{nt} \in R[S]$ for each $n \in Z^+$. If $s \in \mathrm{Supp}(f)$, it follows that $s + nt \in S$ for all n, and therefore t is almost integral over S.

THEOREM 12.5. Assume that T is a unitary integral domain and R is a subring of T containing the identity element of T. Let U be a torsion-free cancellative monoid and let S be a submonoid of U. Assume that $g \in T[U]\backslash(0)$ has canonical form $g = \Sigma_{i=1}^m g_i X^{u_i}$. Then g belongs to the complete integral closure of $R[S]$ if and only if each g_i belongs to the complete integral closure of R and each u_i belongs to the complete integral closure of S.

Proof. Denote by $(R[S])^*$, R^*, and S^* the complete integral closures of $R[S]$, R, and S. Assume first that $g_i \in R^*$ and $u_i \in S^*$ for each i. If $r_i \in R\backslash(0)$ and $s_i \in S$ are chosen so that $r_i g_i^k \in R$ and $s_i + ku_i \in S$ for each $k \in Z^+$, then let $r = r_1 r_2 \ldots r_m$ and let $s = s_1 + \ldots + s_m$. Clearly $rX^s \cdot g^k = rX^s(\Sigma_1^m g_i X^{u_i})^k \in R[S]$ for each $k \in Z^+$, and hence $g \in (R[S])^*$.

Conversely, if $g \in (R[S])^*$, let $f \in R[S]\backslash\{0\}$ be such that $fg^n \in R[S]$ for each $n \in Z^+$. Let $\leq$ be a total order on U compatible with the semigroup operation, and write $f = \Sigma_{i=1}^n f_i X^{s_i}$, where $f_1 \neq 0$, $s_1 < s_2 < \ldots < s_n$, and, without loss of generality, $u_1 < u_2 < \ldots < u_m$. The initial term of fg^k is $f_1 X^{s_1} \cdot (g_1 X^{u_1})^k = f_1 g_1^k X^{s_1 + ku_1}$ for each $k \in Z^+$. Therefore $f_1 g_1^k \in R$ and $s_1 + ku_1 \in S$ for all k. This implies that $g_1 \in R^*$ and $u_1 \in S^*$. Hence $g - g_1 X^{u_1}$ belongs to $(R[S])^* \cap T[U]$. By induction, it follows that

$g_i \in R^*$ and $u_i \in S^*$ for each i .

COROLLARY 12.6. Let D be a unitary integral domain and let G be a torsion-free group.

(1) $D[G]$ is integrally closed if and only if D is integrally closed.

(2) $D[G]$ is completely integrally closed if and only if D is completely integrally closed.

Proof. Let K be the quotient field of D . Since $D[G] \cap K = D$, it follows that D is integrally closed (completely integrally closed) if $D[G]$ has the same property. Assume that D is integrally closed and that $y \in q.f.(D[G])$ is integral over $D[G]$. There exists a finitely generated subgroup H of G such that $y \in q.f.(D[H])$ and y is integral over $D[H]$. Since H is a finitely generated torsion-free group, $H \simeq Z^n$ for some n , and $D[H] \simeq D[X_1,\ldots,X_n][X_1^{-1},\ldots X_n^{-1}]$, a quotient ring of the polynomial ring $D[X_1,\ldots,X_n]$. Since the class of integrally closed domains is closed under polynomial ring formation and under quotient ring formation, it follows that $D[H]$ is integrally closed. Therefore $y \in D[H] \subseteq D[G]$, and $D[G]$ is integrally closed, as asserted.

To prove (2) , we note that $D[G]$ is completely integrally closed in $K[G]$ by Theorem 12.5. Hence, it suffices to prove that $K[G]$ is completely integrally closed. If $\{H_\alpha\}$ is the family of finitely generated subgroups of G , then $K[G]$ is the union of the directed set $\{K[H_\alpha]\}$ of subrings of $K[G]$. Each $K[H_\alpha]$ is Noetherian and integrally closed by (1) , hence completely integrally closed. It follows that $K[G]$ is also completely integrally closed, and this completes the proof of Corollary 12.6.

COROLLARY 12.7. Let D be a unitary integral domain and let S be a torsion-free cancellative monoid with quotient group G. The following conditions are equivalent.

(1) $D[S]$ is completely integrally closed.

(2) D and S are completely integrally closed.

Proof. Theorem 12.5 shows that (1) implies (2). Conversely, if D and S are completely integrally closed, then $D[S]$ is completely integrally closed in $D[G]$, which is completely integrally closed by part (2) of Corollary 12.6. Therefore $D[S]$ is completely integrally closed if (2) is satisfied.

In view of Corollary 12.7, it is natural to conjecture that $D[S]$ is integrally closed if and only if D and S are integrally closed. This conjecture is established in Corollary 12.11, but some preliminary work is required for the proof of (12.11). Basically, the link of the proof that is missing at this point is the result that $D[X;S]$ is integrally closed in $D[X;G]$ if S is integrally closed in G. For this, we prove an analogue of Theorem 12.5; a couple of new concepts are needed.

Let G be a group. By a valuation on G, we mean a homomorphism v of G into a totally ordered group H. We say that v is trivial if $v(G) = \{0\}$; otherwise, v is nontrivial. The set $G_v = \{x \in G \mid v(x) \geq 0\}$ is called the value monoid of v; it is a submonoid of G with quotient group G and it is easy to see that G_v is integrally closed. The next result is the analogue of a well known result for integral domains.

THEOREM 12.8. Let G be a group and let S be a submonoid of G. Let $\{v_\alpha\}_{\alpha \in A}$ be the set of valuations on G such that $G_{v_\alpha} \supseteq S$. Then $\bigcap_\alpha G_{v_\alpha}$ is the integral closure of S in G.

Proof. Let T be the integral closure of S in G. If $t \in T$, then t is integral over each G_{v_α}, and hence $t \in \bigcap_\alpha G_{v_\alpha}$ since each G_{v_α} is integrally closed in G. This proves that $T \subseteq \bigcap_\alpha G_{v_\alpha}$. To prove the reverse inclusion, take $x \in G \setminus T$. Then x is not integral over T, and our aim is to show that $x \notin G_{v_\alpha}$ for some α. Let H be the group of invertible elements of T. We wish to prove three assertions concerning G/H, as follows: (1) G/H is torsion-free, (2) T/H is a positive-subset of G/H, and (3) $x + H$ is not integral over T/H. Thus, if $g + H$ is a torsion element of G/H, then g and $-g$ are integral over T, and hence $g, -g \in T$. Therefore $g, -g \in H$ and G/H is torsion-free, as asserted. Statement (2) follows from the definition of H, and (3) is true because x is not integral over T. Corollary 3.5 implies that there exists a total order $\leq_\alpha$ on G/H such that $0 \leq_\alpha t + H$ for each $t \in T$ and $x + H <_\alpha 0$. Let v_α be the canonical map of G onto G/H, ordered under $\leq_\alpha$. Then $G_{v_\alpha} \supseteq T$, and $x \notin G_{v_\alpha}$ by choice of the ordering $\leq_\alpha$. Therefore $\bigcap_\alpha G_{v_\alpha} \subseteq T$.

THEOREM 12.9. Let v be a valuation on the torsion-free group G and let V be the value monoid of v. For any reduced unitary ring R, the monoid ring $R[V]$ is integrally closed in $R[G]$.

Proof. Assume, to the contrary, that there exists an element f of $R[G]\setminus R[V]$ that is integral over $R[V]$. Without loss of generality, we assume that $v(s) < 0$ for each $s \in \mathrm{Supp}(f)$. If $H = \ker(v)$, then f can be written in the form $\Sigma_{i=1}^{m} X^{s_i} h_i$, where $v(s_1) < \ldots < v(s_m) < 0$ and where each h_i is a nonzero element of $R[H]$. We assume that

$$f_n + a_{n-1} f^{n-1} + \ldots + a_1 f + a_0 = 0 ,$$

where $a_i \in R[V]$ for each i . Since G is torsion-free, $R[G]$ is reduced by Theorem 9.9. Therefore $X^{ns_1} h_1^n \neq 0$. Each element of $\mathrm{Supp}(X^{ns_1} h_1^n)$ has v-value $nv(s_1)$. To obtain a contradiction, we show that $v(s) > nv(s_1)$ for $s \in \mathrm{Supp}(g)$, where

$$g = (f^n - X^{ns_1} h_1^n) + \Sigma_{i=0}^{n-1} a_i f^i .$$

Each such s must belong to $\mathrm{Supp}(f^n - X^{ns_1} h_1^n)$ or to $\mathrm{Supp}(a_i f^i)$ for some $i < n$. Now $s \in \mathrm{Supp}(f^n - X^{ns_1} h_1^n)$ implies s is of the form $(\Sigma_{j=1}^{m} k_j s_j) + h$, where $k_j \geq 0$ for each j , $\Sigma_1^m k_j = n$, $k_1 < n$, and $h \in H$. Applying v , we obtain $v(s) = \Sigma_1^m k_j v(s_j) > \Sigma_1^m k_j v(s_1) = (\Sigma_1^m k_j) v(s_1) = nv(s_1)$. On the other hand, $s \in \mathrm{Supp}(a_i f^i)$ implies $s = w + c$, where $w \in \mathrm{Supp}(a_i)$ and $c \in \mathrm{Supp}(f^i)$. Thus $v(w) \geq 0$, and the argument just given shows that $v(c) \geq iv(s_1)$. Consequently, $v(s) = v(w) + v(c) \geq iv(s_1) > nv(s_1)$. We have proved that $nv(s_1) \in \mathrm{Supp}(f^n + \Sigma_0^{n-1} a_i f^{n-1})$, and hence $f^n + \Sigma_0^{n-1} a_i f^{n-1} \neq 0$. This completes the proof.

The analogues of Theorem 12.5 and Corollary 12.7 follow easily from the two preceding results.

THEOREM 12.10. Assume that T is a unitary domain and R is a subring of T containing the identity element of T. Let U be a torsion-free cancellative monoid and let S be a submonoid of U.

(1) $R[S]$ is integrally closed in $T[U]$ if and only if R is integrally closed in T and S is integrally closed in U.

(2) The integral closure of $R[S]$ in $T[U]$ is $R'[S']$, where R' is the integral closure of R in T and S' is the integral closure of S in U.

Proof. (1): If $R[S]$ is integrally closed in $T[U]$, it is clear that $R = R'$ and $S = S'$. To prove the converse, assume first that U is a group. By Theorem 12.8, $S' = \bigcap_\alpha U_{v_\alpha}$ is an intersection of valuation monoids on U. Since $T[U_{v_\alpha}]$ is integrally closed in $T[U]$ for each α by Theorem 12.9, it follows that the integral closure of $R[S]$ in $T[U]$ is contained in $\bigcap_\alpha T[U_{v_\alpha}] = T[\cap_\alpha U_{v_\alpha}] = T[U']$. To complete the proof in the case where U is a group, it therefore suffices to prove that $R'[U']$ is integrally closed in $T[U']$. Simplifying the notation, this amounts to showing that $R[S]$ is integrally closed in $T[S]$ if R is integrally closed in T. To prove this, take an element $f \in T[S]$ integral over $R[S]$. If G is the quotient group of S, then there exists a finitely generated subgroup H of G such that $f \in T[H]$ and f is integral over $R[H]$. As in the proof of Corollary 12.6, however, $H \simeq Z^n$ for some

n . It is known that R integrally closed in T implies that $R[X_1, \ldots X_n]$ is integrally closed in $T[X_1, \ldots X_n]$, and hence $R[H] \simeq R[Z^n] \simeq R[X_1, \ldots, X_n]_B$ is integrally closed in $T[X_1, \ldots, X_n]_B \simeq T[Z^n] \simeq T[H]$, where B is the multiplicative system generated by $\{X_i\}_{i=1}^n$. Therefore $f \in R[H]$, each coefficient of f is in R , and $f \in R[S]$. This completes the proof of (1) in the case where U is a group. If U is not a group, we let W be the quotient group of U . The case just considered implies that the integral closure of $R[S]$ in $T[W]$ is $R'[M]$, where M is the integral closure of S in W . Therefore, the integral closure of $R[S]$ in $T[U]$ is $R'[M] \cap T[U] = R'[M \cap U] = R'[S']$.

Statement (2) of (12.10) follows easily from (1) .

COROLLARY 12.11. Let D be a unitary integral domain and let S be a torsion-free cancellative monoid.

(1) The integral closure of $D[S]$ is $D'[S']$, where D' is the integral closure of D and S' is the integral closure of S .

(2) $D[S]$ is integrally closed if and only if D and S are integrally closed.

Proof. We need only prove (1) , for (2) then follows. Part (2) of Theorem 12.10 shows that $D'[S']$ is the integral closure of $D[S]$ in $D'[G]$, where G is the quotient group of S . Since $D'[G]$ is an overring of D and since $D'[G]$ is integrally closed by Corollary 12.6, we conclude that $D'[S']$ is the integral closure of $D[S]$.

Having settled the problems of determining conditions

under which a monoid domain $D[S]$ is integrally closed or completely integrally closed, we conclude this section with a description of the integral closure of R in $R[S]$.

THEOREM 12.12. Let R be a unitary ring with nilradical N and let S be a torsion-free cancellative monoid. Assume that $f = \Sigma_{s \in S} f_s X^s \in R[S]$. Then f is integral over R if and only if f_s is nilpotent for each $s \neq 0$. Thus, $R + N[S]$ is the integral closure of R in $R[S]$.

Proof. It is clear that each element of $R + N[S]$ is integral over R . For the converse, assume that f is integral over R . Then $g = f - f_0$ is also integral over R . Let

$$(*) \qquad g^n + r_{n-1}g^{n-1} + \ldots + r_1 g + r_0 = 0$$

be an equation of integral dependence for g over R . If P is a proper prime ideal of R , and if ϕ is the canonical homomorphism of $R[S]$ onto $(R/P)[S]$, then $(*)$ is transformed to an equation of integral dependence for $\phi(g)$ over $(R/P)[S]$. It follows that $\phi(g) = 0$, for if not, take a total order $\leq$ on S . Then by choice of g , either $\deg \phi(g) > 0$ or $\operatorname{ord} \phi(g) < 0$. If, for example, $\deg \phi(g) > 0$, then we conclude that $\deg \phi(g^n) > \deg(\phi(r_{n-1}g^{n-1} + \ldots + g_0))$, a contradiction. Therefore $\phi(g) = 0$ and $g \in P[S]$. Since P is arbitrary, it follows that $g \in N[S]$, and this completes the proof. We note in passing that $N[S]$ is the nilradical of $R[S]$ by Theorem 9.9.

Section 12 Remarks

For polynomial rings, Theorems 12.5 and 12.10 and Corollaries 12.6 and 12.11 are known (see Sections 10 and 13 of [51]). But for rings with zero divisors, equivalent conditions for the polynomial ring to be integrally closed or completely integrally closed are not known. If $R[X]$ is integrally closed, then R is integrally closed and reduced, but the converse fails. For more on this topic, see [3].

Power series rings over integral domains behave like monoid domains in regard to the properties of complete integral closure covered by Theorem 12.5 and Corollary 12.6. But for integral closure, the situation is quite different. For example, for fields F and K with K/F algebraic, $K[[X]]$ need not be integral over $F[[X]]$. Moreover, $D[[X]]$ need not be integrally closed if D is integrally closed. Of course, if D is Noetherian and integrally closed, then D, and hence $D[[X]]$, are completely integrally closed. Again Section 13 of [51] and its references provide additional information on this topic.

§13. Monoid Domains as Prüfer Domains

The closely related classes of Prüfer domains, Bezout domains, Dedekind domains, and principal ideal domains are of fundamental importance in commutative ring theory. In this section we determine necessary and sufficient conditions in order that a given monoid domain $D[S]$ should belong to one of the four classes named. Thus, we assume throughout Section 13 that D denotes a unitary integral domain and that S is a torsion-free cancellative monoid. We begin with the problem of determining conditions under which $D[S]$ is a Prüfer domain.

THEOREM 13.1. Assume that J is an extension domain of D and that each ideal of D is contracted from J. If J is a Prüfer domain, then D is a Prüfer domain.

Proof. It suffices to show that D is integrally closed and that $(a,b)^2 = (a^2,b^2)$ for all $a,b \in D$. Let K be the quotient field of D. Since each principal ideal of D is contracted from J, it follows that $J \cap K = D$, and hence D is integrally closed. Since J is a Prüfer domain, $(a,b)^2J = [(a,b)J]^2 = (a^2,b^2)J$. Contracting to D, we conclude that $(a,b)^2 = (a^2,b^2)$ as we wished to show.

COROLLARY 13.2. Assume that $D[S]$ is a Prüfer domain. If H is a subgroup of S containing 0, then $D[H]$ is a Prüfer domain. In particular, $D = D[\{0\}]$ is a Prüfer domain.

Proof. Theorem 12.2 shows that each ideal of $D[H]$ is contracted from $D[S]$.

THEOREM 13.3. If $S \neq 0$ and if $D[S]$ is a Prüfer domain, then either D is a field or S is a group.

Proof. Assume that S is not a group and choose $s \in S$ such that s is not invertible. If $d \in D\setminus\{0\}$, then $dX^s \in (d^2, X^{2s})$, say $dX^s = d^2f + X^{2s}g$. Since s is not invertible in S and since S is cancellative, it follows that $s \notin \mathrm{Supp}(X^{2s}g)$. Therefore $d = d^2f_i$ for some coefficient f_i of f, and consequently, d is a unit of D since $1 = df_i$. We conclude that D is a field if S is not a group.

Recall that a group G is divisible if the equation $nx = g$ is solvable in G for each $g \in G$ and each $n \in Z^+$. An arbitrary group H can be imbedded in a minimal divisible group H^* called the divisible hull of G. As is implied by the terminology, H^* is uniquely determined up to isomorphism. If H is torsion-free, then considering H as a module over Z, each nonzero element of Z is regular on H, and H^* is merely H_T, where T is the set of nonzero integers. Thus H^* is a vector space over Q, and as a group is isomorphic to a direct sum of copies of the additive group Q. The statement of Theorem 13.4 involves the divisible hull.

THEOREM 13.4. Assume that H is a nonzero torsion-free group with divisible hull H^*. The following conditions are equivalent.

(1) $D[H]$ is a Prüfer domain.

(2) $D[H^*]$ is a Prüfer domain.

(3) D is a field and $H^* \simeq Q$.

(4) $D[H]$ is a Bezout domain.

Proof. (1) <==> (2): Corollary 13.2 shows that (2) implies (1). For the converse, assume that $D[H]$ is a Prüfer domain. For $m \in Z^+$, let $H_m = \{g \in H^* | (m!)g \in H\}$. Then H_m is a subgroup of H^* isomorphic to H and $H = \cup_{m=1}^{\infty} H_m$, where $H_1 \subseteq H_2 \subseteq \ldots$. It follows that each $D[H_m]$ is a Prüfer domain, and hence $D[H] = \cup_1^{\infty} D[H_m]$ is also a Prüfer domain.

(2) ==> (3): We write H^* as $Q \oplus G$ for some subgroup G of H^*. Then $D[H^*] \simeq D[Q][G]$. By Corollary 13.2, $D[Q]$ is a Prüfer domain. Since (2) and (1) are equivalent, $D[Z]$ is also a Prüfer domain. To show that D is a field, take a nonzero element $d \in D$ and let $A = (d^2, (1 - X)^2)$ be the ideal of $D[X]$ generated by d^2 and $(1 - X)^2$. Since A and $XD[X]$ are comaximal, it follows that A is contracted from $D[X]_{\{X^i\}} = D[X, X^{-1}] = D[Z]$, We have $d(1 - X) \in AD[Z]$ since $D[Z]$ is a Prüfer domain. Therefore $d(1 - X) \in AD[Z] \cap D[X] = A$. Write $d(1 - X) = d^2 f + (1 - X)^2 g$, where $f, g \in D[X] = D[1 - X]$. Regarding f and g as polynomials in $1 - X$ with coefficients in D, it follows that $d \in d^2 D$, and this implies that d is a unit of D. Consequently, D is a field. Returning to H^*, we conclude that $G = \{0\}$, for if not, then $G = Q \oplus G_1$ for some G_1. Consequently $D[Q \oplus Q]$ is a Prüfer domain, and this is a contradiction to what was just proved — to wit, $D[Q]$ is not a field because $1 - X$ is a nonzero nonunit of $D[Q]$. Therefore $G = \{0\}$, $H \simeq Q$, and this completes the proof that (2) implies (3).

(3) $\Longrightarrow$ (4): If G_m is the subgroup of Q generated by $1/m!$ for each $m > 0$, then $G_1 \subseteq G_2 \subseteq \ldots$ and $Q = \cup_1^{\infty} G_m$. Therefore $D[Q] = \cup_1^{\infty} D[G_m]$ and $D[G_m] \simeq D[X, X^{-1}]$ is a quotient ring of $D[X]$, a principal ideal domain (PID) since D is a field. Therefore each $D[G_m]$ is a PID, and $D[Q]$ is a Bezout domain.

(4) implies (1) because a Bezout domain is a Prüfer domain.

If G is any abelian group — torsion-free or not — then G is a Z-module and we can consider the group G_T as a vector space over the field $Q = Z_T$, where $T = Z \setminus \{0\}$. The dimension of this vector space is called the torsion-free rank of G and is denoted by $r_0(G)$; $r_0(G)$ is characterized as the supremum of the cardinalities of free (that is, linearly independent) subsets of G. Thus $r_0(G) = 0$ if and only if G is a torsion group, and for a torsion-free group H, $r_0(H) = 1$ if and only if H is a subgroup of Q. For a cancellative semigroup M, we define the torsion-free rank of M to be $r_0(G)$, where G is the quotient group of M. We use $r_0(M)$ to denote the torsion-free rank of M. If $D[S]$ is a Prüfer domain and if G is the quotient group of S, then $D[G]$ is a quotient ring of $D[S]$, and hence $D[G]$ is also a Prüfer domain. By Theorem 13.4, it follows that D is a field and $r_0(S) = 1$ if $D[S]$ is a Prüfer domain. Moreover, if S is not a group, then Theorem 2.9 shows that either $S \subseteq Q_0$ or $-S \subseteq Q_0$, where Q_0 is the set of nonnegative rational numbers. Thus, if $S \neq \{0\}$, then S is isomorphic to a submonoid of Q_0 containing 1. The next result determines conditions under which the monoid ring of such an S, over a field F, is a Prüfer domain.

THEOREM 13.5. Assume that F is a field and that S is a submonoid of Q_0 containing 1. Let G be the quotient group of S. The following conditions are equivalent.

(1) S is integrally closed.

(2) $G \cap Q_0 = S$.

(3) S is the union of an ascending sequence of cyclic monoids.

(4) $F[S]$ is a Bezout domain.

(5) $F[S]$ is a Prüfer domain.

Proof. (1) $\Longrightarrow$ (2): Each element of Q_0 is integral over $Z_0 \subseteq S$. Hence the elements of $G \cap Q_0$ are integral over S, and $G \cap Q_0 = S$ since S is integrally closed.

(2) $\Longrightarrow$ (3): Part (1) of Corollary 2.8 shows that G is the union of a chain of cyclic groups $\{(g_i)\}_{i=1}^{\infty}$. Without loss of generality, we assume that $g_i > 0$ for each i. Then $S = G \cap Q_0 = [\cup_1^{\infty}(g_i)] \cap Q_0 = \cup_1^{\infty}[(g_i) \cap Q_0] = \cup_1^{\infty}\langle g_i\rangle^0$, where $\langle g_i\rangle^0 \subseteq \langle g_{i+1}\rangle^0$ for each i.

(3) $\Longrightarrow$ (4): It follows from (3) that $F[S]$ is the union of a chain of subrings, each isomorphic to $F[Z_0] \simeq F[X]$, a PID. Therefore $F[S]$ is a Bezout domain.

That (4) implies (5) is clear. Moreover, if $F[S]$ is a Prüfer domain, then $F[S]$ is integrally closed, so that S is integrally closed by Corollary 12.11.

We note that condition (3) of Theorem 13.5 is an intrinsic property of S that requires no reference to the assumptions that $S \subseteq Q_0$ and that $1 \in S$. We call a nonzero submonoid T of Q a Prüfer monoid if T is the union of an ascending sequence of cyclic submonoids. Theorem 13.6 is a summary of the results of this section up to this point; the hypotheses of Section 13 concerning D and S are repeated in its statement.

THEOREM 13.6. Assume that D is a unitary integral domain and that S is a nonzero torsion-free cancellative monoid. The following conditions are equivalent.

(1) D[S] is a Prüfer domain.

(2) D[S] is a Bezout domain.

(3) D is a field and to within isomorphism, S is either a subgroup or a Prüfer submonoid of Q .

A Dedekind domain can be characterized as a Noetherian Prüfer domain, and a PID is a Noetherian Bezout domain. In view of these facts, it is easy to determine from Theorem 13.6 conditions under which D[S] is a Dedekind domain or a PID (Theorem 13.8). The proof of (13.8) uses a basic result that is also referred to in Sections 14 and 15.

THEOREM 13.7. If $s,t \in S$ are such that $\langle t\rangle^0$ properly contains $\langle s\rangle^0$, then $(1 - X^t)$ properly contains $(1 - X^s)$ in D[S] . Thus, if D[S] satisfies the ascending chain condition for principal ideals (a.c.c.p.) , then S satisfies the ascending chain condition on cyclic submonoids.

<u>Proof</u>. We have $s = nt$ for some $n > 1$. Therefore $1 - X^s = (1 - X^t)(1 + X^t + \ldots + X^{(n-1)t})$, where $\Sigma_{i=0}^{n-1} X^{it}$ is a nonunit of D[S] by Theorem 11.1. Consequently, $(1 - X^t)$ properly contains $(1 - X^s)$.

THEOREM 13.8. Let D and S be as in the statement of Theorem 13.6. The following conditions are equivalent.

(1) $D[S]$ is a Euclidean domain.

(2) $D[S]$ is a PID .

(3) $D[S]$ is a Dedekind domain.

(4) D is a field and S is isomorphic either to Z or to Z_0 .

Proof. We need only establish the implications (3) ⟹ (4) and (4) ⟹ (1) . Thus, if $D[S]$ is a Dedekind domain, then D is a field by Theorem 13.6. Moreover, S is a submonoid of Q , and S is finitely generated since $D[S]$ is Noetherian (Theorem 7.7). If S is a group, it follows that S is cyclic — that is, $S \simeq Z$. If S is not a group, then Theorem 13.5 shows that S is the union of an ascending sequence of cyclic submonoids. But Theorem 13.7 implies that S satisfies a.c.c. on cyclic submonoids. Consequently, $S = \langle s\rangle^0$ for some s , and $S \simeq Z_0$. This shows that (3) implies (4) .

(4) ⟹ (1) : If D is a field, then $D[Z_0] = D[X]$ is a Euclidean domain. Since a quotient ring of a Euclidean domain is again Euclidean, it follows that $D[Z]$ is also Euclidean.

Section 13 Remarks

A brief account of the theory of Prüfer domains is given in Section I–6 of [83]. For a more detailed account, see Chapter IV of [51]. In particular, Theorem 24.3 of [51] shows that the domain D is a Prüfer domain if and only if D is integrally closed and $(a,b)^2 = (a^2,b^2)$ for all

$a,b \in D$; we have used this characterization of Prüfer domains several times in Section 13.

Our restriction to integral domains in Section 13 will result in some duplication of results in Section 18, where the problems of determining conditions under which $R[S]$ is a Prüfer ring, a Bezout ring, etc. are considered. On the other hand, a more complete treatment of ring-theoretic aspects of the problem are included in Section 18, for the classes of Prüfer and related rings are not as familiar as their domain counterparts.

Another class of domains frequently considered in connection with the classes of Prüfer domains and Dedekind domains is the class of almost Dedekind domains, defined as follows. An almost Dedekind domain is a unitary integral domain D such that D_P is a Noetherian valuation domain for each proper prime ideal P of D. Dedekind domains are almost Dedekind and an almost Dedekind domain is a Prüfer domain. Fxcept for the class of fields, almost Dedekind domains are one-dimensional. A one-dimensional almost Dedekind domain D is Noetherian (hence a Dedekind domain) if and only if no maximal ideal of D is idempotent. Another characterization is that the almost Dedekind domains are the domains in which the cancellation law for ideals holds — that is, $AB = AC$ and $A \neq (0)$ implies $B = C$. Gilmer and Parker in [59] determine necessary and sufficient conditions for $D[S]$ to be an almost Dedekind domain, but their proof requires specialized results about such domains that are outside the scope of this monograph. In Corollary 20.15, however, we determine those group rings $D[G]$ that are almost Dedekind. That result is strong enough to show that D[G] need not be Noetherian if it is almost Dedekind; for example, Q[Q] is such a domain.

For a brief account of the topic of divisible abelian groups, the interested reader should consult Chapter IV of [47].

§14. Monoid Domains as Factorial Domains

As in Section 13, we assume throughout this section that D is a unitary integral domain and that S is a torsion-free cancellative monoid. We seek conditions under which the monoid domain D[S] is factorial. A unitary domain J is factorial if and only if J is a GCD-domain and a.c.c.p. is satisfied in J . Therefore one approach to the problem of determining conditions under which D[S] is factorial would be to determine separately conditions under which D[S] is a GCD-domain or satisfies a.c.c.p. , then mesh these two results together. This is close to the approach we follow. The hedging is on the a.c.c.p. part, but the first step in the proof is to determine necessary and sufficient conditions in order that D[S] should be a GCD-domain. One necessary condition is that S should be a GCD-monoid. Since the operation on S is written as addition, a translation from the notation of Section 6 is required; for example, one equivalent form of the definition of a GCD-monoid, using additive notation, is that for all $s,t \in S$, there exists $u \in S$ such that $(s + S) \cap (t + S) = u + S$. The first two results come easily.

THEOREM 14.1. If D[S] is a GCD-domain, then D is a GCD-domain and S is a GCD-monoid.

<u>Proof</u>. Since D is an algebraic retract of D[S] , Theorem 12.2 shows that D is a GCD-domain. To prove that S is a GCD-monoid, take $s,t \in S$ and let $g \in D[S]$ be such that $(X^s) \cap (X^t) = gD[S]$. Since the monomial X^{s+t} is divisible by g , part (1) of Theorem 11.1 shows that g

is a monomial, necessarily of the form uX^w, where u is a unit of D. Thus $(X^s) \cap (X^t) = (X^w)$. Since (X^v), for $v \in S$, is the semigroup ring $D[v + S]$, it follows that $(s + S) \cap (t + S) = w + S$.

THEOREM 14.2. If D is a GCD–domain and G is a torsion–free group, then $D[G]$ is a GCD–domain.

Proof. Before giving the proof, we observe that each quotient ring J_T of a GCD–domain is again a GCD–domain. To prove this, take principal ideals aJ_T and bJ_T of J_T. Without loss of generality we assume that a and b are in J. Then $aJ \cap bJ = cJ$ for some $c \in J$. Extending to J_T, it follows that $aJ_T \cap bJ_T = (aJ \cap bJ)J_T = cJ_T$, so J_T is a GCD–domain.

To prove that $D[G]$ is a GCD–domain, take nonzero elements $f = \Sigma_1^n f_i X^{s_i}$, $g = \Sigma_1^m g_i X^{t_i}$ of $D[G]$, and let H be the subgroup of G generated by the set $\{s_1,\ldots,s_n,t_1,\ldots,t_m\}$. Then $H \simeq Z^k$ for some k, and $D[H]$ is isomorphic to a quotient ring of the polymonial ring $D[X_1,\ldots,X_k]$. A polynomial ring over GCD–domain is again a GCD–domain, and by the result in the preceding paragraph, $D[H]$ is also a GCD–domain. Therefore $fD[H] \cap gD[H] = hD[H]$ for some $h \in D[H]$. It then follows from Theorems 12.1 and 12.3 that $fD[G] \cap gD[G] = hD[G]$. This completes the proof of Theorem 14.2.

In Theorem 14.5, we establish the converse of Theorem 14.1. The proof of (14.5) uses two preliminary results, plus a couple of new concepts. If A is a nonempty subset of $D\backslash\{0\}$, and if x is a nonzero element of D, we say that

x is prime to A if $aD \cap xD = axD$ for each $a \in A$, equivalently, x is prime to A if $xd:a = xD$ for each $a \in A$.

THEOREM 14.3. Let N be a multiplicative system in D and let T be the set of elements of D that are prime to N. Assume that the following two conditions are satisfied.

(1) Each pair of elements of N has a least common multiple in D.

(2) Each element of D can be expressed in the form nt for some $n \in N$, $t \in T$.

If D_N is a GCD-domain, then D is a GCD-domain.

Proof. Since the elements of T are prime to N, it follows that $tD_N \cap D = tD$ for each $t \in T$. To show D is a GCD-domain, take nonzero elements $b,c \in D$ and write $b = n_1t_1$, $c = n_2t_2$, where $n_1,n_2 \in N$ and $t_1,t_2 \in T$. Let $n = \mathrm{lcm}\{n_1,n_2\}$. We have $bD_N = t_1D_N$ and $cD_N = t_2D_N$. Moreover, $t_1D_N \cap t_2D_N$ is a principal ideal of D_N and is therefore the extension of a principal ideal of D. In fact, $t_1D_N \cap t_2D_N = tD_N$ for some $t \in T$. To complete the proof, we show that $bD \cap cD = ntD$. Since $t \in t_iD_N$, then $t \in t_iD_N \cap D = t_iD$. Hence $nt \in n_it_iD$ for $i = 1,2$, and $ntD \subseteq bD \cap cD$. If $x \in bD \cap cD$, we write x as ms, where $m \in N$ and $s \in T$. Then $n_i | ms$ for $i = 1,2$, and $(n_i) \cap (s) = (n_is)$, so that $n_i | m$; consequently, $n | m$. On the other hand,

$$xD_N \cap D = sD \subseteq bD_N \cap cD_N \cap D = tD ,$$

so that $t | s$. It follows that $nt | ms = x$ and $bD \cap cD = ntD$. as asserted.

Assume that D is a GCD-domain and S is a GCD-monoid. Then S induces a natural grading on the domain $D[S]$, and the homogeneous elements of $D[S]$ under this S-grading are the monomials dX^s. To prove that $D[S]$ is a GCD-domain, we intend to apply Theorem 14.3 to $D[S]$, where N is the set of nonzero homogeneous elements of $D[S]$. The next result paves the way for the application of (14.3) in the setting indicated. If $f = \Sigma_1^n a_i X^{s_i}$ is the canonical form of a nonzero element $f \in D[S]$, then let $d = \gcd\{a_i\}_1^n$ and let $s = \gcd\{s_i\}_1^n$; d is determined up to associates in D and s is determined to within an invertible element of S. We call dX^s the <u>homogeneous content</u> (h-content) <u>of</u> f; it is determined up to a unit factor in $D[S]$. If f has h-content 1, we say that f is <u>homogeneously primitive</u> (h-primitive). Clearly $f = dX^s f^*$, where f^* is h-primitive.

THEOREM 14.4. Assume that D is a GCD-domain and that S is a GCD-monoid. Pick $a, b \in D\backslash\{0\}$, $s, t \in S$, and let $c = \operatorname{lcm}\{a,b\}$, $u = \operatorname{lcm}\{s,t\}$. Assume that $f, g \in D[S]$ are h-primitive. The following statements are valid.

(1) $(a) \cap (b) = (c)$, where parentheses denote the ideal of $D[S]$ generated by the given element.

(2) $(X^s) \cap (X^t) = (X^u)$.

(3) $(a) \cap (X^s) = (aX^s)$.

(4) $(aX^s) \cap (bX^t) = (cX^u)$.

(5) fg is h-primitive.

(6) $(aX^s) \cap (f) = (aX^s f)$.

<u>Proof</u>. Statement (1) follows from Theorem 12.3.

(2): $(X^s) \cap (X^t) = D[s + S] \cap D[t + S] = D[(s + S) \cap (t + S)] = D[u + S] = (X^u)$.

(3): (a) $\cap (X^s) = aD[S] \cap D[s + I] = aD[s + I] = a(X^s) = (aX^s)$.

Statement (4) follows from (1), (2), and (3).

It follows from (4) that each pair of nonzero homogeneous elements of $D[S]$ has a homogeneous least common multiple in $D[S]$. Applying part (4) of Theorem 6.1, we conclude that $\gcd\{aX^s, bX^t\}$ exists in $D[S]$ and is equal to dX^w , where $d = \gcd\{a,b\}$ and $w = \gcd\{s,t\}$. Thus, part (3) of (6.1) then implies that $\gcd\{Y\}$ and $\operatorname{lcm}\{Y\}$ exist for each finite set Y of nonzero homogeneous elements of $D[S]$. We use these observations in proving (5). Thus, let $\leq$ be a total order on S and let $f = \Sigma_1^n a_i X^{s_i}$ and $g = \Sigma_1^m b_j X^{t_j}$ be the canonical forms of f and g , where $s_1 < s_2 < \ldots < s_n$ and $t_1 < t_2 < \ldots < t_m$. To prove that fg is h-primitive, it suffices to show that each nonzero, nonunit homogeneous element w of $D[S]$ fails to divide some homogeneous component of fg . Let $f_i = a_i X^{s_i}$ for $i = 1,2,\ldots,n$ and let $g_j = b_j X^{t_j}$ for $j = 1,2,\ldots,m$. If $\gcd\{w,f_1\} = \gcd\{w,g_1\} = 1$, then $f_1 g_1$ is a homogeneous component of fg not divisible by w . Thus, assume that $\gcd\{w,f_1\} \neq 1$. For $1 \leq i \leq n$, let $w_i = \gcd\{w,f_1,\ldots,f_i\}$. Since f is h-primitive, $w_n = 1$. Choose $i \geq 2$ minimal such that $w_i = 1$. Then w_{i-1} is a nonunit homogeneous divisor of w , and it suffices to show that w_{i-1} fails to divide some homogeneous component of fg . Replacing w by w_{i-1} , we therefore assume that w divides $f_1,\ldots,f_{i-1}$ and that $\gcd\{w,f_i\} = 1$. If $\gcd\{w,g_1\} = 1$, we're finished by a previous case, and if $\gcd\{w,g_1\} \neq 1$, we choose $j \geq 2$ minimal so that $\gcd\{w,g_1,\ldots,g_j\} = 1$. Replacing w by

$\gcd\{w, g_1, \ldots, g_{j-1}\}$, it suffices to consider the case where w divides each of $f_1, \ldots, f_{i-1}, g_1, \ldots, g_{j-1}$ and $\gcd\{w, f_i\} = \gcd\{w, g_j\} = 1$. The $(s_i + t_j)$–component of fg has the form

$$f_i g_j + (\text{a sum of terms } f_v g_y \text{, where either } v < i \text{ or } y < j).$$

Since w divides each such $f_v g_y$ and since $\gcd\{w, f_i g_j\} = 1$, it follows that w fails to divide the $(s_i + t_j)$–component of fg . This completes the proof of (5) .

(6): To prove (6) , we show that if g is a nonzero element of $(aX^s) \cap (f)$, then $g \in (aX^s f)$. Thus, write $g = aX^s k = fq$. Let k_1 be the h–content of k , let q_1 be the h–content of q , and write $k = k_1 k^*$ and $q = q_1 q^*$, where k^* and q^* are h–primitive. Then $aX^s k_1 \cdot k^* = q_1 \cdot q^* f$, where $aX^s k_1$ and q_1 are homogeneous and k^* and $q^* f$ are h–primitive. It follows that k^* and $q^* f$ differ by a unit factor in $D[S]$, so $f | k^*$. Hence $g = aX^s k_1 k^* \in (aX^s f)$.

THEOREM 14.5. If D is a GCD–domain and S is a GCD–monoid, then $D[S]$ is a GCD–domain.

Proof. Under the canonical S–grading on $D[S]$, let N be the set of nonzero homogeneous elements of $D[S]$. Let T be the set of elements of $D[S]$ that are prime to N . We show that the nonzero elements of T are precisely the h–primitive elements of $D[S]$. Part (6) of Theorem 14.4 shows that each h–primitive element of $D[S]$ is in T . And if $f = aX^s f^*$ has nonunit h–content aX^s , then $(f):aX^s = (f^*) \neq (f)$ so that f is not prime to N . We note that conditions (1) and (2) of Theorem 14.3 are satisfied for N and T ; that (2) is satisfied is clear, and part (4) of Theorem 14.4 shows that (1) is satisfied. Moreover $D[S]_N \simeq K[G]$, where K is the quotient field of D and G is the quotient group of S . Theorem 14.2 shows that $D[S]_N$ is a GCD-domain, and hence D[S] is a GCD-domain by Theorem 14.3.

We turn to the problem of determining conditions under which D[S] is factorial. As with the GCD-property, some necessary conditions on D and S are fairly easy to establish. Theorem 14.7 represents an expanded analogue, for factorial domains, of Theorem 14.1. We first prove a result concerning a.c.c.p.

THEOREM 14.6. Let K be the quotient field of D .

(1) If a.c.c.p. is satisfied in D[S] , then D[H] satisfies a.c.c.p. for each subgroup H of S . In particular, D satisfies a.c.c.p.

(2) If D and K[S] satisfy a.c.c.p. , then D[S] satisfies a.c.c.p.

<u>Proof</u>. Statement (1) follows from Theorem 12.2. To prove (2) , take an ascending sequence $(f_1) \subseteq (f_2) \subseteq \ldots$ of nonzero principal ideals of D[S] — say $f_i = g_i f_{i+1}$ for each i . Since K[S] satisfies a.c.c.p. , we assume without loss of generality that $f_1 K[S] = f_2 K[S] = \ldots$. Thus, each g_i is a unit of K[S] . Since units of K[S] are trivial (Theorem 11.1), it follows that each g_i is of the form $d_i X^{s_i}$, where s_i is invertible in S . Let a_1 be a

nonzero coefficient of $f_1 = d_1X^{s_1}f_2$. Then $a_1 = d_1a_2$ for some coefficient a_2 of f_2 . And since $f_2 = d_2X^{s_2}f_3$, then $a_2 = d_2a_3$ for some coefficient a_3 of f_3 , etc. We have $a_1D \subseteq a_2D \subseteq \ldots$, and since a.c.c.p. is satisfied in D , it follows that $a_kD = a_{k+1}D = \ldots$ for some k . Consequently, $d_k, d_{k+1}, \ldots$ are units of D . Therefore $g_i = d_iX^{s_i}$ is a unit of D[S] for $i \geq k$, and $(f_k) = (f_{k+1}) = \ldots$. This completes the proof of (2) .

THEOREM 14.7. Let K be the quotient field of D and let G be the quotient group of S . If D[S] is factorial, then D and S are factorial and S satisfies a.c.c. on cyclic submonoids. Moreover, D[G] and K[S] are factorial domains.

Proof. Since D[S] is factorial, D is factorial by Theorem 12.2. To prove that S is factorial, take a noninvertible element $s \in S$. Write X^s as a finite product of nonunit prime elements of D[S] , say $X^s = \Pi_1 \ldots \Pi_n$. Each Π_i is a monomial, necessarily of the form $u_iX^{p_i}$, where u_i is a unit of D and p_i is not invertible in S . Then $s = p_1 + \ldots + p_n$, and to complete the proof of factoriality of S , we show that each p_i is prime in S . Thus, assume $a, b \in S$ are such that $a + b \in p_i + S$. Then $X^aX^b \in (X^{p_i})$, which implies either $X^a \in (X^{p_i})$ and $a \in p_i + S$ or $X^b \in (X^{p_i})$ and $b \in p_i + S$. Therefore p_i is prime and S is factorial.

Theorem 13.7 implies that S satisfies a.c.c. on cyclic submonoids. Moreover, K[S] and D[G] are factorial because they are quotient rings of D[S] .

COROLLARY 14.8. Under the notation of Theorem 14.7,

$D[S]$ is factorial if and only if D and $K[S]$ are factorial.

Proof. In view of (14.7), we need only show that $D[S]$ is factorial if D and $K[S]$ are factorial. Thus, S is a GCD–monoid by Theorem 14.5, and the same result shows that $D[S]$ is a GCD–domain. Since a.c.c.p. holds in $D[S]$ by part (2) of Theorem 14.6, it follows that $D[S]$ is factorial.

It turns out that the three necessary conditions on D and S given in Theorem 14.7 are also sufficient for factoriality of $D[S]$, but additional work is required for a proof. Assume that D and S are factorial and let H be the group of invertible elements of S. From Section 6 we know that S is isomorphic to $H \oplus M$, where M is the external weak direct sum of α copies of Z_0, with α the cardinality of a complete set of nonassociate prime elements of S. Hence $D[S] \simeq (D[M])[H]$, and $D[M]$ is a polynomial ring over D in a set of indeterminates of cardinality α. Therefore $D[M]$ is factorial, and we seek conditions on H in order that the group ring of H over a factorial domain should be factorial. According to Theorem 14.7, a necessary condition on H is that it satisfies a.c.c. on cyclic submonoids. The next two results are concerned with this condition in an abelian group.

THEOREM 14.9. Assume that G is a group and $g \in G$. Let M be the set of positive integers n such that $nx = g$ is solvable in G (that is, g is divisible by n in G).

(1) The set M is closed under taking positive divisors and under taking least common multiples.

(2) The following conditions are equivalent.

(i) M is finite.

(ii) M contains a largest element (which is is necessarily the least common multiple of the elements of M)

(iii) M contains only finitely many primes p and for each $p \in M$, there is a largest power p^k of p that is in M.

(3) If M is finite, then the element g has infinite order.

<u>Proof</u>. (1): If $n \in M$, it is clear that $d \in M$ for each positive divisor d of n. Thus, if $n, m \in M$, then each prime-power divisor of n and m is in M. Hence, to show that $\mathrm{lcm}\{n,m\}$ is in M, it suffices to consider the case where n and m are relatively prime. Assume that $nx = my = g$ and that $1 = na + mb$. Then $g = nag + mbg = namy + mbnx = mn(ay + bx)$.

(2): The implications (i) $\Longrightarrow$ (ii) and (ii) $\Longrightarrow$ (iii) are patent, and the implication (iii) $\Longrightarrow$ (i) follows from (1).

To prove (3), we note that if g has finite order k, then $(g) = (rg)$ for each positive integer r relatively prime to k so that each such r is in M and M is not finite.

If G and g are as in the statement of (14.9), then the terminology of abelian group theory is that g <u>is of type</u> $(0,0,\ldots)$ if the set M of Theorem 14.9 is finite. To a reader seeing this concept for the first time, the choice of terminology must seem abstruse, but in fact, the type of an arbitrary element $h \in G$ can be defined (see [48, Section 85]); we have no occasion to use the general concept of type. The next result indicates the connection between type $(0,0,\ldots)$ and factoriality of $D[G]$.

THEOREM 14.10. Assume that G is a torsion-free group. The following conditions are equivalent.

(1) G satisfies a.c.c. on cyclic submonoids.

(2) G satisfies a.c.c. on cyclic subgroups.

(3) Each rank-one subgroup of G is cyclic.

(4) Each nonzero element of G is of type $(0,0,\ldots)$.

Proof. $(1) \Longleftrightarrow (2)$: The implication $(2) \Longrightarrow (1)$ follows since G is torsion-free: if $\langle s\rangle^0$ properly contains $\langle t\rangle^0$, then $t = ns$ for some $n > 1$, and hence (s) properly contains (t). For the converse, assume that (1) is satisfied and let $(g_1) \subseteq (g_2) \subseteq \ldots$ be an ascending sequence of nonzero cyclic subgroups of G. We have $g_1 = n_1g_2 = (-n_1)(-g_2)$. Hence by replacing g_2 by $-g_2$ if necessary, we may assume that n_1 is positive. Similarly, after a possible replacement of g_3 by $-g_3$, we may assume that $g_2 = n_2g_3$ with $n_2 > 0$. Without loss of generality, assume that $g_i = n_ig_{i+1}$ for each i, where $n_i > 0$. Then $\langle g_1\rangle^0 \subseteq \langle g_2\rangle^0 \subseteq \ldots$, so (1) implies that there exists $k \in Z^+$ such that $n_i = 1$ for each $i \geq k$. Thus $(g_k) = (g_{k+1}) = \ldots$, and (2) is satisfied.

Corollary 2.8 shows that each rank-one group is the union of an ascending sequence of cyclic subgroups. Therefore (2) implies (3).

(3) ⟹ (4): Let g be a nonzero element of G and let $H = \{x \in G | nx \in (g)$ for some $n \in Z^+\}$. H is a rank-one subgroup of G since $\{g\}$ is a maximal free subset of H. Since (3) is satisfied, $H = (h)$ is cyclic. Write $g = mh$. Replacing h by $-h$ if necessary, we assume that m is positive. If k is any positive integer such that $kx = g$ is solvable, then a solution x_0 of this equation is in (h). Write $x_0 = qh$. Then $g = mh = kqh$, and since G is torsion-free, $m = kq$ and $k \le m$. Therefore g is of type $(0,0,\ldots)$.

(4) ⟹ (2): If (2) fails, then there exists a strictly ascending sequence $(g_1) < (g_2) < \ldots$ of cyclic subgroups of G. It is clear that each g_i fails to be of type $(0,0,\ldots)$. This completes the proof of Theorem 14.10.

Suppose that D is factorial and G is a group such that each nonzero element of G is of type $(0,0,\ldots)$. Our goal is to prove that $D[G]$ is factorial. Corollary 14.8 shows that it suffices to consider the case where D is a field, and Theorem 14.15 is that result: $D[G]$ is factorial if D is a field and each nonzero element of G is of type $(0,0,\ldots)$. In preparation for the proof of (14.15), we first record a result that follows from the proof of part (2) of Theorem 14.6.

THEOREM 14.11. Assume that J is an extension domain of D and that T is a torsion-free cancellative monoid containing S as a submonoid. Assume that noninvertible elements of S are noninvertible in T. If a.c.c.p. is satisfied in D and in $J[T]$, then it is satisfied in $D[S]$.

COROLLARY 14.12. Assume that F is a subfield of the field K and that H is a subgroup of the group G. If $K[G]$ is factorial, then $F[H]$ is factorial.

Proof. Theorem 14.2 shows that $F[H]$ is a GCD–domain, and $F[H]$ satisfied a.c.c.p. by Theorem 14.11. Therefore $F[H]$ is factorial.

A subgroup H of an abelian group G is said to be pure in G if G/H is torsion–free; this is equivalent to the condition that each solution x_0 of an equation of the form $nx = h$, where $n \in Z^+$ and $h \in H$, is in H. If L is a subgroup of G, and if L^*/L is the torsion subgroup of G/L, then L^* is called the pure subgroup of G generated by L. Obviously L^* is the smallest pure subgroup of G containing L, and it consists of all elements $g \in G$ such that $ng \in L$ for some $n \in Z^+$. The condition that each nonzero element of a torsion–free group G is of type $(0,0,\ldots)$ is equivalent to the condition that $(g)^*$ is cyclic for each $g \in G$. If H is a pure subgroup of G and if G/H is finitely generated, then H is a direct summand of G. This is the main result about pure subgroups that we use, and the result itself follows since G/H is torsion–free and finitely generated, hence free.

THEOREM 14.13. Let H be a pure subgroup of the torsion–free group G. If F is a field, then prime elements of $F[H]$ are prime in $F[G]$.

Proof. Let p be a prime element of $F[H]$, and assume that p divides a product fg in $F[G]$ — say $fg = ph$, where $h \in F[G]$. Let M be the subgroup of G generated by $H \cup \mathrm{Supp}(f) \cup \mathrm{Supp}(g) \cup \mathrm{Supp}(h)$. Then H is pure in M and M/H is finitely generated, so $M = H \oplus M_1$ for some subgroup M_1 of M . The element p is prime in $F[H]$, and hence in $(F[H])[M_1] \simeq F[M]$. Since p divides fg in $F[M]$, it follows that p divides f or g in $F[M]$, and hence in $F[G]$. This proves that p is prime in $F[G]$.

Most of the work required for the proof of Theorem 14.15 is contained in the proof of the next result.

THEOREM 14.14. Let F be an algebraically closed field, let G be a torsion-free group such that each nonzero element of G is of type $(0,0,\ldots)$, and let H be a finitely generated subgroup of G such that G/H is a torsion group. Then each prime element of $F[H]$ can be expressed as a finite product of prime elements of $F[G]$.

<u>Proof</u>. Let p be a prime element of $F[H]$ and let $p = \Sigma_{i=1}^{n} r_i X^{h_i}$ be the canonical form of p , where $h_1 < h_2 < \ldots < h_n$ under a fixed total order $\leq$ on G . Since X^{h_1} is a unit of $F[H]$, we assume without loss of generality that $h_1 = 0$. Let k be the largest positive integer by which h_n is divisible in G . We prove that p is a finite product of prime elements of $F[G]$ by showing that if $f_1, f_2, \ldots, f_t$ are nonunits of $F[G]$ such that $p = f_1 f_2 \ldots f_t$, then $t \leq k$. Once this is proved, it follows that p is a finite product of irreducible elements of $F[G]$, and since $F[G]$ is a GCD-domain, irreducible elements of $F[G]$ are prime (Theorem 6.7).

Thus, assume that there exist nonunits $f_1, \ldots, f_t$ of

$F[G]$, with $t > k$, such that $p = f_1 f_2 \ldots f_t$. Let M be the subgroup of G generated by H and $\cup_1^t \mathrm{Supp}(f_i)$. Then $M \simeq Z^w$ for some w , and hence $F[M]$ is factorial. Consider the following diagram:

$$\begin{array}{ccc} F[M] & \text{———} & F(X;M) \\ | & & | \\ F[H] & \text{———} & F(X;H) \end{array}$$

In the diagram, $F(X;H)$ and $F(X;M)$ denote the quotient fields of $F[H]$ and $F[M]$, respectively. We prove that $F(X;M)$ is finite and normal over $F(X;H)$, that $F[H]$ is integrally closed, and that $F[M]$ is the integral closure of $F[H]$ in $F(X;M)$. Theorem 14.2 implies that $F[H]$ is integrally closed. Let $\{g_i\}_1^m$ be a finite set of generators of M . Then $F(X;M) = F(X;H)(\{X^{g_i}\}_1^m)$. Moreover, for each i between 1 and m , there is a positive integer n_i such that $n_i g_i \in H$. It follows that X^{g_i} is a root of the pure equation $Y^{n_i} - X^{n_i g_i}$ over $F(X;H)$. Since F is algebraically closed, F contains the n_ith roots of unity, and $Y^{n_i} - X^{n_i g_i}$ splits into linear factors in $F(X;M)[Y]$. Therefore $F(X;M)/F(X;H)$ is finite and normal. We have shown that $F[M]$ is integral over $F[H]$, and since $F[M]$ is integrally closed, $F[M]$ is the integral closure of $F[H]$ in $F(X;M)$. Moreover, we have also shown that if $g \in M$, then the conjugates of X^g over $F(X;H)$ are of the form αX^g , where α is a root of unity in F ; in particular, if $f = \Sigma_1^r a_i X^{g_i} \in F[M]$ and if σ is an element of the Galois group T of $F(X;M)$ over $F(X;H)$, then $\sigma(f) = \Sigma_1^r a_i \sigma(X^{g_i})$, and hence f and $\sigma(f)$ have the same support.

Since p can be factored in $F[M]$ as a product of t nonunits, where $t > k$, it follows that the prime factorization of p in $F[M]$ is of the form $p = wp_1^{e_1}\dots p_r^{e_r}$, where w is a unit of $F[M]$ and where $e_1 + e_2 + \dots + e_r \geq t$. We write p_1 as $X^{g_0}q_1$, where q_1 has order 0 and degree g_v. It is clear that the ideal (q_1) of $F[M]$ generated by q_1 is a minimal prime ideal of (p). Moreover, $(q_1) \cap F[H] = pF[H]$ by the lying-under theorem, and because $F(X;M)/F(X;H)$ is normal and $F[H]$ is an integrally closed domain, it follows that $\{\sigma((q_1)) = (\sigma(q_1)) \mid \sigma \in T\}$ is the set of prime ideals of $F[M]$ lying over $pF[H]$ in $F[H]$. Thus, each (p_i) is of the form $(\sigma_i(q_1))$ for some $\sigma_i \in T$. Therefore

$$(p) = (p_1^{e_1}\dots p_r^{e_r}) = ([\sigma_1(q_1)]^{e_1}\dots[\sigma_r(q_1)]^{e_r}),$$

so that there is a unit uX^g of $F[M]$ such that

$$p = uX^g[\sigma_1(q_1)]^{e_1}\dots[\sigma_r(q_1)]^{e_r}.$$

Our previous observations show that each $\sigma_i(q_1)$ has order 0 and degree g_v. Since p has order 0 and degree h_n, it follows that $g = 0$ and $h_n = (e_1 + \dots + e_r)g_v$. Because $e_1 + \dots + e_r \geq t > k$, this contradicts the choice of k. Hence $t \leq k$ as asserted, and p is a finite product of prime elements of $F[G]$.

THEOREM 14.15. Assume that F is a field and G is a torsion-free group such that each nonzero element of G is of type $(0,0,\dots)$. Then $F[G]$ is factorial.

<u>Proof</u>. In view of Corollary 14.12, we assume without

loss of generality that F is algebraically closed. Take a nonzero nonunit f of $F[G]$. Let H be the subgroup of G generated by $\mathrm{Supp}(f)$ and let H^* be the pure subgroup of G generated by H. Since H is finitely generated, $F[H]$ is factorial and f is a finite product of prime elements of $F[H]$. Theorem 14.14 shows that each prime element of $F[H]$ is a finite product of primes of $F[H^*]$; thus, f is a finite product of primes of $F[H^*]$. But Theorem 14.13 shows that prime elements of $F[H^*]$ are also prime in $F[G]$, and hence f is a finite product of primes in $F[G]$. Consequently, $F[G]$ is a factorial domain.

The next theorem summarizes the main results of this section. Most of the work required to prove Theorem 14.16 has already been done.

THEOREM 14.16. Assume that D is a unitary integral domain and that S is a torsion-free cancellative monoid. Let H be the group of invertible elements of S. The following conditions are equivalent.

(1) $D[S]$ is factorial.

(2) D and S are factorial and S satisfies a.c.c. on cyclic submonoids.

(3) D and S are factorial and each nonzero element of H is of type $(0,0,\ldots)$.

<u>Proof</u>. Theorem 14.7 shows that (1) implies (2). If (2) is satisfied, then clearly H satisfies a.c.c. on cyclic submonoids, so each nonzero element of H is of type $(0,0,\ldots)$ by Theorem 14.10. Therefore (2) implies (3). If (3) is satisfied, then since S is factorial, $D[S]$ is

isomorphic to $J[H]$, where J is a polynomial ring over D . Therefore J is factorial. Let K be the quotient field of J . Theorem 14.15 shows that $K[H]$ is factorial, and hence $J[H]$ is factorial by Corollary 14.8. This completes the proof.

As indicated in the introduction of this section, we do not determine conditions under which $D[S]$ satisfies a.c.c.p. But for a group S , such conditions are given in the next result.

THEOREM 14.17. Let G be a torsion-free group. The group ring $D[G]$ satisfies a.c.c.p. if and only if D satisfies a.c.c.p. and each nonzero element of G is of type $(0,0,\ldots)$.

Proof. Theorems 12.2, 13.7, and 14.8 show that D satisfies a.c.c.p. and each nonzero element of G is of type $(0,0,\ldots)$ if $D[G]$ satisfies a.c.c.p. For the converse, let K be the quotient field of D . Theorem 14.15 implies that $K[G]$ satisfies a.c.c.p. Since D also satisfies a.c.c.p. , it then follows from Theorem 14.11 that a.c.c.p. is satisfied in $D[G]$.

Theorem 14.15 can be used to give examples of finite-dimensional non-Noetherian factorial domains of any dimension $r \geq 2$. (A one-dimensional factorial domain is a PID , hence Noetherian.) For each $r \geq 2$, there exists an indecomposable — hence non-finitely generated — torsion-free abelian group L_r such that each nonzero element of L_r is of type $(0,0,\ldots)$ [48, Section 88]. If F is a field, then Theorem 14.15 shows that $F[L_r]$ is factorial. This domain is non-Noetherian since L_r is not finitely generated, and Theorem 17.1 shows that $\dim F[L_r] = r$.

Section 14 Remarks

Many of the results of Section 14 were originally proved in the paper [58]. There is some discussion at the end of Section 7 of [58] of the problem of determining conditions under which a.c.c.p. is satisfied in $D[S]$. One troublesome aspect of a.c.c.p. is that it doesn't behave well with respect to localization in either direction. For example, if $D = Q[Q]$, then D is an almost Dedekind domain (see the remarks at the end of Section 13) that does not satisfy a.c.c.p. , but D_P is Noetherian for each maximal ideal P of D . On the other hand, if S is the additive semigroup consisting of 0 and the rational numbers $q \geq 1$, then an elementary degree argument shows that $F[S]$ satisfies a.c.c.p. But S has quotient group Q , $F[Q]$ is a localization of $F[S]$, and $F[Q]$ does not satisfy a.c.c.p.

Another condition frequently considered in connection with a.c.c.p. and factorization properties is the condition that D should be <u>atomic</u>, which is defined to mean that each nonzero element of D is a finite product of irreducible elements (atoms) of D . If D satisfies a.c.c.p. , then D is atomic. The converse fails, but examples are hard to come by (see [64] and [135]). The question of determining conditions under which $D[S]$ is atomic is open (see, however, Section 8 of [97]).

§15. Monoid Domains as Krull Domains

We continue in Section 15 to denote by D a unitary integral domain, and by S a torsion-free cancellative monoid. Krull domains are frequently treated from two points of view; one approach is in terms of valuation overrings, and the other is in terms of divisorial ideals and the v-operation. Our treatment follows this separation of approaches. In this section we regard a Krull domain as the intersection of a family of rank-one discrete valuation overrings of finite character; in Section 16, where we determine the divisor class group of a monoid domain that is a Krull domain, the topics of divisorial ideals and the v-operation naturally arise.

Throughout this section we generally use various modifications of the letter w to denote valuations on groups, while forms of v are used for valuations on fields. (There is an exception to this practice in the statement of Theorem 15.7, where w^* is used for a valuation on a field.) If $w: G \longrightarrow \Gamma$ is a valuation from the group G into the ordered group Γ, then $w(G)$ is an ordered subgroup of Γ called the <u>value group</u> of w. We say that w is <u>discrete</u> if $w(G)$ is discrete, and the <u>rank of</u> w is defined as the rank of $w(G)$, as an ordered group. Thus, w is discrete of rank one if and only if $w(G)$ is isomorphic to Z. A family $\{w_\alpha\}_{\alpha \in A}$ of valuations on G is said to be <u>of finite character</u> if, for each $g \in G$, the set $\{\alpha \in A \mid w_\alpha(g) \neq 0\}$ is finite.

Assume that G is the quotient group of S. We say that S is a <u>Krull monoid</u> if there exist a family $\{w_\alpha\}_{\alpha \in A}$

of rank-one discrete valuations on G such that $\{w_\alpha\}_{\alpha \in A}$ has finite character and $S = \{g \in G \mid w_\alpha(g) \geq 0$ for each $\alpha \in A\}$; the latter condition is equivalent to the assertion that S is the intersection of the valuation monoids of the valuations w_α . In Theorem 15.5 we show that $D[S]$ is a Krull domain if D is a Krull domain, S is a Krull monoid, and each element of G is of type $(0,0,\ldots)$. The proof of (15.5) uses several preliminary results, but a proof of the converse of (15.5) can be given at once.

THEROEM 15.1. Let H be the group of invertible elements of S . If $D[S]$ is a Krull domain, then D is a Krull domain, S is a Krull monoid, and each nonzero element of H is of type $(0,0,\ldots)$.

Proof. If K is the quotient field of D , then $D = K \cap D[S]$, and hence D is a Krull domain. Since a Krull domain satisfied a.c.c.p. , it follows that S satisfies a.c.c. on cyclic submonoids, and hence each nonzero element of H is of type $(0,0,\ldots)$. Let G be the quotient group of S , and let $\{v_\alpha\}_{\alpha \in A}$ be a family of rank-one discrete valuations defining $D[S]$ as a Krull domain. For $\alpha \in A$, define $w_\alpha : G \longrightarrow Z$ by $w_\alpha(g) = v_\alpha(X^g)$, and let $B = \{\alpha \in A \mid w_\alpha$ has rank one$\}$. It is clear that the family $\{w_\alpha\}_{\alpha \in B}$ has finite character. Moreover, for $g \in G$,

$$w_\alpha(g) \geq 0 \text{ for each } \alpha \in B \iff w_\alpha(g) \geq 0 \text{ for each } \alpha \in A$$

$$\iff v_\alpha(X^g) \geq 0 \text{ for each } \alpha \in A \iff X^g \in D[S] \iff g \in S .$$

Therefore S is a Krull monoid.

If H is the group of invertible elements of S, then S is factorial if and only if S is of the form $H \dot{\oplus} (Z_0)^{\mu}$ for some cardinal μ. Theorem 15.2 provides an analogue of this result for Krull monoids. The statement of (15.2) uses some new notation. Let $F = \Sigma_{i \in I} Ze_i$ be a free abelian group with free basis $\{e_i\}_{i \in I}$. For $j \in I$, the mapping $\Pi_j : F \longrightarrow Z$ defined by $\Pi_j(\Sigma n_i e_i) = n_j$ is called the jth <u>projection map on</u> F. It is, of course, a rank-one discrete valuation on F. The family $\{\Pi_i\}_{i \in I}$ is of finite character, and we denote by F_+ the Krull monoid determined by this family; thus $F_+ = \{\Sigma n_i e_i \mid n_i \geq 0$ for each $i \in I\}$, the positive cone of F under the cardinal order.

THEOREM 15.2. Let H be the group of invertible elements of S. The following conditions are equivalent.

(1) S is a Krull monoid.

(2) S is of the form $H \oplus T$, where T is of the form $M \cap F_+$ for some free group F and some subgroup M of F.

(3) S is of the form $H \oplus T$, with T of the form $G \cap F_+$, where F is a free group and G is the quotient group of T.

<u>Proof</u>. (1) $\Longrightarrow$ (2): Let U be the quotient group of S and let $\{w_\alpha\}_{\alpha \in A}$ be a family of rank-one discrete valuations on U defining S as a Krull monoid. Let $F = \Sigma Ze_\alpha$ be the free abelian group with free basis $\{e_\alpha\}$ indexed by A, and let $f : G \longrightarrow F$ be the homomorphism defined by $f(g) = \Sigma w_\alpha(g) e_\alpha$. Since F is free, the group G splits as $\ker(f) \oplus G_1$, where $G_1 \simeq f(G)$ under the restriction of f to G_1. We show that $H = \ker(f)$. By definition,

$\ker(f) = \{g \in G \mid w_\alpha(g) = 0$ for each $\alpha \in A\}$. Therefore $\ker(f)$ is a subgroup of H. If $h \in H$, then for each $\alpha \in A$, $w_\alpha(h) \geq 0$ and $w_\alpha(-h) = -w_\alpha(h) \geq 0$, so $w_\alpha(h) = 0$ and $h \in \ker(f)$. Therefore $G = H \oplus G_1$ and $S = H \oplus (G_1 \cap S)$. On the other hand, $G_1 \cap S$ is isomorphic under f to $f(G_1) \cap f(S) = f(G) \cap f(S) = f(S) = f(G) \cap F_+$. Taking $M = f(G)$, this establishes (2).

The implication (2) $\Longrightarrow$ (3) is clear. To prove that (3) implies (1), we regard S and G as subgroups of $H \oplus F$, where $F = \Sigma Z e_i$ is free abelian with free basis $\{e_i\}_{i \in I}$. For each i, let $\mu_i : H \oplus F \longrightarrow Z$ be the mapping defined by $\mu_i(h + \Sigma n_j e_j) = n_i$. We note that the family $\{\mu_i\}_{i \in I}$ has finite character. Let w_i be the restriction of μ_i to G and let $J = \{i \in I \mid w_i$ has rank one$\}$. Then for $g = h + \Sigma n_j e_j \in G$, we have

$$w_i(g) \geq 0 \text{ for each } i \in J \iff w_i(g) \geq 0 \text{ for each } i \in I$$

$$\iff g - h \in G \cap F_+ = T \iff g \in H + T = S .$$

Therefore $\{w_i\}_{i \in J}$ is a family of valuations on G that determines S as a Krull monoid.

Assume that $S = H \oplus T$, where $T = G \cap F_+$ is as in part (3) of Theorem 15.2. Assume that D is a Krull domain and that each nonzero element of H is of type $(0, 0, \ldots)$. Then $D[S] \simeq D[H][T]$, and hence to prove that $D[S]$ is a Krull domain, it suffices to prove that $D[H]$ is a Krull domain and that the monoid ring of T over a Krull domain is again a Krull domain. The next result, which concerns extension of valuations on fields, will take care of the case of D[H] , and that result can be used to dispose of the general case.

If V is a valuation overring of D and if M is the maximal ideal of V , recall that $P = M \cap D$ is called the <u>center</u> of V on D , and that V is <u>essential</u> for D if V is a quotient ring of D , in which case V is necessarily D_P .

THEOREM 15.3. Assume that V is a valuation overring of D with center P on D . Let v be a valuation on the quotient field K of D associated with V . Let Γ be the value group of v and define the function $v^*: D[S]\backslash\{0\} \longrightarrow \Gamma$ as follows. If $f = \Sigma_1^n a_i X^{s_i}$ is the canonical form of the nonzero element f of D[S] , then $v^*(f) = \inf\{v(a_i)\}_1^n$.

(1) v^* determines a valuation on the quotient field L of D[S] ; the value group of v^* is Γ . The valuation ring V^* of v^* is an overring of D[S] with center P[S] on D[S] .

(2) v is essential for D if and only if v^* is essential for D[S] .

(3) If $\{V_\alpha\}_{\alpha\in A}$ is a family of valuation overrings of D such that $D = \cap_{\alpha\in A} V_\alpha$, and if v_α is a valuation on K associated with V_α for each $\alpha \in A$, then $D[S] = K[S] \cap (\cap V_\alpha^*)$. Moreover, if $\{v_\alpha\}$ has finite character, then $\{v_\alpha^*\}$ also has finite character.

<u>Proof</u>. (1): Let $\geq$ be a total order on G compatible with the group operation. Let $f = \Sigma_1^n a_i X^{s_i}$ and $g = \Sigma_1^m b_i X^{t_i}$ be the canonical forms of nonzero elements $f,g \in D[S]$, where

$s_1 < s_2 < \ldots < s_n$ and $t_1 < t_2 < \ldots < t_m$. Choose k and u minimal so that $v^*(f) = v(a_k)$ and $v^*(g) = v(b_u)$. If $f + g \neq 0$ and if $a_j + b_j$ is a nonzero coefficient of $f + g$, then $v(a_j + b_j) \geq \inf\{v(a_j), v(b_j)\} \geq \inf\{v(a_k), v(b_u)\} = \inf\{v^*(f), v^*(g)\}$. Therefore $v^*(f + g) \geq \inf\{v^*(f), v^*(g)\}$. The inequality $v^*(fg) \geq v^*(f) + v^*(g)$ is clear. Moreover, the coefficient of $X^{s_k + t_u}$ in fg has the form

$$a_k b_u + (\text{a sum of terms } a_x b_y \text{, where either } x < k \text{ or } y < u).$$

The v-value of this coefficient is $v(a_k) + v(b_u) = v^*(f) + v^*(g)$. Therefore $v^*(fg) = v^*(f) + v^*(g)$. We have proved that v^* determines a valuation on L, and it is clear that Γ is the value group of v^*. The inclusion $D[S] \subseteq V^*$ holds since v is nonnegative on D, and the very definitions of P and v^* imply that $P[S]$ is the center of v^* on $D[S]$.

(2): The equality $D[S]_{P[S]} \cap K = D_P$ implies that v is essential for D if v^* is essential for $D[S]$. We prove the converse. Since $D_P[S]$ is a quotient ring of $D[S]$, it suffices to show that V^* is essential for $D_P[S]$. Thus, we assume without loss of generality that $D = V$. Let $f = \Sigma_1^n a_i X^{s_i}$ be an arbitrary nonzero element of $D[S]$. If a is a generator of the ideal $(a_1, \ldots, a_n)D$ of D, then f can be written as af^*, where $v^*(f) = v(a)$ and $f^* \in D[S] \backslash P[S]$. Thus, an arbitrary nonzero element of V^* is of the form af^*/bg^*, where $v(a) \geq v(b)$, and hence $a/b \in D$. Consequently, $af^*/bg^* = (a/b)f^*/g^* \in D[S]_{P[S]}$, and $V^* = D[S]_{P[S]}$, as asserted.

(3): For a fixed α, it is clear that $V_\alpha^* \cap K[S] = V_\alpha[S]$. Thus,

$$K[S] \cap (\cap_{\alpha \in A} V_\alpha^*) = \cap_{\alpha \in A}(V_\alpha[S]) = (\cap V_\alpha)[S] = D[S] .$$

Assume that $\{v_\alpha\}$ has finite character and let $f = \Sigma_1^n a_i X^{s_i}$ be the canonical form of a nonzero element $f \in D[S]$. The set $B_i = \{\alpha \in A \mid v_\alpha(a_i) \neq 0\}$ is finite for each i between 1 and n, and hence $B = \cup_{i=1}^n B_i = \{\alpha \in A \mid v_\alpha(a_i) \neq 0$ for some i between 1 and $n\}$ is also finite. For $\alpha \in A \backslash B$, $v_\alpha^*(f) = 0$; hence $\{v_\alpha^*\}$ has finite character. This completes the proof.

THEOREM 15.4. Assume that D is a Krull domain and H is a group such that each nonzero element of H is of type $(0,0,\ldots)$. Then $D[H]$ is a Krull domain.

Proof. Let K be the quotient field of D. If $K = D$, then $D[H]$ is factorial, hence a Krull domain. If $K \neq D$, let $\{v_\alpha\}_{\alpha \in A}$ be a family of rank-one discrete valuations on K defining D as a Krull domain. Let L be the quotient field of $D[H]$ and for each $\alpha \in A$, let v_α^* be the rank-one discrete valuation on L determined as in part (1) of Theorem 15.3. If V_α^* is the valuation ring of v_α^*, then part (3) of Theorem 15.3 shows that $\{v_\alpha^*\}_{\alpha \in A}$ has finite character. Moreover, $D[H] = K[H] \cap (\cap V_\alpha^*)$, and since $K[H]$ is a Krull domain, $D[H]$ is also a Krull domain.

THEOREM 15.5. Let H be the group of invertible elements of S. Assume that D is a Krull domain, that S is a Krull monoid, and that each nonzero element of H is of type $(0,0,\ldots)$. Then $D[S]$ is a Krull domain.

Proof. We have already observed that it suffices to consider the case where $H = (0)$. Then S is of the form $G \cap F_+$, where F is a free group and G is the quotient group of S . Since $G \subseteq F$, each nonzero element of G is of type $(0,0,\ldots)$. Therefore $D[G]$ is a Krull domain by Theorem 15.4. Moreover, $D[F_+]$ is a Krull domain since it is isomorphic to a polynomial ring over D . It follows that $D[S] = D[G] \cap D[F_+]$ is also a Krull domain.

Combining Theorems 15.1 and 15.5, we have proved the following summary result.

THEOREM 15.6. Assume that D is a unitary integral domain and that S is a torsion-free cancellative monoid. Let H be the group of invertible elements of S . The following conditions are equivalent.

(1) $D[S]$ is a Krull domain.

(2) D is a Krull domain, S is a Krull monoid, and each nonzero element of H is of type $(0,0,\ldots)$.

If D is a Krull domian, it is known that the set of nontrivial essential valuation overrings of D is $\{D_{P_\alpha}\}_{\alpha\in A}$ where $\{P_\alpha\}_{\alpha\in A}$ is the set of minimal primes of D . Each D_{P_α} is rank-one discrete, the family $\{D_{P_\alpha}\}$ is of finite character, and $D = \cap_{\alpha\in A} D_{P_\alpha}$. The family $\{D_{P_\alpha}\}$ is called the defining family for the Krull domain D ; it is contained in each family F of rank-one discrete valuation overrrings of D such that F has finite character and $D = \cap F$, and hence is the unique family of essential valuation overrings of D satisfying the three properties listed. Theorem 15.3 and the proof of Theorem 15.4 determine the defining family for $D[H]$, for D and H as in the statement of (15.4), as follows. If K is the quotient field of D , then $K[H]$ is factorial and the defining family for $K[H]$ is $\{K[H]_{(q_i)}\}_{i\in I}$ where $\{q_i\}_{i\in I}$ is complete set of nonassociate prime elements of $K[H]$. Each of these valuation rings $K[H]_{(q_i)}$ is essential for $D[H]$ since $K[H]$ is a quotient ring of $D[H]$. Theorem 15.3 and the proof of (15.4) show that

$$D[H] = K[H] \cap (\cap_{\alpha\in A} D[H]_{P_\alpha[H]}) ,$$

where $\{P_\alpha\}_{\alpha\in A}$ is the family of minimal primes of D . This equality implies that the defining family for $D[H]$ is

$$\{K[H]_{(q_i)}\}_{i\in I} \cup \{D[H]_{P_\alpha[H]}\}_{\alpha\in A} .$$

For a general Krull domain $D[S]$, where $S = H \oplus (G \cap F_+)$ as in part (3) of Theorem 15.2, the proof of Theorem 15.5 shows that the defining family for $D[S]$ consists of the defining family for $D[H][G]$, together with those members of the defining family for $D[H][F_+]$ that are essential for $D[S]$. We note that $D[H][G] \simeq D[H \oplus G]$, where $H \oplus G$ is isomorphic to the quotient group of S ; the defining family for $D[H][G]$ is determined in the preceding paragraph. Let L be the quotient field of $D[H]$. Using the fact that $L[F_+]$ is factorial, the defining family for $D[H][F_+]$ can be determined from part (3) of Theorem 15.3 as

$$\{L[F_+]_{(h_j)}\}_{j\in J} \cup \{D[H][F_+]_{Q_\beta[F_+]}\}_{\beta\in B} ,$$

where $\{h_j\}_{j\in J}$ is a complete set of nonassociate prime elements of $L[F_+]$ and where $\{Q_\beta\}_{\beta\in B}$ is the set of minimal prime ideals of $D[H]$. We pursue this line of reasoning no further, for there is a better approach to determining the essential valuations for $D[S]$ that are not in the defining family for $D[H][G]$. The approach alluded to involves proving an analogue of Theorem 15.3 for valuations on a group.

THEOREM 15.7. Let G be the quotient group of S and let w be a valuation on G whose associated valuation monoid W contains S . Let Γ be the value group of w and define the function $w^*: D[S]\setminus\{0\} \longrightarrow \Gamma$ as follows. If $f = \Sigma_1^n a_i X^{s_i}$ is the canonical form of the nonzero element f of S , then $w^*(f) = \inf\{w(s_i)\}_1^n$.

(1) w^* determines a valuation on the quotient field L of $D[S]$; the value group of w^* is Γ . The valuation ring W^* is an overring of D with center $D[S\setminus T]$ on $D[S]$, where T is the submonoid of S consisting of all $s \in S$ such that $w(s) = 0$.

(2) w^* is essential for $D[S]$ if and only if W is the quotient monoid of S with respect to T .

(3) Assume that $\{w_\alpha\}_{\alpha \in A}$ is a family of valuations on G such that $S = \cap_\alpha W_\alpha$, where W_α is the valuation monoid of w_α . Then $D[S] = D[G] \cap (\cap W_\alpha^*)$. Moreover, if $\{w_\alpha\}$ has finite character, then $\{w_\alpha^*\}$ also has finite character.

<u>Proof</u>. (1) : Let $\geq$ be a total order on G compatible with the group operation. Take nonzero elements $f,g \in D[S]$, and let $f = \Sigma_1^n a_i X^{s_i}$ and $g = \Sigma_1^m b_i X^{t_i}$ be the canonical forms of f and g , where $s_1 < s_2 < \ldots < s_n$ and $t_1 < \ldots < t_m$. Choose k and u minimal such that $w^*(f) = w(s_k)$ and $w^*(g) = w(t_u)$. If $f \neq -g$ and if $s \in \mathrm{Supp}(f + g) \subseteq \mathrm{Supp}(f) \cup \mathrm{Supp}(g)$, then either

$w(s) \geq w(s_k)$ or $w(s) \geq w(t_u)$. Either way, $w(s) \geq \inf\{w^*(f), w^*(g)\}$, and hence $w^*(f+g) \geq \inf\{w^*(f), w^*(g)\}$. The inequality $w^*(fg) \geq w^*(f) + w^*(g)$ is clear, and the reverse inequality holds since $s_k + t_u \in \mathrm{Supp}(fg)$; to wit $a_k b_u$ is the coefficient of $X^{s_k+t_u}$ in fg by choice of k and u . Therefore w^* determines a valuation on L with value group Γ . The inclusion $D[S] \subseteq W^*$ holds since w is nonnegative on S , and the assertion concerning the center of w^* is immediate from the definition of w^* .

(2) : Assume that w^* is essential for $D[S]$. For $g \in W$, write $X^g = f/h$ for some $f \in D[S]$, $h \notin D[S \setminus T]$. There exists $t \in \mathrm{Supp}(h) \cap T$. From the equality $X^g h = f$, it follows that $g + t \in S$, and hence $g = (g+t) - t$ belongs to the quotient monoid Y of S with respect to T . Thus $W \subseteq Y$, and the reverse inclusion is clear. For the converse, we note that $U = \{X^t \mid t \in T\}$ is a multiplicative system in $D[S]$ consisting of units of W^* . Thus, to show that W^* is essential for $D[S]$, it suffices to show that it is essential for $D[S]_U = D[W]$. Each nonzero element f of $D[W]$ is expressible in the form $X^a f^*$, where $w^*(f) = w(a)$, $f^* \in D[W]$, and $w^*(f^*) = 0$. Thus, an arbitrary element of W^* is of the form $X^a f^*/X^b g^*$, where $w(a) \geq w(b)$ and where f^* and g^* are not in M_0 , the center of W^* on $D[W]$. It follows that $X^a f^*/X^b g^* = X^{a-b} f^*/g^* \in D[W]_{M_0}$, and hence $W^* = D[W]_{M_0}$.

(3) : For a fixed α , we have $D[G] \cap W_\alpha^* = D[W_\alpha]$. Therefore

$$D[G] \cap (\cap_\alpha W_\alpha^*) = \cap_\alpha D[W_\alpha] = D[\cap W_\alpha] = D[S] .$$

Assume that $\{w_\alpha\}$ has finite character and let $f = \Sigma_1^n a_i X^{s_i}$ be the canonical form of a nonzero element $f \in D[S]$. As in the proof of part (3) of Theorem 15.3, the set $B = \{\alpha \in A \mid w_\alpha(s_i) \neq 0$ for some i between 1 and $n\}$ is finite, and $w_\alpha^*(f) = 0$ for each $\alpha \in A \setminus B$. Thus $\{w_\alpha^*\}$ is also of finite character.

If G is the quotient group of S and if w is a valuation on G whose valuation monoid W contains S, then we say that w is <u>essential for</u> S (and that W is <u>essential for</u> S) if W is the quotient monoid of S with respect to $T = \{s \in S \mid w(s) = 0\}$. The next result follows from Theorem 15.7.

THEOREM 15.8. Assume that S is a Krull monoid with quotient group G and that each nonzero element of G is of type $(0,0,\ldots)$. Let $\{W_\alpha\}_{\alpha \in A}$ be the family of rank-one discrete valuation monoids on G that are essential for S. Then $S = \cap_{\alpha \in A} W_\alpha$, and $\{W_\alpha\}$ has finite character. If $\{U_\beta\}_{\beta \in B}$ is any family of rank-one discrete valuation monoids on G of finite character such that $S = \cap_{\beta \in B} U_\beta$, then $\{W_\alpha\} \subseteq \{U_\beta\}$.

<u>Proof</u>. Let w_α be a valuation associated with W_α, u_β a valuation associated with U_β, and let K be a field. If F is the defining family for $K[G]$, then part (3) of Theorem 15.7 shows that $F \cup \{U_\beta^*\}_{\beta \in B}$ is a defining family for the Krull domain $K[S]$. The defining family for $K[S]$ is $F \cup \{U_\gamma^*\}_{\gamma \in C}$, where $C = \{\beta \in B \mid U_\beta^*$ is essential for $S\}$. We have $K[S] = K[G] \cap (\cap_{\gamma \in C} U_\gamma^*) = K[\cap_\gamma U_\gamma]$ (see the proof of Theorem 15.7), so that $S = \cap_\gamma U_\gamma$. The inclusion

$\{U_\gamma\} \subseteq \{W_\alpha\}$ is clear. On the other hand, each W_α^* is an essential rank-one discrete valuation overring of $D[S]$, and hence belongs to the defining family $F \cup \{U_\gamma^*\}$. The center of W_α^* on $D[S]$ is $D[S \setminus T_\alpha]$, where $T_\alpha = \{s \in S \mid w_\alpha(s) = 0\}$. This ideal extends to $K[G]$ in $K[G]$, for X^s is a unit of $K[G]$ for each $s \in S \setminus T_\alpha$. It follows that $W_\alpha^* \notin F$, and hence $W_\alpha^* = U_\gamma^*$ for some $\gamma \in C$. Thus $K[G] \cap W_\alpha^* = K[W_\alpha] = k[G] \cap U_\gamma^* = K[U_\gamma]$, so $W_\alpha = U_\gamma$. We conclude that $\{U_\gamma\} = \{W_\alpha\}$, and this completes the proof.

COROLLARY 15.9. Assume that $D[S]$ is a Krull domain. The defining family for $D[S]$ is

$$\{K[G]_{(q_i)}\}_{i\in I} \cup \{D[S]_{P_\alpha[S]}\}_{\alpha\in A} \cup \{W_\beta^*\}_{\beta\in B},$$

where the notation is as follows: K is the quotient field of D, G is the quotient group of S, $\{q_i\}_{i\in I}$ is a complete set of nonassociate prime elements of $K[G]$, and $\{P_\alpha\}_{\alpha\in A}$ is the set of minimal primes of D; finally, $\{W_\beta\}_{\beta\in B}$ is the family of rank-one discrete valuation monoids on G that are essential for S.

Several remarks on the statement of Theorem 15.8 and Corollary 15.9 seem appropriate. First, (15.8) is stated for valuation monoids on G, rather than valuations, for the following reasons. If w is a rank-one discrete valuation on G with value group dZ, where $d > 0$, then the valuation monoid W of w is $\ker(w) \oplus \langle g\rangle^0$, where $w(g) = d$. Note that $\ker(w)$ is precisely the group of invertible elements of W. Thus W does not determine w uniquely, but rather up to the value of w on g. This wee difficulty can be avoided by assuming that all rank-one discrete valuations dealt with have value group Z (such a valuation is said to be normed). Working with normed valuations, Theorem 15.8 can be rephrased by replacing $\{W_\alpha\}$ and $\{U_\beta\}$ by $\{w_\alpha\}$ and $\{u_\beta\}$, respectively. In the statement of Corollary 15.9, neither of the families $\{K[G]_{(q_i)}\}_{i\in I}$ or $\{W^*_\beta\}$ is described implicitly in terms of the domain $D[S]$; in the other direction, $D[S]_{P_\alpha[S]}$ could be written as $D[G]_{P_\alpha[G]}$. With $K[G]_{(q_i)}$, a description in terms of $D[S]$ is easily obtained: $K[G]_{(q_i)} = D[S]_{(q_i)\cap D[S]}$. Similarly, $W^*_\beta = D[S]_{D[S\setminus T_\beta]}$, and in Theorem 16.10 of Section 16 we show that $\{S\setminus T_\beta\}_{\beta\in B}$ is the set of minimal prime ideals of S.

If S is finitely generated, then the quotient group G of S is also finitely generated. The converse need not hold in general, but we show in Theorem 15.11 that the converse holds if S is a Krull monoid. An ancillary result is used in the proof of (15.11).

THEOREM 15.10. Let A be an infinite subset of F_+, where $F = \Sigma_1^k Ze_i$ is a free group of finite rank k. Denote by $\leq$ the cardinal order on F. Then A contains an infinite strictly ascending sequence under $\leq$.

Proof. We use induction on k. The case where $k = 1$ is patent. In the case where $k > 1$, let $M = \Pi_1(A)$ be the set of first coordinates of elements of A. If $\Pi_1^{-1}(m)$ is infinite for some $m \in M$, then the induction hypothesis implies that $\Pi_1^{-1}(m)$, and hence A, contains an infinite strictly increasing sequence. On the other hand, if $\Pi_1^{-1}(m)$ is finite for each $m \in M$, then there exists an infinite subset B of A such that distinct elements of B have distinct first coordinates. Thus, we may assume that $B = \{b_i\}_1^\infty$, where $\Pi_1(b_1) < \Pi_1(b_2) < \ldots$. Let $c_i = b_i - \Pi_1(b_i)e_1$ for each i. If $\{c_i\}_{i=1}^\infty$ is a finite set, then there exists $c \in \{c_i\}_1^\infty$ such that the set $I = \{i \in Z^+ \mid c_i = c\}$ is infinite, and $\{b_i\}_{i \in I}$ is an infinite strictly ascending sequence in A. In the contrary case, where $\{c_i\}_1^\infty$ is infinite, the induction hypothesis yields an infinite strictly ascending sequence $c_{i_1} < c_{i_2} < \ldots$. Let $j_1 = i_1$. Only finitely many elements of the set $\{\Pi_1(b_i)\}_1^\infty$ are less than $\Pi_1(b_{j_1})$. Let j_2 be the smallest integer i_t such that $\Pi_1(b_{i_t}) > \Pi_1(b_{j_1})$. Then $b_{j_1} < b_{j_2}$ and we can repeat the argument on b_{j_2}. To wit, only finitely many elements of $\{\Pi_1(b_i)\}_1^\infty$ are less than $\Pi_1(b_{j_2})$. We let j_3 be the smallest i_u such that $\Pi_1(b_{i_u}) > \Pi_1(b_{j_2})$. Then $b_{j_1} < b_{j_2} < b_{j_3}$, and a continuation of the process yields an infinite strictly ascending sequence $\{b_{j_r}\}$ in A. The result follows by the principle of mathematical induction.

THEOREM 15.11. Assume that S is a Krull monoid with quotient group G. If G is finitely generated as a group, then S is finitely generated as a monoid.

<u>Proof</u>. Assume that G is generated by $\{g_i\}_1^n$. A family $\{w_\alpha\}_{\alpha \in A}$ of nontrivial valuations on G of finite character is necessarily finite since $\{\alpha \in A \mid w_\alpha(g_i) \neq 0\}$ is finite for each i between 1 and n. In particular, a defining family for S is finite. The proof of Theorem 15.2

shows that S is of the form $H \oplus T$, where H is the group of invertible elements of S and T is of the form $M \cap F_+$, where F is a finitely generated free group and M is a subgroup of F. Since $H \subseteq G$, H is finitely generated both as a group and as a monoid. Hence, it suffices to show that T is finitely generated as a monoid. Consider F as a partially ordered group under the cardinal order $\leq$. Theorem 15.10 shows that the set B of nonzero minimal elements of T is finite. Let U be the submonoid of T generated by B. To complete the proof, we show that $T \subseteq U$. Thus, take a nonzero element $t \in T$. Only finitely many elements of F_+ are less than t, so $t \geq b_1$ for some element $b_1 \in B$. If $t = b_1$, then $t \in U$. If $t \neq b_1$, then $t > t - b_1$ and $t - b_1 \in M \cap F_+ = T$. Repeating the process, there exists $b_2 \in B$ such that $b_2 \leq t - b_1$. If $t - b_1 = b_2$, we're finished, and in the other case, $t > t - b_1 > t - b_1 - b_2$ is a decreasing sequence in T. Again because only finitely many elements of F_+ are less than t, it follows that $t = b_1 + \ldots + b_k \in U$ for some k. Therefore $T = U$, and this completes the proof.

If T is of the form $M \cap F_+$, where $F = \Sigma_{\alpha \in A} Ze_\alpha$ is free on $\{e_\alpha\}_{\alpha \in A}$, then the monoid domain $D[T]$ can be regarded as a subring of the polynomial ring $D[\{X_\alpha\}_{\alpha \in A}]$ over D. Moreover, $D[T]$ is generated as a ring over D by "pure monomials" $X_{\alpha_1}^{e_1} X_{\alpha_2}^{e_2} \ldots X_{\alpha_n}^{e_n}$, with $e_i \geq 0$ for each i. Conversely, each ring $D[\{m_\beta\}_{\beta \in B}]$, where each m_β is a pure monomial in the indeterminates X_α, is of the form $D[U]$, where U is a submonoid of F_+. Rings of the form $D[\{m_\beta\}]$ arise as objects of natural interest in various parts of commutative algebra. Thus, it seems worthwhile to state the following consequence of Theorem 15.11 as a separate result.

COROLLARY 15.12. Assume that D is a Noetherian integrally closed domain and that $\{X_i\}_{i=1}^n$ is a finite set of indeterminates over D. Let $\{m_\alpha\}_{\alpha\in A}$ be a set of pure monomials in the indeterminates X_i, let T be the monoid generated by $\{m_\alpha\}$, and let $J = D[\{m_\alpha\}]$. The following conditions are equivalent.

(1) T is finitely generated and integrally closed.

(2) J is Noetherian and integrally closed.

(3) J is a Krull domain.

We remark that under the notation of Corollary 15.12, integral closure of T does not, in general, imply that T is finitely generated. For example, $D[\{XY^i\}_{i=0}^{\infty}] = D + XD[X,Y]$ is a non-Noetherian integrally closed domain for each integrally closed domain D.

Section 15 Remarks

Three general references on Krull domains are [20, Ch. VII], [44], and [51]. A number of other classes of domains that are related to Krull domains have been considered in the literature. Among classes whose definitions involve certain conditions on the family of valuation overrings of the given domain, there are the classes of domains of finite character, domains of finite real character, domains of finite rational character, domains of Krull type, and generalized Krull domains (for the definitions, see Section 43 of [51]). Each of these concepts has a natural extension to monoids, so that one could consider the classes of (always torsion-free and cancellative) monoids of finite character, etc. Using Theorems 15.3 and 15.7 and other techniques from this section, it can be shown that the monoid domain D[S] is in one of the named classes if and only if D[G] belongs to the class, and S belongs to the monoid class of the same name; here G denotes the quotient group of S . For D of characteristic 0 , Matsuda in [97] has shown that D[G] is in a given class if and only if D is in the class and each element of G is of type (0,0,...) . For D of characteristic $p > 0$, the conditions are almost the same; the difference in this case is that divisibility of nonzero elements of G by arbitrary powers of p does not affect whether or not D[G] belongs to the class (see [100] for details).

Another class of domains related to Krull domains are the Π-domains, where D is defined to be a Π-domain if each principal ideal of D is a finite product of prime ideals of D . According to Theorem 46.7 of [51], D is a Π-domain iff D is a Krull domain whose minimal primes are invertible iff D_M is factorial for each maximal ideal M of D and minimal primes of D are finitely generated. Anderson and Anderson [5] have shown that D[S] is a Π-domain if and only if D is a Π-domain, S is factorial, and each element of the quotient group of S is of type (0,0,...) .

§16. The Divisor Class Group of a Krull Monoid Domain

Theorem 15.6 gives necessary and sufficient conditions for a monoid domain $D[S]$ to be a Krull domain, and subsequent results in Section 15, such as Corollary 15.9, determine the defining family for $D[S]$ and the family of minimal primes of $D[S]$. Using heavily these results from Section 15, we determine in Corollaries 16.7 and 16.9 the divisor class group of $D[S]$. To begin, we review some of the basic terminology, results, and notation concerning the v-operation on an integral domain, and in particular on a Krull domain. Later in the section we replicate much of this development for cancellative monoids, assuming no familiarity on the part of the reader with the results in the context of monoids.

Let D be a unitary integral domain with quotient field K and denote by $F(D)$ the set of nonzero fractional ideals of D. For $F \in F(D)$, F^{-1} is defined to be $D:F = \{x \in K \mid xF \subseteq D\}$, and the fractional ideal $(F^{-1})^{-1}$ is denoted by F_v. Equivalently, F_v is the intersection of the family of principal fractional ideals of D that contain F. The mapping $F \longrightarrow F_v$ is called the v-operation on D, and a fractional ideal F is said to be divisorial (or a **v-ideal**) if $F = F_v$. The v-operation on D induces an equivalence relation $\sim$ on $F(D)$ defined by setting $A \sim B$ if $A_v = B_v$. The equivalence classes under $\sim$ are called divisor classes of D, the class of $A \in F(D)$ is denoted by $\operatorname{div}(A)$, and the set of all divisor classes of D is denoted by $\mathcal{D}(D)$. Under the operation $\operatorname{div}(A) + \operatorname{div}(B) = \operatorname{div}(AB)$, $\mathcal{D}(D)$ is a commutative monoid with zero element $\operatorname{div}(D)$, and $\mathcal{D}(D)$ is a group if and only if D is completely integrally closed. The set $\mathcal{P}(D) = \{\text{div}(xD) \mid x \in K, x \neq 0\}$ is a subgroup of the group of invertible elements of $\mathcal{D}(D)$, and $\mathcal{C}(D) = \mathcal{D}(D)/\mathcal{P}(D)$ is called the <u>divisor class monoid</u> of D; if D is completely integrally closed, then $\mathcal{C}(D)$ is the <u>divisor class group</u> of D. For $F \in \mathcal{F}(D)$, we denote by $[\text{div}(F)]$ the element $\text{div}(F) + \mathcal{P}(D)$ of $\mathcal{C}(D)$ determined by $\text{div}(F)$.

We now specialize the discussion of the preceding paragraph to the case of a Krull domain. Thus, assume that D is a Krull domain, that $\{P_i\}_{i \in I}$ is the family of minimal prime ideals of D, and that v_i is a normed valuation on K associated with the valuation domain $V_i = D_{P_i}$ for each $i \in I$. In this case it is known that $\{\text{div}(P_i)\}_{i \in I}$ is a free basis for the group $\mathcal{D}(D)$. In fact, if $F \in \mathcal{F}(D)$, then $F_v = \cap_{i \in I} FV_i$, and $FV_i = V_i$ for all but a finite set of subscripts i. Moreover, for an integral ideal A we have

$$A_v = \cap_{j=1}^{k} P_j^{(n_j)} = (\cap_1^k P_j^{n_j})_v = (\Pi_1^k P_j^{n_j})_v$$

is a finite intersection of symbolic powers of the minimal primes P_j of D that contain A. The first result of the section relates the divisor class group of D to that of a quotient overring D_N of D.

THEOREM 16.1. Let N be a multiplicative system in the Krull domain D. The mapping $F \longrightarrow FD_N$ of $\mathcal{F}(D)$ into $\mathcal{F}(D_N)$ induces a homomorphism ϕ of $\mathcal{C}(D)$ onto $\mathcal{C}(D_N)$. The kernel of ϕ is generated by $\{[\text{div}(P)] \mid P$ is a minimal prime of D that meets $N\}$.

<u>Proof</u>. Consider the mapping $\sigma : \mathcal{D}(D) \longrightarrow \mathcal{D}(D_N)$ defined

by $\sigma(\mathrm{div}(F)) = \mathrm{div}(FD_N)$ for $F \in \mathcal{F}(D)$. To show that σ is well-defined, we first show that $(FD_N)_v = F_v D_N$. There exists an integral ideal A of D and a nonzero element d of D such that $F = d^{-1}A$. Hence $F_v = d^{-1}A_v$ and $(FD_N)_v = d^{-1}(AD_N)_v$. Thus, it suffices to prove the equality $(AD_N)_v = A_v D_N$. If each minimal prime of D containing A meets N, then it is clear that $(AD_N)_v = D_N = A_v D_N$. Thus, assume that the set $\{P_j\}_1^r$ of minimal primes of D containing A is labeled so that P_j does not meet N for $1 \le j \le k$, while P_j meets N for $k + 1 \le j \le r$. Then

$$A_v = \cap_{j=1}^r P_j^{(n_j)} , \text{ where } AD_{P_j} = P_j^{n_j} D_{P_j}$$

for each j . We have

$$A_v D_N = \cap_1^r (P_j^{(n_j)} D_N) = \cap_1^k (P_j^{(n_j)} D_N) = \cap_1^k (P_j D_N)^{(n_j)} .$$

Moreover, $\{P_j D_N\}_1^k$ is the set of minimal primes of D_N that contain AD_N and

$$(AD_N)(D_N)_{P_j D_N} = AD_{P_j} = P_j^{n_j} D_{P_j} = (P_j D_{P_j})^{n_j} (D_N)_{P_j D_N} \text{ for each } j .$$

Therefore $(AD_N)_v = \cap_{j=1}^k (P_j D_N)^{(n_j)} = A_v D_N$. Thus, if $\mathrm{div}(F) = \mathrm{div}(G)$, then $F_v = G_v$, and hence $\mathrm{div}(FD_N) = \mathrm{div}((FD_N)_v) = \mathrm{div}(F_v D_N) = \mathrm{div}(G_v D_N) = \mathrm{div}(GD_N)$. This shows that σ is well-defined. It is clear that σ preserves addition, and σ is surjective since each integral ideal of D_N is extended from D . Since $\mathcal{D}(D)$ is free on $\{\mathrm{div}(P)\}$ and $\mathcal{D}(D_N)$ is free on $\{\mathrm{div}(PD_N) \mid P$ does not meet $N\}$, it follows that $\{\mathrm{div}(P) \mid P$ meets $N\}$ generates the kernel of σ .

Since $\sigma(P(D)) \subseteq P(D_N)$, then σ induces a surjective homomorphism $\phi: C(D) \longrightarrow C(D_N)$ defined by $\phi([\mathrm{div}(F)]) = [\mathrm{div}(FD_N)]$. Suppose $[\mathrm{div}(F)]$ belongs to the kernel of ϕ . Then $F_vD_N = yD_N$ is principal, so $\mathrm{div}(y^{-1}F) \in \ker \sigma$, and hence $\mathrm{div}(y^{-1}F) = \mathrm{div}(P_1^{k_1}\ldots P_m^{k_m})$ for some set $\{P_i\}_1^m$ of minimal primes of D meeting N and some set of integers $\{k_i\}_1^m$. It follows that $[\mathrm{div}(F)] = [\mathrm{div}(P_1^{k_1}\ldots P_m^{k_m})] = \Sigma_1^m k_i[\mathrm{div}(P_i)]$, and hence $\{[\mathrm{div}(P)] \mid P$ is a minimal prime of D that meets N$\}$ generates $\ker \phi$, as asserted.

There are strong similarities between the proofs of Theorem 16.1 and the next result, which relates the class group of a Krull domain $D[S]$ to that of D .

THEOREM 16.2. Assume that the monoid domain $D[S]$ is a Krull domain.

(1) If A is a nonzero integral ideal of D , then $(A[S])_v = A_v[S]$.

(2) The mapping $F \longrightarrow FD[S]$ of $F(D)$ into $F(D[S])$ induces an injective homomorphism ϕ of $C(D)$ into $C(D[S])$.

(3) If S is a group, then ϕ is surjective and $C(D[S]) \simeq C(D)$.

Proof. (1): Let $\{P_i\}_1^k$ be the set of minimal primes of D that contain A and for $1 \le i \le k$, let v_i be a normed valuation on K , the quotient field of D, associated with the valuation ring $V_i = D_{P_i}$. Corollary 15.9 shows that $\{P_i[S]\}_1^k$ is the set of minimal primes of $D[S]$ that contains $A[S]$. Let v_i^* be the valuation associated with $D[S]_{P_i[S]}$, defined as in the statement of Theorem 15.3 We have

$A_v = \cap_{i=1}^{k} P_i^{(m_i)}$, where $m_i = \inf\{v_i(a) \mid a \in A\}$, and

$(A[S])_v = \cap_1^k (P_i[S])^{(n_i)}$, where $n_i = \inf\{v_i^*(b) \mid b \in A[S]\}$.

It is clear from the definition of v_i^* that $n_i = m_i$ for each i . Since

$$A_v D[S] = (\cap_1^k P_i^{(n_i)}) D[S] = \cap_1^k (P_i^{(n_i)}[S]) ,$$

to complete the proof of (1) , it suffices to show that $P_i^{(n_i)}[S] = (P_i[S])^{(n_i)}$ for each i . This follows, however, from part (2) of Corollary 8.7: since S is torsion-free and cancellative, $P_i^{(n_i)}[S]$ is $P_i[S]$-primary , and hence is the contraction to $D[S]$ of

$$P_i^{(n_i)}[S] \cdot D[S]_{P_i[S]} = (P_i[S]D[S]_{P_i[S]})^{n_i} .$$

This completes the proof of (1) .

(2): The mapping ϕ is defined by $\phi([\mathrm{div}(F)]) = [\mathrm{div}(F \cdot D[S])]$. Using (1) , the proof that ϕ is well-defined is the same as the proof of the same assertion for σ in Theorem 16.1. Clearly ϕ is a homomorphism. Suppose $[\mathrm{div}(F)] \in \ker \phi$. Without loss of generality we assume that F is an integral ideal of D . Then $F_v D[S] = fD[S]$ is principal. Since $fD[S]$ contains monomials, $f = aX^s$ is a monomial, and since $F_v \subseteq aX^s D[S]$, it follows that s is invertible in S , and hence $F_v D[S] = aD[S]$. Contraction to D then yields $F_v = aD$, so that $[\mathrm{div}(F)] = 0$ and ϕ is injective, as asserted.

(3): Suppose S is a group. Theorem 15.6 shows that each nonzero element of S is of type $(0,0,\ldots)$. Hence, if $N = D\backslash\{0\}$, then $(D[S])_N = K[S]$ is factorial. Thus

$K[S]$ has trivial class group, and Theorem 16.1 shows that $C(D[S])$ is generated by $\{[\mathrm{div}(Q)] \mid Q$ is a minimal prime of $D[S]$ that meets $N\}$. Applying Corollary 15.9, it follows that $D[S]$ is generated by $\{[\mathrm{div} P_\alpha[S]] \mid P_\alpha$ is a minimal prime of $D\}$. Since each $[\mathrm{div} P_\alpha[S]]$ belongs to the range of ϕ, we conclude that ϕ is an isomorphism of $C(D)$ onto $C(D[S])$.

Suppose $D[S]$ is a Krull domain. Part (3) of Theorem 16.2 determines $C(D[S])$ in terms of D and S for a group S. Using this result, Theorem 16.3 advances such a determination in the case where S need not be a group.

THEOREM 16.3. Assume that $D[S]$ is a Krull domain. Let K be the quotient field of D, let G be the quotient group of S, and let H be the group of invertible elements of S.

(1) $C(D[S]) \simeq C(D) \oplus C(K[S])$.

(2) As in the statement of Theorem 15.2, express S as $H \oplus T$, where $T = G \cap F_+$, with F a free group. Then $C(K[S]) \simeq C(L[T])$, where L is the quotient field of $K[H]$.

<u>Proof</u>. (1): Let $\phi: C(D) \longrightarrow C(D[S])$ be defined as in the statement of Theorem 16.2. Since $D[G]$ is a quotient ring of $D[S]$, Theorem 16.1 shows that the mapping $\alpha: C(D[S]) \longrightarrow C(D[G])$ defined by $\alpha([\mathrm{div} F]) = [\mathrm{div} FD[G]]$ is a homomorphism. Moreover, part (3) of Theorem 16.2 shows that the map $\mu: C(D) \longrightarrow C(D[G])$ defined by $\mu([\mathrm{div} F]) = [\mathrm{div}\, FD[G]]$ is an isomorphism. Since $\mu^{-1}\alpha\phi$ is the identity map on $C(D)$, it follows that $\phi(C(G))$ is a direct summand of $C(D[S])$ — say $C(D[S]) = \phi(C(D)) \oplus M$. If

$N = D \setminus \{0\}$, then as in the proof of Theorem 16.2, there exists a surjective homomorphism τ from $C(D[S])$ onto $C(K[S])$ with kernel generated by $\{[\mathrm{div} P_\alpha[S]] \mid P_\alpha$ is a minimal prime of $D\}$; since this latter subgroup is precisely $\phi(C(D))$, it follows that $C(K[S]) \simeq C(D[S])/\phi(C(D)) \simeq M$. Therefore $C(D[S]) = \phi(C(D)) \oplus M \simeq C(D) \oplus C(K[S])$, where the isomorphism $C(D) \simeq \phi(C(D))$ follows from part (2) of Theorem 16.2. This completes the proof of part (1) .

(2) : We have $K[S] \simeq K[H][T]$, so part (1) shows that $C(K[S]) \simeq C(K[H]) \oplus C(L[T])$. But Theorems 14.15 and 15.1 show that $K[H]$ is factorial, so $C(K[H]) = \{0\}$, and $C(K[S]) \simeq C(L[T])$, as asserted.

In view of Theorem 16.3, we are left with the problem of determining $C(L[T])$, where L is a field and T is of the form $G_1 \cap F_+$, where G_1 is the quotient group of T . We handle this problem in two stages. First, we define the v—operation on an arbitrary cancellative monoid S . For S completely integrally closed, we define the divisor class group $C(S)$ of S . This development is analogous to the usual treatment for integral domains, and hence will provide a review of the material concerning divisorial ideals, class groups, etc. in domains that comprises most of the introduction to this section. If K is a field and $K[S]$ is a Krull domain, we prove in Theorem 16.6 that $C(K[S]) \simeq C(S)$. This result can be viewed as the counterpart of part (3) of Theorem 16.2. Then for S of the form $G \cap F_+$, we prove in Theorem 16.8 that $C(S) \simeq F/G$. When combined with Theorem 16.3, these results provide the description of $C(D[S])$, for $D[S]$ Krull, that we seek. To begin, we develop the theory of the v—operation and related concepts for a cancellative monoid.

Let S be a cancellative monoid with quotient group G. A nonempty subset I of G is called a fractional ideal of S if (1) $S + I \subseteq I$ and (2) there exists $s \in S$ such that $s + I \subseteq S$. We remark that a fractional ideal of S need not be a subsemigroup of G. A fractional ideal I is said to be principal if $I = x + S$ for some $x \in G$. Denote by $F(S)$ the set of fractional ideals of S. Under ordinary addition of subsets of G — that is, $A + B = \{a + b \mid a \in A$ and $b \in B\}$ — $F(S)$ is a commutative monoid with zero element S. If $I, J \in F(S)$, then $I:J$ is defined to be $\{x \in G \mid x + J \subseteq I\}$. We show in part (1) of Theorem 16.4 that $I:J \in F(S)$. The fractional ideal $S:(S:I)$ is denoted by I_v and is called the divisorial ideal associated with I; if $I = I_v$, then I is divisorial.

THEOREM 16.4. Let the notation and hypothesis be as in the preceding paragraph.

(1) If $I, J \in F(S)$, then $I:J \in F(S)$.

(2) $I:(x + J) = -x + (I:J)$ for each $x \in G$. In particualr, $S:(S:(x + S)) = S:(-x + S) = x + S$, so $x + S$ is divisorial.

(3) If $J_1 \subseteq J_2$, then $I:J_1 \supseteq I:J_2$. Hence $(J_1)_v \subseteq (J_2)_v$.

(4) I_v is the intersection of the family of all principal fractional ideals of S that contain I.

(5) $(I_v)_v = I_v$.

(6) $(x + I)_v = x + I_v$ for each $x \in G$, $I \in F(S)$.

(7) $(I + J)_v = (I_v + J_v)_v$.

Proof. (1): Take $L \in F(S)$. We first observe that $L \cap S \neq \phi$ and that $(L \cap S) + S \subseteq L$. The second assertion is clear. For the first, take $y \in L$ and write it as $u - v$, where $u, v \in S$. Then $u = v + y \in S + L \subseteq L$, and hence $u \in L \cap S$. To show that $I:J$ is nonempty, choose $u \in I \cap S$ and $a \in S$ such that $a + J \subseteq S$. Then $u + a + J \subseteq u + S \subseteq I$, so $u + a \in I:J$. That $S + (I:J) \subseteq I:J$ is clear from the definitions. Finally, take $v \in J \cap S$ and take $b \in S$ such that $b + I \subseteq S$. If $x \in I:J$, then $b + v + x \in b + I \subseteq S$, so $b + v + (I:J) \subseteq S$, and this completes the proof of (1).

The proofs of (2) and (3) are routine, and hence are omitted. To prove (4), note that $I \subseteq x + S$ implies $I_v \subseteq (x + S)_v = x + S$, so I_v is contained in the intersection of the family of principal fractional ideals of S that contain I; this family is nonempty since $s + I \subseteq S$ implies $I \subseteq -s + S$ for some $s \in S$. Conversely, if $y \in G$ is such that $y \notin x + S$ for some $x + S$ containing I, then $-x \in S:I$ but $y - x \notin S$, so $y \notin S:(S:I) = I_v$.

Statement (5) follows from (4) and from the fact that $x + S \supseteq I$ if and only if $x + S \supseteq I_v$. To prove (6), we note that

$$I_v = \cap\{a + S \mid I \subseteq a + S\}, \text{ and hence}$$

$$x + I_v = x + \cap\{a + S\} = \cap\{x + a + S \mid I \subseteq a + S\} =$$

$$\cap\{x + a + S \mid x + I \subseteq x + a + S\} =$$

$$\cap\{b + S \mid x + I \subseteq b + S\} = (x + I)_v .$$

(7) The inclusion $I + J \subseteq I_v + J_v$ implies that $(I + J)_v \subseteq (I_v + J_v)_v$. Moreover, since $I_v + J_v \subseteq (I + J)_v$,

then $(I_v + J_v)_v \subseteq ((I + J)_v)_v = (I + J)_v$, and hence the equality $(I + J)_v = (I_v + J_v)_v$ holds, as asserted.

If S is a cancellative monoid with quotient group G , then the v-operation induces an equivalence relation ~ on $F(S)$ defined by $I \sim J$ if $I_v = J_v$. For $I \in F(S)$, div(I) denotes the equivalence class of I under ~ , and $\mathcal{D}(S)$ denotes the set of all divisor classes of S . Part (7) of Theorem 16.4 shows that the operation + on $\mathcal{D}(S)$ defined by div(I) + div(J) = div(I + J) is well-defined. Under + , the set $\mathcal{D}(S)$ forms a commutative monoid with zero element div(S) . Moreover, the set $P(S) = \{div(x + S) \mid x \in G\}$ is a subgroup of the group of invertible elements of $\mathcal{D}(S)$. The factor monoid $C(S) = \mathcal{D}(S)/P(S)$ is called the <u>divisor class monoid</u> of S . As in the case of integral domains, [div(I)] denotes the element div(I) + $P(S)$ of $C(S)$. Under what conditions is $\mathcal{D}(S)$ a group? Based on the domain case, the answer should be "if and only if S is completely integrally closed". Theorem 16.5 shows that, indeed, this is the appropriate condition. Recall from Section 12 that S is <u>completely integrally closed</u> if S contains each element x of G such that $s + nx \in S$ for some fixed element $s \in S$ and for each positive integer n .

THEOREM 16.5. Let S be a cancellative monoid with quotient group G . Then $\mathcal{D}(S)$ is a group if and only if S is completely integrally closed.

<u>Proof</u>. Assume that S is completely integrally closed. To show that $\mathcal{D}(S)$ is a group, we show that [div(I)] has an inverse in $\mathcal{D}(S)$ for each $I \in F(S)$. Without loss of

generality we assume that I is a divisorial ideal contained in S. It suffices to show that $\mathrm{div}(I + (S:I)) = \mathrm{div}(S)$ — that is, that $(I + (S:I))_v = S$. Since $I + (S:I) \subseteq S$, it's enough to show that S is contained in each principal fractional ideal $x + S$ of S containing $I + (S:I)$. Thus $x + S \supseteq I + (S:I)$ implies that $-x + I \subseteq S:(S:I) = I_v = I$. Therefore $-2x + I \subseteq -x + I \subseteq I$, and $-nx + I \subseteq I$ for each positive integer n. Hence $-x$ is almost integral over S, so $-x \in S$ and $S \subseteq x + S$, as asserted.

Conversely, assume that $\mathcal{D}(S)$ is a group. Let $y \in G$ be almost integral over S — say $s + ny \in S$ for each $n \in Z^+$, where $s \in S$. Let I be the fractional ideal $\cup_{n=0}^{\infty}(ny + S)$ of S. Then $I + I = I$, so $\mathrm{div}(I) = \mathrm{div}(S)$ in $\mathcal{D}(S)$. Since S is divisorial, it follows that $I \subseteq S$ so $y \in S$ and S is completely integrally closed.

Suppose K is a field and $K[S]$ is a Krull domain. In order to prove that $\mathcal{C}(K[S])$ and $\mathcal{C}(S)$ are isomorphic, one preliminary result is needed. The statement of Theorem 16.5 contains an aberration in notation: for a fractional ideal I of S, we write $K[I]$ for the family of elements $f \in K[G]$ such that $\mathrm{Supp}(f) \subseteq I$. This is a variance from usual notation because I need not itself be a semigroup; it's easy to show, however, that $K[I]$ is a fractional ideal of $K[S]$. As usual, $K[I]:K[J]$ denotes the set of elements y of the quotient field of $K[S]$ such that $yK[J] \subseteq K[I]$.

THEOREM 16.6. Assume that S is a torsion-free cancellative monoid with quotient group G. Let K be a field.

(1) $K[I]:K[J] = K[I:J]$ for $I,J \in \mathcal{F}(S)$.

(2) $(K[I])_v = K[I_v]$.

(3) The mapping $\mu: \mathcal{D}(S) \longrightarrow \mathcal{D}(K[S])$ defined by $\mu(\mathrm{div}(I)) = \mathrm{div}(K[I])$ is an injective homomorphism.

Proof. (1): The inclusion $K[I:J] \subseteq K[I]:K[J]$ is immediate. For the reverse inclusion, take $y \in K[I]:K[J]$. If $s \in J$, then $yX^s \in K[I]$ implies $y \in X^{-s}K[I] \subseteq K[G]$. For $y \in K[G]$, however, it is clear that $y \in K[I]:K[J]$ iff $yX^s \in K[I]$ for each $s \in J$ iff $t + s \in I$ for each $t \in \mathrm{Supp}(y)$ and each $s \in J$ iff $\mathrm{Supp}(y) \subseteq I:J$ iff $y \in K[I:J]$.

(2) It follows from (1) that $(K[I])_v = K[S]:(K[S]:K[I]) = K[S:(S:I)] = K[I_v]$.

(3): For $I,J \in \mathcal{F}(S)$, we have $\mathrm{div}(I) = \mathrm{div}(J)$ iff $I_v = J_v$ iff $K[I_v] = K[J_v]$ iff $(K[I])_v = (K[J])_v$ iff $\mathrm{div}(K[I]) = \mathrm{div}(K[J])$. Therefore μ is both well-defined and injective. Moreover, $\mu(\mathrm{div}(I) + \mathrm{div}(J)) = \mu(\mathrm{div}(I + J)) = \mathrm{div}(K[I + J]) = \mathrm{div}(K[I] \cdot K[J]) = \mathrm{div}(K[I]) + \mathrm{div}(K[J]) = \mu(\mathrm{div}(I)) + \mu(\mathrm{div}(J))$, so μ is a homomorphism.

THEOREM 16.7. Assume that K is a field and $K[S]$ is a Krull domain. The mapping $\phi: \mathcal{C}(S) \longrightarrow \mathcal{C}(K[S])$ defined by $\phi([\mathrm{div}(I)]) = [\mathrm{div}(K[I])]$ is an isomorphism of $\mathcal{C}(S)$ onto $\mathcal{C}(K[S])$.

Proof. Let $\mu: \mathcal{D}(S) \longrightarrow \mathcal{D}(K[S])$ be defined as in part (3) of Theorem 16.6. Since $\mu(\mathrm{div}(y + S)) = \mathrm{div}(K[y + S]) = \mathrm{div}(X^yK[S])$, it follows that $\mu(\mathcal{P}(S)) \subseteq \mathcal{P}(K[S])$, so μ induces the homomorphism ϕ . We show that ϕ is both injective and surjective. If $[\mathrm{div}(I)] \in \ker\phi$, then $K[I_v]$ is principal and $K[I_v]$ contains monomials. Therefore

$K[I_v]$ is generated by X^a for some $a \in G$. The equality $K[I_v] = X^a K[S] = K[a + S]$ implies, however, that $I_v = a + S$, so $[\mathrm{div}(I)] = 0$ and ϕ is injective.

To complete the proof, we show that ϕ is surjective. The equality $K[G] = K[S]_N$, where $N = \{X^s \mid s \in S\}$, implies that $K[G]$ is a Krull domain, and then Theorems 15.1 and 14.15 show that $C(K[G]) = \{0\}$. Therefore Theorem 16.1 shows that $C(K[S])$ is generated by $\{[\mathrm{div}(Q_\alpha)] \mid Q_\alpha$ is a minimal prime of $K[S]$ that meets $N\}$. Consulting Corollary 15.9, we conclude that each Q_α is of the form $K[I_\alpha]$ for an appropriate proper prime ideal I_α of S. Since $[\mathrm{div}(K[I_\alpha])] = \phi([\mathrm{div}(I_\alpha)])$ for each α, it follows that range $\phi \supseteq C(K[S])$, and hence ϕ is surjective. This completes the proof of Theorem 16.7.

A combination of Theorems 16.3 and 16.7 yields the following corollary.

COROLLARY 16.8. If the monoid domain $D[S]$ is a Krull domain, then $C(D[S]) \simeq C(D) \oplus C(S)$.

Assume that S is a Krull monoid with quotient group G and assume that $H = \{0\}$ is the set of invertible elements of S. Let $\{W_\alpha\}_{\alpha \in A}$ be the family of rank-one discrete valuation monoids on G that are essential for S, let w_α be a normed valuation on G associated with W_α, and let P_α be the center of w_α on S. Theorem 15.8 shows that $S = \cap_{\alpha \in A} W_\alpha$. Let $F = \Sigma_{\alpha \in A} Z_\alpha$ be the weak direct sum of the family $\{Z_\alpha\}_{\alpha \in A}$, where $Z_\alpha = Z$ for each α, and let $\{e_\alpha\}_{\alpha \in A}$ be the canonical free basis for F. Since $H = \{0\}$, the mapping $\sigma : G \longrightarrow F$ defined by $\sigma(x) = \Sigma_\alpha w_\alpha(x) e_\alpha$ is an imbedding of G in F. Part (3) of Theorem 15.2 shows that $G \cap F_+ = S$ in this case, and it is an S of this form to which Theorem 16.3 reduces the problem of determining the divisor class group of a Krull monoid domain.

THEOREM 16.9. Let the notation and hypothesis be as in the preceding paragraph.

(1) $\{\mathrm{div}(P_\alpha)\}_{\alpha \in A}$ is a free basis for $\mathcal{D}(S)$.

(2) Let $y \in G\setminus\{0\}$ and let $\{w_i\}_1^n$ be the finite subset of $\{w_\alpha\}$ consisting of those valuations with nonzero value on y. Write P_i for the center of w_i on S. Then $\mathrm{div}(y + S) = \mathrm{div}(\Sigma_1^n w_i(y)P_i)$.

(3) $\mathcal{C}(S) \simeq F/G$.

Proof. We remark that in the statement of (2), $w_i(y)P_i$ means $-w_i(y)(S:P_i)$ if $w_i(y) < 0$. In proving each of (1) and (2), we consider the monoid domain $K[S]$ of S over a field K, and we transfer properties of the Krull domain $K[S]$ to S.

(1): That $\{\mathrm{div}(P_\alpha)\}$ is a free subset of $\mathcal{D}(S)$ follows from part (3) of Theorem 16.6 and the fact that $\{\mathrm{div}(K[P_\alpha])\}$ is a free subset of $\mathcal{D}(K[S])$. To prove that $\{\mathrm{div}(P_\alpha)\}$ generates $\mathcal{D}(S)$, it suffices to show that $\mathrm{div}(I)$ belongs to the subgroup generated by $\{\mathrm{div}(P_\alpha)\}$ for each integral divisorial ideal $I \neq S$ of S. Part (2) of Theorem 16.6 shows that $K[I]$ is a proper divisorial ideal of $K[S]$, and Corollary 15.9 shows that the set of minimal primes of $K[S]$ containing $K[I]$ is $\{K[Q_j]\}_{j=1}^m$, where $\{Q_j\}_1^m$ is the subset of $\{P_\alpha\}$ consisting of elements P_α that contain I. Hence $\mathrm{div}(K[I])$ belongs to the subgroup of $\mathcal{D}(K[S])$ generated by $\{div(K[Q_j])\}_1^m$, and since each of these elements is in the range of the mapping μ of Theorem 16.6, it follows that $div(I)$ belongs to the subgroup of $\mathcal{D}(S)$ generated by $\{div(Q_j)\}_1^m$.

(2): To prove (2), it suffices to consider the case where $y \in S$. The proof of (1) shows that there exist positive integers $k_1,\ldots,k_n$ such that $div(y + S) = \Sigma_1^n k_i div(P_i) = div(\Sigma_1^n k_i P_i)$ in this case. Moreover, since $K[y + S] = X^y K[S]$ is principal, the integer k_i is determined as $w_i^*(X^y) = w_i(y)$, where w_i^* is the valuation on q.f.$(K[S])$ associated with w_i as in the proof of Theorem 15.7. Therefore $div(y + S) = div(\Sigma_1^n w_i(y)P_i)$, as asserted.

(3): Let $\{e_\alpha\}_{\alpha\in A}$ be the canonical free basis for F and consider the homomorphism $\tau: F \longrightarrow \mathcal{C}(S)$ determined by $\tau(e_\alpha) = [div(P_\alpha)]$. Part (1) implies that τ is surjective. We show that $\sigma(G)$ is the kernel of τ. Thus, if $\Sigma_1^n m_i e_i$ is a nonzero element of F, then $\tau(\Sigma m_i e_i) = \Sigma m_i[div(P_i)] = [div(\Sigma m_i P_i)] = 0$ if and only if $(\Sigma m_i P_i)_v$ is principal. According to parts (2) and (1), however, this occurs if and only if there exists $y \in G\backslash\{0\}$ such that $w_i(y) = m_i$ for each i and $w_\alpha(y) = 0$ for $\alpha \notin \{1,\ldots,n\}$. But this is precisely the condition that $\Sigma m_i e_i = \sigma(y)$. Thus $\ker\tau = \sigma(G)$ as asserted, and $\mathcal{C}(S) \simeq F/\sigma(G) \simeq F/G$. This completes the proof of Theorem 16.9.

Since Theorem 16.9 provides an alternate description of $\mathcal{C}(S)$ in the case of a Krull monoid S with no nonzero invertible element, its combination with Theorems 16.3 and 16.7 yields the following result.

COROLLARY 16.10. Assume that $D[S]$ is a Krull domain. Let H be the group of invertible elements of S, and express S as $H \oplus T$. Let G be the quotient group of T, let $\{w_\alpha\}_{\alpha \in A}$ be the family of rank-one discrete valuations on G that are essential for T, let $F = \Sigma_{\alpha \in A} Z_\alpha$ be the weak direct sum of groups $Z_\alpha = Z$, and let σ be the canonical imbedding of G in F. Then $C(D[S]) \simeq C(D) \oplus (F/\sigma(G))$.

We conclude the section with a result concerning Krull monoids that follows from Theorem 16.9; the statement of Theorem 16.11 is an expected result.

THEOREM 16.11. Let S be a Krull monoid with quotient group G, let $\{W_\alpha\}_{\alpha \in A}$ be the family of rank-one discrete valuation monoids on G that are essential for S, and let P_α be the center of W_α on S. Then $\{P_\alpha\}_{\alpha \in A}$ is the set of minimal prime ideals of S.

Proof. Clearly each P_α is prime in S, and there is no inclusion relation between P_α and P_β for $\alpha \neq \beta$. Thus, to prove (16.11), it's enough to show that each prime ideal P of S contains some P_α. Express S as $H \oplus T$, where H is the group of invertible elements of S. By definition, each P_α is contained in T, and in fact, it's straightforward to show that $\{P_\alpha\}$ is the family of centers of rank-one discrete valuation monoids on the quotient group of T that are essential for T, a Krull monoid. Replacing P by $P \cap T$, we therefore assume without loss of generality that $S = T$. Choose $x \in P$. Part (1) of Theorem 16.9 implies that there exists a finite subset $\{P_i\}_1^n$ of $\{P_\alpha\}$ and positive integers $k_1, k_2, \ldots, k_n$ such that

$x + S = (\Sigma_1^n k_i P_i)_v$. Therefore $P \supseteq x + S \supseteq \Sigma_1^n k_i P_i$, and hence P contains some P_i .

Section 16 Remarks

Detailed treatments of the material cited in the introduction to this section concerning the v–operation on a Krull domain can be found in [20; Ch. VII, Sect. 1], [44; Ch. I and II], and [51; Sect. 34, 44]. In particular, (16.1) is a special case of Theorem 7.1 of [44]. Other references for the divisor class group of a Krull monoid domain are [95], [7], [28], and [21]. More specifically, part (3) of (16.2) is Proposition 5.3 of [95], part (1) of of (16.3) is part (1) of Proposition 7.3 of [7], and Corollaries 16.8 and 16.10 appear in [28]. For some specific examples of calculations of $C(D[S])$, see [6–8], [28], and [21].

CHAPTER IV

RING–THEORETIC PROPERTIES OF MONOID RINGS

Chapter IV continues the main theme of Chapter III — that of determining, for various ring–theoretic properties E , necessary and sufficient conditions on a unitary ring R and a monoid S in order that R[S] should have property E . The main difference in the two chapters is that in Chapter IV, R[S] is not normally assumed to be a domain; hence R may contain zero divisors and S is not uniformly taken to be torsion–free and cancellative. In Section 17, E is the property of being (von Neumann) regular, and certain chain conditions, including d.c.c. , are considered in Section 20. It is trivial to determine conditions under which a monoid domain is regular or Artinian (that is, a field). On the other hand, the properties E considered in Sections 18 and 19, including the conditions of being a Prüfer ring or an arithmetical ring, are closely related to properties considered in Section 13 of Chapter III. One new wrinkle arises in the case of rings with zero divisors, however. While a unitary integral domain is arithmetical if and only if it is a Prüfer domain, in Section 19 we show that if S is not torsion–free, then R[S] may be arithmetical, but not a Prüfer ring. Similar situations exist in regard to the properties Bezout ring-arithmetical ring and the properties of being a principal ideal ring or a ZPI–ring.

§17. Monoid Rings as von Neumann Regular Rings

Throughout this section we use the term regular ring for a ring that is regular in the sense of von Neumann. The definition, for a ring R that need not be commutative, is that each $a \in R$ is expressible in the form axa for some $x \in R$. For R commutative, R is regular iff each principal ideal of R is idempotent iff R is a zero-dimensional reduced ring.

We assume throughout the section that R is a unitary (commutative) ring and that S is a monoid. Our first results are directed toward a proof that $R[S]$ is zero-dimensional if and only if R is zero-dimensional and S is periodic. Since R is a homomorphic image of $R[S]$, the inequality $\dim R \leq \dim R[S]$ is always satisfied.

THEOREM 17.1. Assume that G is a group. If G has torsion-free rank α, then $\dim R[G] = \dim R[\{X_\lambda\}_{\lambda \in \Lambda}]$, where $|\Lambda| = \alpha$.

Proof. First, a comment on the statement of the theorem seems appropriate. We do not distinguish among different infinite cardinalities in dealing with Krull dimension, so if α is infinite, the equality in the statement of (17.1) is interpreted as meaning that $R[G]$ is infinite-dimensional.

Let $\{y_i\}_{i \in I}$ be a maximal free subset of G, and let $H = \Sigma_{i \in I} Z y_i$ be the subgroup of G generated by $\{y_i\}$. Then G/H is a torsion group so $R[G]$ is integral over $R[H]$ and $\dim R[G] = \dim R[H]$. Moreover, the torsion-free rank of H is $|I| = \alpha$, so we assume without loss of generality that $G = H$. In that case, $R[G]$ is isomorphic to

$R[\{X_\lambda\}_{\lambda\in\Lambda}, \{X_\lambda^{-1}\}]$. The latter ring is integral over $R[\{X_\lambda + X_\lambda^{-1}\}]$, for each of X_λ and X_λ^{-1} is a root of the monic polynomial $Y^2 - (X_\lambda + X_\lambda^{-1})Y + 1$ over $R[\{X_\lambda + X_\lambda^{-1}\}]$. A straightforward degree argument shows, however, that $\{X_\lambda + X_\lambda^{-1}\}$ is algebraically independent over R . Consequently, $\dim R[G] = \dim R[\{X_\lambda\}_{\lambda\in\Lambda}]$, as asserted.

COROLLARY 17.2. If $\dim R = \dim R[S] = n < \infty$, then $S/\sim$ is periodic, where $\sim$ is the cancellative congruence on S .

<u>Proof</u>. We have $\dim R \le \dim R[S/\sim] \le \dim R[S]$, the latter inequality holding since $R[S/\sim]$ is a homomorphic image of $R[S]$. Therefore $\dim R[S/\sim] = n$. Let G be the quotient group of $S/\sim$. Then $\dim R \le \dim R[G] \le \dim R[S/\sim]$ since $R[G]$ is a quotient ring of $R[S/\sim]$. The equality $\dim R = \dim R[G] = n$ then implies, by Theorem 17.1, that G has torsion-free rank 0 . Consequently, G is a torsion group and $S/\sim$ is periodic.

THEOREM 17.3. Assume that R has finite dimension k . Then $\dim R[S] = k$ if and only if S is periodic.

<u>Proof</u>. If S is periodic, then $R[S]$ is integral over R and $\dim R[S] = \dim R$. Conversely, assume that $\dim R[S] = \dim R$ and choose $s \in S$. Corollary 17.2 shows that there exist distinct positive integers m and n such that $ms \sim ns$, where $\sim$ is the cancellative congruence on S . Thus $ms + c = ns + c$ for some $c \in S$. Let $I = \{t \in S \mid k_1 s + t = k_2 s + t$ for some $k_1, k_2 \in Z^+, k_1 \ne k_2\}$; I is an ideal of S , and s is a periodic element of S if and only if I meets the semigroup $\langle s\rangle$ of S . Assume

that $I \cap \langle s \rangle = \phi$. By Theorem 11.8, there exists a prime ideal J of S containing I such that $J \cap \langle s \rangle = \phi$. Let U be the submonoid $S \backslash J$ of S . Then $R[S] = R[U] + R[J]$, $R[J]$ is an ideal of $R[S]$, and $R[U] \cap R[J] = (0)$. Consequently, $R[U] \simeq R[S]/R[J]$ is a homomorphic image of $R[S]$ so that $k = \dim R \le \dim R[U] \le \dim R[S] = k$, and $\dim R[U] = k$. Applying Corollary 17.2 to the monoid ring $R[U]$, we see that $U/\sim$ is periodic. Since $s \in \langle s \rangle \subseteq U$, it follows that $k_1 s + u = k_2 s + u$ for distinct positive integers k_1 and k_2 and some element $u \in U$. But then $u \in J \cap U$, which contradicts the fact that $J \cap U = \phi$. Consequently, s is periodic and S is periodic, as asserted.

Theorem 9.17 shows that $R[S]$ is a reduced ring if and only if R is reduced, S is free of asymptotic torsion, and p is regular on R for each p such that S is not p-torsion-free. This result and Theorem 17.3 enable us to determine the monoid rings that are regular.

THEOREM 17.4. Let R be a unitary ring and let S be a monoid. The ring $R[S]$ is regular if and only if the following four conditions are satisfied.

(1) R is regular.

(2) S is free of asymptotic torsion.

(3) p is regular on R for each prime p such that S is not p-torsion-free.

(4) S is periodic.

Proof. If $R[S]$ is regular, then R is regular since R is a homomorphic image of $R[S]$. Thus $\dim R = \dim R[S] = 0$, and Theorem 17.3 implies that S is periodic. Moreover,

$R[S]$ reduced implies that (2) and (3) are satisfied. Therefore conditions (1)–(4) are satisfied if $R[S]$ is regular. The converse follows by similar reasoning.

Since regular elements of a von Neumann regular ring are invertible, condition (3) of Theorem 17.4 is equivalent to the condition that p is a unit of R for each prime p such that S is not p-torsion-free. If $\{M_\lambda\}_{\lambda \in \Lambda}$ is the set of maximal ideals of R and if p_λ is the characteristic of R/M_λ for each $\lambda \in \Lambda$, then p is a unit of R if and only if $p \notin \{p_\lambda\}_{\lambda \in \Lambda}$. Thus, stated in terms of $\{p_\lambda\}_{\lambda \in \Lambda}$, (3) can be replaced by the condition that S is p_λ-torsion-free for each λ, where "0-torsion-free" is interpreted as "free of asymptotic torsion". Note that if some $p_\lambda \neq 0$, then condition (3) implies condition (2); in particular, this is the case if R has nonzero characteristic. To obtain other equivalent forms of the conditions in Theorem 17.4, we consider some consequences of the combination of conditions (2) and (4) of that result.

THEOREM 17.5. If the semigroup T is free of asymptotic torsion, then $\langle t \rangle$ is a group for each periodic element t of T. Moreover, if T is periodic, then T is p-torsion-free for a given prime p if and only if p divides the order of no element of T.

Proof. As in Section 2, let m and r be the period and index, respectively, of t. It suffices to prove that $r = 1$. If $r > 1$, then $(r-1)t \neq (m+r-1)t$ by definition of r. But for $k \geq 2$, we have $k(r-1)t = k(m+r-1)t$ since $k(r-1) \equiv k(m+r-1) \pmod{m}$ and since $k(r - 1) \geq r$. This contradiction to the fact that T is free of asymptotic torsion shows that $\langle t \rangle$ is a group, as asserted.

Since any two elements of a p-group are p-equivalent, it follows that if T is p-torsion-free, then T contains no nontrivial p-group, and hence no element of order divisible by p . Conversely, if T is not p-torsion-free, then there exist distinct elements $b, c \in T$ such that $pb = pc$. If p does not divide the order of c , then c and pc generate the same subgroup of T ; hence b and c are elements of the subgroup $G = \langle b \rangle$ of T . If e is the identity of G and d is the inverse of c in G , then $b + d \neq e$ since $b \neq c$, but $p(b + d) = e$. Therefore, p divides the order of an element of S .

In view of Theorem 17.5 and the remarks following Theorem 17.4, we can state a variety of conditions on R and S that are equivalent to the condition that $R[S]$ is regular.

COROLLARY 17.6. Assume that R is regular, that $\{M_\lambda\}_{\lambda \in \Lambda}$ is the set of maximal ideals of R , and that $\text{char}(R/M_\lambda) = p_\lambda$ for each $\lambda \in \Lambda$. Assume that S is periodic and free of asymptotic torsion. The following conditions are equivalent.

(1) p is regular on R for each prime p such that S is not p-torsion-free.

(2) p is a unit of R for each prime p that divides the order of an element of S .

(3) No p_λ divides the order of an element of S .

If the semigroup T is periodic and free of asymptotic torsion, then Theorem 17.5 implies that T is a union of periodic subgroups. We embark on a brief consideration of this and some related conditions in a semigroup T. Thus, the material in the rest of the section is primarily concerned with some of the theory of commutative semigroups that was not covered in Chapter 1. In particular, we abandon for the rest of the section the restriction to consideration of monoids that has been the rule for several sections. We use the term semilattice to describe a semigroup, each of whose elements is idempotent. The results concerning the Archimedean decomposition of a semigroup established in Theorems 17.9 and 17.10 will subsequently be used in Section 23.

THEOREM 17.7. Assume that the semigroup T is the union of a family $\{G_\alpha\}_{\alpha\in A}$ of subgroups of T. Let E be the semilattice of idempotents of T, and for $e \in E$, let $H_e = \cup\{G_\alpha \mid e$ is the identity element for $G_\alpha\}$. Each H_e is a subgroup of G, the family $\{H_e\}_{e\in E}$ is a partition of G, and $H_e + H_f \subseteq H_{e+f}$ for $e,f \in E$. Moreover, if each G_α is periodic, then each H_e is also periodic.

Proof. If $t \in T$, then $t \in G_\alpha$ for some α. Let e_α be the identity element of G_α. Then there exists $x \in G_\alpha$ such that $t + x = e_\alpha$. We observe that e_α is the unique idempotent of T with the properties that $t + e_\alpha = t$ and $e_\alpha \in t + T$. To prove this, assume that $e \in E$ is such that $t + e = t$ and $e = t + y$ for some $y \in T$. Then $e = t + y = (e_\alpha + t) + y = e_\alpha + (t + y) = e_\alpha + e = (x + t) + e = x + (t + e) = x + t = e_\alpha$.

Thus, H_e may be alternately described as $\{t \in T | t + e = t$ and $e \in t + T\}$. Since each G_α contains a unique idempotent and since $T = \cup G_\alpha$, it follows easily that $\{H_e\}_{e \in E}$ is a partition of G. To prove that H_e is a subgroup of T, take $a, b \in H_e$. Then $(a + b) + e = a + (b + e) = a + b$, and $e = e + e \in (a + T) + (b + T) \subseteq a + b + T$; hence H_e is closed. It is clear that e is the identity element of H_e, and if $t \in G_\alpha \subseteq H_e$, there exists $s \in G_\alpha$ such that $t + s = e$; since $s \in H_e$, it follows that each H_e is a subgroup of T. To prove that $H_e + H_f \subseteq H_{e+f}$, take $a \in H_e$, $b \in H_f$. Then $a = a + e$, $b = b + f$, $e \in a + T$, and $f \in b + T$. Consequently, $a + b = a + b + e + f$ and $e + f \in (a + T) + (b + T) \subseteq a + b + T$. This implies that $a + b \in H_{e+f}$, and hence $H_e + H_f \subseteq H_{e+f}$, as asserted. Finally, it is clear that each H_e is periodic if each G_α is periodic.

COROLLARY 17.8. The following conditions are equivalent in a semigroup T.

(1) T is periodic and is free of asymptotic torsion.

(2) T is a union of periodic subgroups.

(3) T is a disjoint union of periodic subgroups.

Proof. In view of Theorems 17.5 and 17.7, it suffices to prove that (3) implies (1). Thus, assume that $T = \cup_{\alpha \in A} G_\alpha$ is a partition of G into periodic subgroups G_α. It is clear that T is periodic. Assume that $s, t \in T$ are asymptotically equivalent. If $s \in G_\alpha$ and $t \in G_\beta$, then $\alpha = \beta$ since $\langle s \rangle \subseteq G_\alpha$, $\langle t \rangle \subseteq G_\beta$, and $\langle s \rangle \cap \langle t \rangle \neq \phi$. Choose relatively prime positive integers n and m such that $ns = nt$ and $ms = mt$, and write $1 = nx + my$ for integers x and y. Since G_α is a group, we have $s = nxs + mys = xnt + ymt = t$. Therefore T is free of asymptotic torsion.

Assume that T is a periodic semigroup that is free of asymptotic torsion. We remark that if Theorem 17.7 is applied to T, where $\{G_\alpha\}_{\alpha \in A}$ is the family of cyclic subgroups of T, then elements $a, b \in T$ belong to the same subgroup H_e of T if and only if the identity element for $\langle a \rangle$ is the same as the identity element for $\langle b \rangle$.

Theorem 17.7 is a special case of a more general decomposition theorem for commutative semigroups that we proceed to establish. To partially motivate the development, we note that in Theorem 17.7, the set E of idempotents of T is a semilattice and that the partition $T = \cup_{e \in E} H_e$ is compatible with the operation on E in the sense that $H_e + H_f \subseteq H_{e+f}$ for all $e, f \in E$. In general, we say that a semigroup U is a <u>semilattice of subsemigroups</u> $\{U_\alpha \mid \alpha \in A\}$ if A is a semilattice, $U = \cup_{\alpha \in A} U_\alpha$ is a partition of U, and $U_\alpha + U_\beta \subseteq U_{\alpha+\beta}$ for all $\alpha, \beta \in A$. The decomposition theorem referred to above states that each commutative semigroup T admits a unique decomposition as a semilattice of Archimedean subsemigroups, where a semigroup U is <u>Archimedean</u> if $\mathrm{rad}(x + U) = \mathrm{rad}(y + U)$ for all $x, y \in U$. (The terminology Archimedean is suggested by the theory of ordered groups, for such a group is Archimedean, as customarily defined, if and only if its semigroup of positive elements is Archimedean as defined above.) To obtain the decomposition theorem, we introduce a new congruence on a semigroup T.

THEOREM 17.9. For a semigroup T, define a relation ρ on T by $s \rho t$ if $\mathrm{rad}(s + T) = \mathrm{rad}(t + T)$.

(1) ρ is a congruence on T, and the factor semigroup $U = T/\rho$ is a semilattice. If μ is any congruence on T such that T/μ is a semilattice, then $\mu \geq \rho$.

(2) Let $\{T_u\}_{u \in U}$ be the set of equivalence classes of T under ρ, indexed by the elements of U in such a way that T_u maps to u under the canonical map $f: T \longrightarrow U$. Then $T = \cup_{u \in U} T_u$ is the unique representation of T as a semilattice of Archimedean semigroups.

Proof. (1): That ρ is an equivalence relation on T is clear. To prove that ρ is compatible with the semigroup operation on T, assume that $s \rho t$ and that $x \in T$. Then $ns \in t + T$ for some $n \in Z^+$, so $n(s + x) \in t + nx + T \subseteq t + x + T$. Consequently, $s + x \in \mathrm{rad}(t + x + T)$, and hence $\mathrm{rad}(s + x + T) \subseteq \mathrm{rad}(t + x + T)$. Similarly, $\mathrm{rad}(t + x + T) \subseteq \mathrm{rad}(s + x + T)$ so that $(s + x)\rho(t + x)$. We have $t \rho 2t$ for each $t \in T$, so T/ρ is a semilattice. Assume that T/μ is a semilattice and that $s, t \in T$ are such that $s \rho t$. This implies that $ms = t + x$ and $nt = s + y$ for some $m, n \in Z^+$ and some $x, y \in T$. Thus

$$s \mu ms \mu (t + x) \text{ and } t \mu nt \mu (s + y), \text{ so that}$$

$$(s + t)\mu(2t + x)\mu(t + x)\mu s \text{ and}$$

$$(s + t)\mu(2s + y)\mu(s + y)\mu t.$$

Therefore $s \mu t$ and $\mu \geq \rho$. This establishes (1).

To prove (2), we first show that each equivalence class T_u of T under ρ is an Archimedean subsemigroup of T. Thus, take $a,b \in T_u$. Since $a \mu b$, then $2a \mu (a+b)$. Moreover, $a \mu 2a$ and hence $a \mu (a+b)$. Therefore $a + b \in T_u$, and T_u is a subsemigroup of T. To show that T_u is Archimedean, choose $m,n \in Z^+$ and $x,y \in T$ such that $ma = b + x$ and $nb = a + y$. Then $(m+1)a = b + a + x$, where $a \in \text{rad}(a + x + T) \subseteq \text{rad}(a + T)$. It follows that $\text{rad}(a+T) = \text{rad}(a + x + T)$, and hence $a + x \in T_u$. The equation $(m+1)a = b + (a + x)$ then implies that a belongs to the radical of $b + T_u$ in T_u; by the same argument, b belongs to $\text{rad}(a + T_u)$ in T_u. Therefore $\text{rad}(a + T_u) = \text{rad}(b + T_u)$, so each T_u is Archimedean. Clearly $T = \cup_{u \in U} T_u$ is a partition of T, and the indexing is such that to show $T_u + T_v \subseteq T_{u+v}$, we need only show that $f(a+b) = u + v$ for each $a \in T_u$, $b \in T_v$. This is clear, however, since $f(a) = u$, $f(b) = v$, and $f(a+b) = f(a) + f(b)$.

To prove uniqueness of the representation $T = \cup_{u \in U} T_u$, take any representation $T = \cup_{c \in C} W_c$ of T as a semilattice of Archimedean subsemigroups W_c. For $a,b \in T$, it suffices to show that $a,b \in W_c$ if and only if $a \rho b$. Since W_c is Archimedean, it is clear that $a \rho b$ if $a,b \in W_c$. Conversely, assume that $a \rho b$ and that $ma = b + x$ and $nb = a + y$. Choose $c,d,e,f \in C$ such that $a \in W_c$, $b \in W_d$, $x \in W_e$, and $y \in W_f$. Then $ma \in W_{mc} = W_c$ and $b + x \in W_{d+e}$, so $c = d + e$ and hence $c + d = 2d + e = d + e = c$. Considering the equation $nb = a + y$, we conclude by similar reasoning that $c + d = d$. Hence $c = d$, and a and b belong to the same set W_c of the partition. This completes the proof of Theorem 17.9.

With the notation as in Theorem 17.9, the semigroups T_u are called the Archimedean components of T. Any group G is Archimedean, for G is the only ideal of G. Thus, uniqueness of the Archimedean components implies that the subgroups H_e of T in the statement of Theorem 17.7 are the Archimedean components of T. It is in this sense that Theorem 17.7 is a special case of (17.9). For T free of asymptotic torsion, Corollary 17.11 shows that a given Archimedean component U of T is a group if and only if U contains an idempotent.

THEOREM 17.10. The Archimedean components of T are cancellative if and only if T is free of asymptotic torsion.

Proof. Assume that the Archimedean components of T are cancellative. If $a,b \in T$ are asymptotically equivalent, then it is clear that $\text{rad}(a + T) = \text{rad}(b + T)$, so a and b belong to the same Archimedean component U of T. Choose $n \in Z^+$ such that $na = nb$ and $(n + 1)a = (n + 1)b$. Then $na + a = nb + b = na + b$. Since U is cancellative, it follows that $a = b$, so T is free of asymptotic torsion.

Conversely, assume that T is free of asymptotic torsion and that a,b,c are elements of an Archimedean component U of T such that $a + c = b + c$. Choose $n \in Z^+$ such that both na and nb belong to $c + U$, say $na = c + x$ and $nb = c + y$. Then $(n + 1)a = a + c + x = b + c + x = na + b$; hence $(n + 2)a = (n + 1)a + b = na + b + b = na + 2b$. By induction, it follows that if $0 \le i \le k$, then $(n + k)a = (n + i)a + (k - i)b$. Similarly, we have $(n + k)b = (n + i)b + (k - i)b$ for $0 \le i \le k$. In particular,

$$2na = na + nb = 2nb \text{ and } (2n+1)a = (n+1)a + nb = (2n+1)b .$$

Therefore a and b are asymptotically equivalent, and $a = b$. It follows that U is cancellative, as asserted.

COROLLARY 17.11. Assume that T is free of asymptotic torsion and that U is an Archimedean component of T. Then U is a group if and only if U contains an idempotent.

Proof. We need only show that the existence of an idempotent $e \in U$ implies that U is a group. Theorem 17.10 implies that U is cancellative, and hence e is an identity element for U. If $a \in U$, then $e \in \text{rad}(e + u) = \text{rad}(a + U)$ implies that $e \in a + U$, and hence a is invertible in U. Therefore U is a group.

To conclude this section, we show that if T is free of asymptotic torsion, then T can be imbedded in a semigroup that is a semilattice of subgroups.

THEOREM 17.12. Assume that T is free of asymptotic torsion and let $T = \cup_{u \in U} T_u$ be the representation of T as a semilattice of its Archimedean components. Let G_u be the quotient group of T_u for each $u \in U$. Then T can be imbedded in a semigroup W such that $W = \cup_{u \in U} W_u$ is a semilattice of subgroups W_u, where $W_u \simeq G_u$ for each $u \in U$.

Proof. We represent the elements of G_u in the form used in Section 1 — that is, as equivalence classes $[a,b]$, where $a,b \in T_u$ and where $[a,b] = [c,d]$ if and only if $a + d = b + c$. Since the sets T_u are disjoint, the groups

G_u are also disjoint. Let $W = \cup_{u \in U} G_u$. We define an operation $+$ on W as follows. If $[a,b] \in G_u$ and $[c,d] \in G_v$, then $[a,b] + [c,d]$ is the element $[a + c, b + d]$ of G_{u+v}. This operation is well-defined, for if $[a,b] = [a',b']$ and $[c,d] = [c',d']$, then $a + b' = a' + b$ and $c + d' = c' + d$ so that $(a + c) + (b' + d') = (a' + c') + (b + d)$ and $[a + c, b + d] = [a' + c', b' + d']$. Associativity of the operation follows because $+$ is associative on T. The operation on W agrees with the operation on G_u for each $u \in U$, and hence each G_u is a subgroup of W. Since $G_u + G_v \subseteq G_{u+v}$ by the very definition of $+$ on W, it follows that $W = \cup_{u \in U} G_u$ is a representation of W as a semilattice of subgroups. Finally, we show that the natural inclusion mapping of T into W, where $a \in T_u$ is identified with $[2a,a] \in G_u$, is a semigroup homomorphism. This is true since $[2a,a] + [2b,b] = [2(a + b), a + b]$. Hence, we can take $W_u = G_u$ for each $u \in U$ in the statement of Theorem 17.12.

Section 17 Remarks

The problem of determining conditions under which a semigroup ring $T[U]$ is regular has been investigated in cases other than where T is a commutative unitary ring and U is a commutative monoid. For example, Gilmer and Teply show in [61] that the statement of Theorem 17.4 remains valid if the hypothesis that R is unitary is dropped; this is one reason we have chosen to state condition (3) of (17.4) in terms of p being regular <u>on</u> R. If U is a group, then Connell [36], using previous work of several authors, shows that T[U] is regular if and only if T is regular, U is locally finite, and the order of each element of U is a unit of T ; here neither T nor U is required to be commutative. Finally, Weissglass [131] has considered the problem of determining conditions under which T[U] is regular, where neither T nor U is required to be commutative and U is not necessarily a monoid.

While Boolean rings form a subclass of the class of regular rings, the problem of determining the semigroup rings T[U] that are Boolean can be solved using only first principles. The result is that T[U] is Boolean if and only if T is Boolean and U is a semilattice. Necessity of these conditions follows from the equality $(tX^u)^2 = t^2X^{2u}$ and the fact that a Boolean ring is commutative; sufficiency follows easily from the given conditions and the observation that T[U] has characteristic 2 if T is Boolean.

The material in Section 17 concerning the representation of a commutative semigroup as a semilattice of its Archimedean components has been applied in work on questions concerning semigroup rings; see, for example, [132], [128–129], and [30].

§18. Monoid Rings as Prüfer Rings

In Section 13, conditions on a monoid domain $D[S]$ have been given in order that it should be a Prüfer domain, a Bezout domain, a Dedekind domain, or a PID. In this section we consider certain analogues of these classes of domains for rings with zero divisors. The analogues of the classes just named are, respectively, the classes of Prüfer rings, Bezout rings, general ZPI-rings, and principal ideal rings (PIR's). At the same time, we consider here the classes of arithmetical rings and multiplication rings; an arithemetical integral domain is the same as a Prüfer domain and the concepts of multiplication domain and Dedekind domain agree, but these equivalences do not carry over to the case of rings. Each of the classes named is treated to some degree in [51], except for the class of Prüfer rings. On the other hand, the concepts may be unfamiliar to some readers, so we develop in this section enough of the theory to facilitate our treatment of the monoid ring characterization problems. More detailed references are given in the Section 18 Remarks.

Throughout the section, R denotes a unitary ring and S denotes a torsion-free cancellative monoid. We say that R is a <u>Prüfer ring</u> if each finitely generated regular ideal of R is invertible. An invertible ideal $A = (a_1,\ldots,a_k)$ has the property that $A^n = (a_1^n,\ldots,a_k^n)$ for each $n \in Z^+$. Thus, it is clear that a Prüfer ring R satisfies the following condition.

(18.1) If $a,b \in R$ and if at least one of a or b is regular, then $ab \in (a^2,b^2)$.

For the sake of brevity, we use the ad hoc term P-*ring* in this section for a unitary ring R satisfying condition (18.1). Theorem 18.9 shows that the monoid ring $R[S]$ is a Prüfer ring if and only if it is a P-ring. Theorem 18.2 lists three elementary properties of P-rings.

THEOREM 18.2. Let N be a regular multiplicative system in R, and assume that $\{R_i\}_{i=1}^{\infty}$ is an ascending sequence of subrings of R such that $R = \cup_{i=1}^{\infty} R_i$.

(1) If R is a P-ring, then R_N is a P-ring.

(2) If each R_i is a P-ring, then R is a P-ring.

(3) If $R[S]$ is a P-ring, then R is a P-ring.

Proof. (1): If $\alpha, \beta \in R_N$ with β regular, then there exist $a, b \in R$ such that b is regular, $\alpha R_N = aR_N$, and $\beta R_N = bR_N$. Then $ab \in (a^2, b^2)$ implies that $\alpha\beta \in (a^2, b^2)R_N = (\alpha^2, \beta^2)$.

(2): Choose $a, b \in R$ with b regular. There exists $i \in Z^+$ such that $a, b \in R_i$, and b is of necessity regular in R_i. Thus $ab \in \{a^2, b^2\}R_i$, which implies that $ab \in (a^2, b^2)$.

(3): If $a, b \in R$ with b regular, then b is regular in $R[S]$, and hence $ab \in \{a^2, b^2\}R[S]$. But Theorem 12.2 shows that each ideal of R is contracted from $R[S]$. In particular, $ab \in (a^2, b^2) = \{a^2, b^2\}R[S] \cap R$.

COROLLARY 18.3. Assume that G is the quotient group of S and that G^* is the divisible hull of G. If $R[S]$ is a P-ring, then $R[G]$ and $R[G^*]$ are P-rings.

Proof. It is immediate from Theorem 18.2 that $R[G]$ is

a P–ring. Moreover, the proof of Theorem 13.4 shows that G^* can be expressed as the union of an ascending sequence $\{G_m\}_{m=1}^{\infty}$ of subgroups, where $G_m \simeq G$ for each m. Hence each $R[G_m]$ is a P–ring, so Theorem 18.2 implies that $R[G^*]$ is a P–ring.

The next result is a generalization of Theorem 13.3.

THEOREM 18.4. If $S \neq \{0\}$ and if $R[S]$ is a P–ring, then either R is a regular ring or S is a group.

Proof. Assume that S is not a group and choose $s \in S$ such that s is not invertible. Choose $r \in R\backslash\{0\}$. Since S is cancellative, X^s is a regular element of $R[S]$. Thus $rX^s \in (r^2, X^{2s})$, say $rX^s = r^2 f + X^{2s} g$. Since s is not invertible in S and since S is cancellative, it follows that $s \notin \mathrm{Supp}(X^{2s} g)$. Therefore $r = r^2 f_i$ for some coefficient f_i of f, and this implies that $(r) = (r^2)$ is idempotent. We conclude that R is regular if S is not a group.

We subsequently show that $R[S]$ a P–ring implies that R is a regular ring in any case. The key to this result is a treatment of the case of the rational group ring $R[Q]$.

THEOREM 18.5. If $R[Q]$ is a P–ring, then R is a regular ring.

Proof. Choose $r \in R\backslash\{0\}$. Corollary 8.6 shows that $1 - X$ is regular in $R[Q]$, so $r(1 - X) \in (r^2, (1 - X)^2)$. Since $R[Q]$ is the union of the ascending sequence $\{R[Z/n!]\}_{n=1}^{\infty}$ of subrings, it follows that

$r(1 - X) \in \{r^2, (1 - X)^2\}R[X^{\pm 1/m}]$ for some $m \in Z^+$. By a change of variable we can assume that $r(1 - X^m) \in (r^2, (1 - X^m)^2)$ in $R[X, X^{-1}] = R[X]_{\{X^i\}}$. The ideal $A = \{r^2, (1 - X^m)^2\}R[X]$ is comaximal with $XR[X]$ in $R[X]$, and hence A is prime to the multiplicative system $\{X^i\}$. Therefore A is contracted from $R[X]_{\{X^i\}_1^\infty}$, so $r(1 - X^m) \in A$. We next observe that $R[X]$ is a free $R[X^m]$–module with a free basis consisting of $\{1, X, \ldots, X^{m-1}\}$. Thus the ideal $B = \{r^2, (1 - X^m)^2\}R[X^m]$ is contracted from $R[X]$ by Theorem 12.2, so $r(1 - X^m) \in B$. There exists an R–automorphism of $R[X^m]$ mapping $1 - X^m$ to X^m, and under this mapping we obtain $rX^m \in \{r^2, X^{2m}\}R[X^m]$. An easy computation then shows that $r \in r^2R$, and thus R is regular, as asserted.

The results up to this point enable us to give equivalent conditions for the group ring $R[H]$ of a torsion–free group H to be a P–ring or a Prüfer ring (Theorem 18.8). But in order to broaden the statement of that result, we consider briefly the notion of an arithmetical ring. We recall the definition: R is <u>arithmetical</u> if $A \cap (B + C) = (A \cap B) + (A \cap C)$ for all ideals A, B, C of R; this condition is equivalent to distributivity of $+$ over $\cap$ in the lattice of ideals of R, and it is also equivalent to the validity of the Chinese Remainder Theorem in R. It is well known that a unitary integral domain is arithmetical if and only if it is a Prüfer domain. We call a unitary ring R a <u>chained ring</u> (or <u>valuation ring</u>) if the set of ideals of R is linearly ordered under inclusion.

THEOREM 18.6. The ring R is arithmetical if and only if R_M is a chained ring for each maximal ideal M of R. A Bezout ring is arithmetical and an arithmetical ring is a Prüfer ring.

Proof. Assume that R is arithmetical. Since each ideal of R_M is extended from R and since extension of ideals of R to R_M distributes over both sum and intersection, it follows that R_M is also arithmetical. Thus, we assume without loss of generality that R is quasi-local with maximal ideal M, and we prove that R is a chained ring. It suffices to prove that the set of principal ideals of R is linearly ordered under inclusion. Thus, take $a,b \in R$ and assume that $b \notin (a)$. We have

$$(a) = (a) \cap (b, a-b) = [(a) \cap (b)] + [(a) \cap (a-b)] .$$

Thus, write $a = u + v$, where $u = ca = db \in (a) \cap (b)$ and $v = ra = s(a-b) \in (a) \cap (a-b)$. If c is a unit of R, then $a = c^{-1}db \in (b)$. Otherwise, $c \in M$. Moreover, $sb = (s-r)a$ with $b \notin (a)$ implies that $s \in M$. Hence $1-c-s$ is a unit of R, and since $(1-c-s)a = -sb \in (b)$, we conclude that $a \in (b)$ and that R is chained. For the converse, let $\{M_\lambda\}_{\lambda \in \Lambda}$ be the set of maximal ideals of R and let e_λ and c_λ denote extension and contraction of ideals of R to R_{M_λ} for each $\lambda \in \Lambda$. Since R_{M_λ} is chained, its lattice of ideals is distributive under $+$ and $\cap$. Thus, for ideals A,B,C of R we have

$$A \cap (B + C) = \cap_{\lambda \in \Lambda}[A \cap (B + C)]^{e_\lambda c_\lambda} =$$

$$\cap_\lambda [A^{e_\lambda} \cap (B^{e_\lambda} + C^{e_\lambda})]^{c_\lambda} = \cap_\lambda[(A^{e_\lambda} \cap B^{e_\lambda}) + (A^{e_\lambda} \cap C^{e_\lambda})]^{c_\lambda} =$$

$$\cap_\lambda [(A \cap B) + (A \cap C)]^{e_\lambda c_\lambda} = (A \cap B) + (A \cap C) .$$

This completes the proof that R is arithmetical if and only if each localization R_M is a chained ring.

Assume that R is a Bezout ring. Then R_M is Bezout for each maximal ideal M of R, and hence to prove that R is arithmetical, it suffices to prove that a quasi-local Bezout ring T is chained. Let M be the maximal ideal of T and choose nonzero elements $a,b \in T$. If $(c) = (a,b)$, then there exist elements $r,s,x,y \in T$ such that $c = ar + bs$, $a = cx$, and $b = cy$. Then $c = crx + csy$, so $c(1 - rx - sy) = 0$. Since $c \neq 0$, $1 - rx - sy$ is a nonunit of T, so $x \notin M$ or $y \notin M$. If $x \notin M$, for example, then $c \in (a)$ and $(b) \subseteq (a)$. Thus, a Bezout ring is arithmetical.

Finally, assume that R is arithmetical and let A be a finitely generated regular ideal of R. Choose a regular element $b \in A$, and let $\{M_\lambda\}$, e_λ, and c_λ be as in the first paragraph of this proof. Since R_{M_λ} is chained, A^{e_λ} is principal for each $\lambda \in \Lambda$, and hence $(b)^{e_\lambda} = A^{e_\lambda}[(b)^{e_\lambda}:A^{e_\lambda}]$. Because A is finitely generated, $(b)^{e_\lambda}:A^{e_\lambda} = [(b):A]^{e_\lambda}$. Therefore

$$(b) = \cap_\lambda (b)^{e_\lambda c_\lambda} = \cap_\lambda (A^{e_\lambda}[(b):A]^{e_\lambda})^{c_\lambda} =$$

$$\cap_\lambda (A[(b):A])^{e_\lambda c_\lambda} = A[(b):A] .$$

Since (b) is invertible, this proves that A is invertible, and hence R is a Prüfer ring.

THEOREM 18.7. If R is a regular ring, then the polynomial ring $R[X]$, the integral group ring $R[Z]$, and the rational group ring $R[Q]$ are Bezout rings.

Proof. If $R[X]$ is a Bezout ring, then its quotient ring $R[Z]$ is also a Bezout ring. Moreover, since $R[Q] = \cup_1^{\infty} R[Z/n!]$ with $R[Z/n!] \approx R[Z]$ for each n, it also follows that $R[Q]$ is a Bezout ring. Hence, we prove that the polynomial ring $R[X]$ is a Bezout ring. It suffices to prove that if f and g are nonzero elements of $R[X]$, then (f,g) is principal. We assume that $\deg f \leq \deg g$ and we use induction on $\deg f$. If $\deg f = 0$ — that is, if $f \in R$ — then we can assume that f is idempotent. Then $(f,g) = (h)$, where $h = f + (1 - f)g$, for $f = fh$ and $g = (1 - f + fg)h$. We assume that (f,g) is principal if $\deg f \leq n$, and we consider the case where $\deg f = n + 1$. Let t be the leading coefficient of f and let e be an idempotent generator of the ideal tR of R. Then $R = eR \oplus (1 - e)R$, $R[X] = eR[X] \oplus (1 - e)R[X]$, and $(f,g) = (ef,eg) \oplus ((1 - e)f, (1 - e)g)$. The ideal $((1 - e)f, (1 - e)g)$ is principal by the induction hypothesis. And as an element of $eR[X]$, the leading coefficient et of ef is a unit since $et \cdot eR = eR$. Hence $eg = q \cdot ef + r$ for some $q,r \in eR[X]$ with $r = 0$ or $\deg r < \deg ef$. It follows that $(ef,eg) = (ef,r)$. If $r \neq 0$, the induction hypothesis implies that (ef,eg) is principal, and if $r = 0$, it is clear that (ef,eg) is principal. Since each of the summands of the ideal (f,g)

in the decomposition

$$(f,g) = (ef,eg) \oplus ((1-e)f, (1-e)g)$$

is principal, (f,g) is also principal. This completes the proof.

THEOREM 18.8. Assume that H is a nonzero torsion-free group with divisible hull H^*. The following conditions are equivalent.

(1) $R[H]$ is a P-ring.

(2) $R[H^*]$ is a P-ring.

(3) R is a regular ring and $H^* \simeq Q$.

(4) $R[H]$ is a Bezout ring.

(5) $R[H]$ is arithmetical.

(6) $R[H]$ is a Prüfer ring.

Proof. Corollary 18.3 shows that (1) implies (2), Theorem 18.7 shows that (3) implies (4), and the implications (4) $\Longrightarrow$ (5) and (5) $\Longrightarrow$ (6) follow from Theorem 18.6. Since each Prüfer ring is a P-ring, we need only prove that (2) implies (3). The group H^* is a direct sum of copies of Q, and in view of Theorem 18.5, it suffices to show that the assumption that there is more than one Q-summand leads to a contradiction. Thus, assume that $H^* = Q \oplus Q \oplus K$, where K is a subgroup of H^*. Part (3) of Theorem 18.2 implies that $R[Q \oplus Q] \simeq R[Q][Q]$ is a P-ring. Hence $R[Q]$ is a regular ring by Theorem 18.5, and this contradicts Theorem 17.4, for Q is not periodic as a semigroup. We conclude that $H^* \simeq Q$, and hence (2) implies (3).

We are able to extend Theorem 18.8 to monoid rings without further ado.

THEOREM 18.9. Assume that R is a unitary ring and that S is a nonzero torsion-free cancellative monoid. The following conditions are equivalent.

(1) $R[S]$ is a Bezout ring.

(2) $R[S]$ is arithmetical.

(3) $R[S]$ is a Prüfer ring.

(4) $R[S]$ is a P-ring.

(5) R is a regular ring, and to within isomorphism, S is either a subgroup of Q or a Prüfer submonoid of Q.

Proof. In view of Theorems 18.6 and 18.9, we need only prove that (4) implies (5) and (5) implies (1). Thus, assume that $R[S]$ is a P-ring and let H be the quotient group of S. Theorem 18.2 implies that $R[H]$ is a P-ring. Hence R is regular and $H^* \simeq Q$ by Theorem 18.8. We consider S as a submonoid of Q. If S contains both positive and negative rationals, then S is a subgroup of Q by Theorem 2.9. In the contrary case, S is isomorphic to a submonoid U of Q_0 containing 1; without loss of generality we assume that $S = U$. To show that S is a Prüfer monoid, it suffices, by Theorem 13.5, to show that $H \cap Q_0 \subseteq S$. Take $h = a - b \in H \cap Q_0$, where $a,b \in S$. We assume that $h \neq 0$ so that $a > b$. Since X^a and X^b are regular elements of $R[S]$, a P-ring, we have $X^{a+b} \in (X^{2a}, X^{2b})$. This implies that either $a + b \in 2a + S$ or $a + b \in 2b + S$. As each element of $2a + S$ is greater than $a + b$. we have $a + b \in 2b + S$ and $h = a - b \in S$.

Thus $H \cap Q_0 \subseteq S$, and (4) implies (5), as asserted.

(5) $\Longrightarrow$ (1): If R is regular and S is a subgroup of Q, then Theorem 18.8 shows that $R[S]$ is a Bezout ring. On the other hand, if S is a Prüfer submonoid of Q, then write $S = \cup_1^\infty S_i$, where each S_i is a cyclic monoid and $S_i \subseteq S_{i+1}$ for each i. Then $R[S] = \cup_1^\infty R[S_i]$, where each $R[S_i]$ is isomorphic to $R[X]$, a Bezout ring by Theorem 18.7 Therefore $R[S]$ is also a Bezout ring, and this completes the proof of Theroem 18.9.

The ring-theoretic analogue of a Dedekind domain is the notion of a *general* ZPI-*ring*, defined as a ring in which each ideal is a finite product of prime ideals. Theorem 39.2 of [51] shows that a unitary ring R is a general ZPI-ring if and only if R is a finite direct sum of Dedekind domains and special primary rings, where a *special primary ring* is, by definition, a local PIR with nilpotent maximal ideal. (The term *special principal ideal ring* (SPIR) is sometimes used instead of special primary ring.) For our purposes, the equivalence in terms of Dedekind domains and SPIR's could be taken as the definition of a unitary general ZPI-ring. Note in particular that a unitary general ZPI-ring is Noetherian. An ideal A of a ring T is said to be a *multiplication ideal* if A is a factor of each ideal of T that it contains — that is, if $\{AB_\alpha \mid B_\alpha$ is an ideal of $T\}$ is the set of ideals of T contained in A; the ring T is a *multiplication ring* if each ideal of T is a multiplication ideal. It is clear that a regular ideal of a unitary ring is a multiplication ideal if and only if it is invertible, and hence a unitary integral domain is a multiplication ring if and only if it is a Dedekind domain. Moreover, each unitary multiplication ring is a Prüfer ring. On the other hand, a regular ring is a multiplication ring (for $A \leq B$ in a regular ring implies $A = A^2 = AB$), and hence a multiplication ring with zero divisors need not be Noetherian. Theorem 18.10 determines necessary and sufficient conditions in order that $R[S]$ should be a multiplication ring or a general ZPI–ring. While Theorem 18.10 is a summary result, we do not repeat our standing hypotheses concerning R and S in its statement.

THEROEM 18.10. The following conditions are equivalent.

(1) $R[S]$ is a PIR.

(2) $R[S]$ is a general ZPI–ring.

(3) $R[S]$ is a multiplication ring.

(4) R is a finite direct sum of fields and S is isomorphic to Z or to Z_0.

Proof. (1) $\Longrightarrow$ (2): Each PIR is a finite direct sum of PID's and SPIR's, and hence is a general ZPI–ring.

(2) $\Longrightarrow$ (3): Dedekind domains and SPIR's are multiplication rings, and a finite direct sum of unitary multiplication rings is again a multiplication ring. Therefore each general ZPI–ring is a multiplication ring.

(3) $\Longrightarrow$ (4): Assume that $R[S]$ is a multiplication ring. Then $R[S]$ is a Prüfer ring, and hence R is a regular ring and S is either a subgroup of Q or a Prüfer submonoid of Q. We show that R is Noetherian by showing that each proper prime ideal P of R is finitely generated. By Corollary 8.6, $1 - X$ is not a zero divisor in $R[S]$, so

$(P, 1 - X)$ is invertible, and hence finitely generated. Since P is the image of $(P, 1 - X)$ under the augmentation map, it follows that P is also finitely generated. A Noetherian regular ring is a finite direct sum of fields — say $R = F_1 \oplus \ldots \oplus F_n$. Thus $R[S] = F_1[S] \oplus \ldots \oplus F_n[S]$, and each $F_i[S]$ is an integral domain that is a multiplication ring, hence a Dedekind domain. Applying Theorem 13.8, we conclude that either $S \simeq Z$ or $S \simeq Z_0$.

(4) $\Longrightarrow$ (5): If $R = F_1 \oplus \ldots \oplus F_k$ is a direct sum of fields F_i and if S is Z or Z_0, then each $F_i[S]$ is a PID, and hence $R[S] = \Sigma_1^k \oplus F_i[S]$ is a PIR.

Section 18 Remarks

Some detailed references on topics treated in Section 18 are the following. Prüfer rings: [27], [65]; arithmetical rings [45], [82]; general ZPI-rings [107], [13]; multiplication rings [106], [108], [56], [66], [4]. Moreover, [91, Chapters IX and X] is a general reference for these topics. We remark that a ZPI-*ring*, as opposed to a general ZPI-ring, is a ring T in which each nonzero ideal is uniquely expressible as a finite product of prime ideals (if T is unitary, then factors of T are disregarded in considering uniqueness). A unitary ring T is a ZPI-ring if and only if it is either a Dedekind domain or a SPIR. The choice of terminology comes from the German *Zerlegung Primideale*.

Much of the material in Section 18 stems from [59], but new proofs have been necessary for most of the results of that paper. This is because Gilmer and Parker in [59] made strong use of a result of Griffin in [65] stating that a unitary ring R is a Prüfer ring if and only if R is an integrally closed P–ring. While it is true that a Prüfer ring is an integrally closed P–ring, the status of the converse is in question; Griffin's proof shows only that in an integrally closed P–ring R, each ideal of the form $(a_1,\ldots,a_n)$, with $a_1,\ldots,a_{n-1}$ regular, is invertible. A P–ring need not be integrally closed, and hence need not be a Prüfer ring. For example, if K is the Galois field with 4 elements and if F is the prime subfield of K, then the domain $D = F + XK[[X]]$ is a P–ring, but is not integrally closed. For details of this example, as well as related results, see [112], [27], and [53].

§19. Monoid Rings as Arithmetical Rings — the Case Where S is not Torsion-Free

Results of Section 18 show that if R is unitary and S is torsion-free and cancellative, the monoid ring $R[S]$ is a Prüfer ring iff it is arithmetical iff it is a Bezout ring. Moreover, $R[S]$ is a general ZPI-ring iff it is a multiplication ring iff it is a PIR. An examination of the proofs in Section 18 suggests that the assumption that S is torsion-free is used sparingly, and that it may be possible to extend some of the results to the case where S is any cancellative monoid. This section is devoted to that end. The key concept in this development turns out to be that of an arithmetical ring, rather than that of a Prüfer ring. Our results show that even for finite groups, a group ring that is arithmetical need not be a Bezout ring, and that $R[G]$ a general ZPI-ring does not, in general, imply that $R[G]$ is a PIR.

The first four results of the section are devoted to a determination of conditions under which a monoid ring is quasi-local (that is, has a unique maximal ideal) or a chained ring; these results do not require that the monoid in question is cancellative. They are related to the problem of determining conditions under which $R[S]$ is arithmetical in two ways: first, a chained ring is arithmetical, and second, $R[S]$ is arithmetical if and only if each localization $(R[S])_M$ of $R[S]$ at a maximal ideal M is chained. In connection with the latter condition, we remark that for a multiplicative system N in R, the rings $R[S]_N$ and $R_N[S]$ are canonically isomorphic for any unitary ring R and any monoid S.

Throughout the section, R denotes a unitary ring and S denotes a cancellative monoid.

THEOREM 19.1. Assume that U is a nonzero monoid. The monoid ring $R[U]$ is quasi-local if and only if R is quasi-local with maximal ideal M, $\operatorname{char}(R/M) = p \neq 0$, and U is a p-group.

Proof. Assume first that $R[U]$ is quasi-local. Since R is a homomorphic image of $R[U]$, R is also quasi-local. Let M be the maximal ideal of R and let $F = R/M$. Then $F[U] \simeq R[U]/M[U]$ is quasi-local. Pick $u \in U$. Then $1 - X^u$ is a nonunit of $F[U]$ and $(1 - X^u) + (X^u) = F[U]$, so X^u is a unit of $F[U]$ and u is invertible in U. Therefore U is a group. Let H be a subgroup of U. Theorem 12.2 shows that each ideal of $F[H]$ is contracted from $F[U]$; in particular, each maximal ideal of $F[H]$ is contracted from a maximal ideal of $F[U]$, and hence $F[H]$ is quasi-local. Since $F[Z] \simeq F[X,X^{-1}]$ is not quasi-local, it follows that U is a torsion group. Let p be a prime that divides the order of an element of U. We show that $p = \operatorname{char}(F)$; this will imply that U is a p-group. Thus, choose $u \in U$ of order p. Then $F[(u)] \simeq F[X]/(X^p - 1)$ is quasi-local. This implies that $X - 1$ is the only prime factor of $X^p - 1$ in $F[X]$, and therefore $X^p - 1 = b(X - 1)^p$ for some unit b of F. Since $X^p - 1$ is monic, $b = 1$ and $X^p - 1 = (X - 1)^p$. Therefore F has characteristic p.

To prove the converse, let I be the augmentation ideal of $R[U]$, where R is quasi-local with maximal ideal M,

$\mathrm{char}(R/M) = p \neq 0$, and U is a p–group. Then $R[U]/(M[U] + I) \simeq R/M$, so $M[U] + I$ is maximal in $R[U]$. Let P be maximal in $R[U]$. Since $R[U]$ is integral over R , then $P \cap R$ is maximal in R . Therefore $P \cap R = M$ and $P \supseteq M[U]$. The ideal I is generated by $\{1 - X^u \mid u \in U\}$ and each $1 - X^u$ is nilpotent modulo $M[U]$ since R/M has characteristic p . Consequently, $I \subseteq P$, so $P = M[U] + I$. This completes the proof of Theorem 19.1.

COROLLARY 19.2. If $U \neq \{0\}$, then the monoid ring $R[U]$ is a local ring if and only if R is a local ring with residue field of characteristic $p \neq 0$ and U is a finite p–group.

Proof. The result follows immediately from Theorem 19.1 and the fact that $R[U]$ is Noetherian if and only if R is Noetherian and U is finitely generated (Theorem 7.7).

In order to determine conditions under which $R[U]$ is a chained ring, we use an auxillary result from group theory. For a prime p , the p–_quasicyclic group_ is the multiplicative group of complex p^nth roots of unity, taken over all $n \in Z^+$. If G is quasicyclic or cyclic of prime-power order, then it is well known that the set of subgroups of G is linearly ordered under inclusion. Corollary 19.3 establishes the converse.

THEOREM 19.3. If the set of subgroups of the group G is linearly ordered under inclusion, then G is either cyclic of prime-power order or a quasicyclic group.

Proof. To obtain the result, we need not assume that G

is abelian in advance; in fact, if $a_1,\dots,a_n \in G$, then the subgroup of G generated by $\{a_i\}_1^n$ is one of the cyclic groups (a_i). Since the subgroups of Z are not linearly ordered, G is a torsion group. Hence G is the direct sum of its primary components. It is clear, however, that G is indecomposable, so that G is a p-group for some prime p. Let (x_1) be a subgroup of G of order p. If $G = (x_1)$, we're finished. Otherwise, choose $y_2 \in G\backslash(x_1)$. Then $(x_1) \subseteq (y_2)$, and in fact, $(x_1) = (p^k y_2)$, where y_2 has order p^{k+1}. It follows that there exists $x_2 \in (y_2)$ of order p^2 such that $x_1 = px_2$. If $G = (x_2)$, the proof is complete, and if $G \neq (x_2)$, we can continue the process. Hence, if G is not cyclic, then there exists a sequence $\{x_i\}_1^\infty$ in G such that x_i has order p^i and $px_{i+1} = x_i$ for each $i \geq 1$. Let $H = \cup_1^\infty (x_i)$. No cyclic subgroup of G contains H, and hence each cyclic subgroup of G is contained in H. Therefore $G = H$, and G is the p-quasicyclic group in this case.

THEOREM 19.4. Assume that U is a nonzero monoid. The monoid ring $R[U]$ is a chained ring if and only if R is a field of characteristic $p \neq 0$, U is a group, and U is either cyclic of order p^n or p-quasicyclic.

Proof. Assume that $R[U]$ is a chained ring. Theorem 19.1 implies that R is quasi-local with maximal ideal M and that U is a p-group, where $p = \text{char}(R/M)$. We show that R is a field. Thus, pick $m \in M$ and $u \in U\backslash\{0\}$. The ideals (m) and $(1 - X^u)$ of $R[U]$ are comparable. Since $1 - X^u \notin (m)$, it follows that $m \in (1 - X^u)$. Under the augmentation map, $(1 - X^u)$ goes to (0) and m is mapped to itself. Hence $m = 0$, $M = (0)$, and R is a field of characteristic p. Since the set of kernel ideals of $R[U]$ is linearly ordered, the set of congruences on U is chained, and hence the set of subgroups of U forms a chain. By Theorem 19.3, we conclude that U is either cyclic of order p^n or p–quasicyclic.

If R is a field of characteristic p and U is p–quasicyclic, then $U = \cup_{i=1}^{\infty} U_i$, where U_i is the cyclic subgroup of U of order p^i. If $f, g \in R[U]$, then $f, g \in R[U_n]$ for some n. Hence, to prove that $R[U]$ is chained, it suffices to consider the case where $U = U_n$ is cyclic of order p^n. In that case,

$$R[U] \simeq R[X]/(X^{p^n} - 1) = R[X]/(X - 1)^{p^n}$$

is a SPIR with maximal ideal $(X - 1)/(X - 1)^{p^n}$, hence a chained ring. This completes the proof.

We turn to the problem of determining conditions under which $R[S]$ is arithmetical (recall that S is assumed to be cancellative). Theorem 19.5 lists some basic properties of arithmetical rings.

THEOREM 19.5. Let N be a multiplicative system in R, and assume that $\{R_i\}_{i \in I}$ is a directed family of subrings of R such that $R = \cup_{i \in I} R_i$.

(1) If R is arithmetical, then so is R_N.

(2) If R is arithmetical, then R is integrally closed.

(3) If each R_i is arithmetical, then R is

arithmetical.

(4) Assume that T is an extension ring of R with a free R-module basis containing 1. If T is arithmetical, then R is also arithmetical.

Proof. (1) follows from the facts that each ideal of R_N is extended from R and that extension of ideals of R to R_N distributes over both $+$ and $\cap$.

Since R is a Prüfer ring if R is arithmetical, it suffices to prove the statement in the Section 18 Remarks to the effect that a Prüfer ring R is integrally closed. Take an element y in the total quotient ring of R that is integral over R. If y is a root of a monic polynomial over R of degree $n+1$, then the fractional ideal $F = (1,y,\ldots,y^n)$ of R is both regular and idempotent. Since R is a Prüfer ring, F is invertible, so $F = F^2$ implies that $F = R$. Hence $y \in R$ and R is integrally closed.

(3) The proof of Theorem 18.6 shows that R is arithmetical if $a \in [(a) \cap (b)] + [(a) \cap (a-b)]$ for all $a,b \in R$. Fix $a,b \in R$. Then $a,b \in R_i$ for some i. Since R_i is arithmetical, $a \in [\{a\}R_i \cap \{b\}R_i] + [\{a\}R_i \cap \{a-b\}R_i]$, and this implies that the corresponding relation also holds in R.

(4): If $\{x_i\}_{i\in I}$ is a free R-module basis for T, then $AT = \Sigma_{i\in I}Ax_i$ for each ideal A of R. Because $\{x_i\}$ is free, it follows that if A and B are ideals of R, then $AT \cap BT = (\Sigma Ax_i) \cap (\Sigma Bx_i) = \Sigma(A \cap B)x_i = (A \cap B)T$. Now take ideals A,B,C are ideals of R. Since each ideal of R is contracted from T, it suffices to prove that $[A \cap (B + C)]T = [(A \cap B) + (A \cap C)]T$. Using the fact that T is arithmetical, this is straightforward:

$$[A \cap (B + C)]T = AT \cap (BT + CT) = (AT \cap BT) + (AT \cap CT)$$

$$= (A \cap B)T + (A \cap C)T = [(A \cap B) + (A \cap C)]T$$

COROLLARY 19.6. Assume that S has quotient group G and that H is a subgroup of S. If $R[S]$ is arithmetical, then so are $R[G]$ and $R[H]$. In particular, R is arithmetical.

In giving conditions under which $R[S]$ is arithmetical, where S is not torsion-free, we separate the cases where S is, or is not, periodic.

THEOREM 19.7. Assume that S is neither torsion-free nor periodic. Let G be the quotient group of S and let H be the torsion subgroup of G.

(1) If $R[S]$ is arithmetical, then $H \subseteq S$, $R[S/H]$ is arithmetical, R is a regular ring, and $R[H]$ is regular.

(2) Conversely, if the conditions in (1) are satisfied, then $R[S]$ is a Bezout ring, hence arithmetical.

<u>Proof</u>. (1): If $R[S]$ is arithmetical, then (19.6) shows that $R[S]$ is integrally closed. Thus S is integrally closed by Theorem 12.10. Since H is integral over S, it follows that $H \subseteq S$. Moreover, $R[S/H]$ is arithmetical since it is a homomorphic image of $R[S]$. Because $S/H \subseteq G/H$ is torsion-free and cancellative, Theorem 18.9 shows that R is a regular ring. To show that $R[H]$ is regular, we must show that the order of each element $h \in H$ is a unit of R. Pick $g \in G \setminus H$. The subgroup K of G generated by $\{h,g\}$ is $(h) \oplus (g)$. Part (1) of Theorem 19.5 shows that $R[G]$ is arithmetical, and Corollary 19.6

implies that $R[K] \approx R[(h)][Z]$ is also arithmetical. Applying either Theorem 18.8 or the part of (1) already proved, we conclude that $R[(h)]$ is regular, and hence the order of h is a unit of R (Theorem 17.4). This establishes (1).

To prove (2), we show that $R[S]$ is the union of an ascending sequence of Bezout rings. Since S/H is torsion-free and cancellative, Theorem 18.9 shows that either S/H is a subgroup of Q or a Prüfer submonoid of Q. In the first case, S is a group and S/H is the union of an ascending sequence $\{(s_n + H)\}_{n=1}^{\infty}$ of nonzero subgroups. It follows that $S = \cup_{n=1}^{\infty}[(s_n) + H]$. Moreover, $(s_n) + H = (s_n) \oplus H$ since $(s_n) \cap H = \{0\}$. Therefore $R[S] = \cup_{n=1}^{\infty}R[(s_n) \oplus H]$, where $R[(s_n) \oplus H] \simeq R[H][Z]$ is a Bezout ring for each n. This proves that $R[S]$ is a Bezout ring if S/H is a group. The proof in the other case is similar; there $S/H = \cup_{n=1}^{\infty}\langle s_n + H\rangle^0$ is the union of an ascending sequence of cyclic monoids and $S = \cup_{n=1}^{\infty}[\langle s_n\rangle^0 \oplus H]$. Since $R[\langle s_n\rangle^0 \oplus H] \simeq R[H][Z_0]$ is a Bezout ring in this case as well, this establishes (2).

The conditions in (1) of Theorem 19.7 could be stated more explictly in terms of R, S, and H, but we choose not to do so, for Theorem 18.9 gives conditions on S/H in order that $R[S/H]$ should be arithmetical for R regular, and Theorem 17.4 provides necessary and sufficient conditions on R and H for $R[H]$ to be regular.

A cancellative periodic monoid is a group, and hence the case remaining from Theorem 19.7 in the problem of determining when $R[S]$ is arithmetical is that in which S is a torsion group. Theorem 19.9 provides a reduction in this case.

THEOREM 19.9. Let $\{M_\lambda\}_{\lambda\in\Lambda}$ be the set of maximal ideals of R, and assume that S is a torsion group. Then $R[S]$ is arithmetical if and only if each $R_{M_\lambda}[S]$ is arithmetical.

Proof. If $R[S]$ is arithmetical, then $R[S]_{R\setminus M_\lambda} \simeq R_{M_\lambda}[S]$ is arithmetical for each λ. For the converse, let M be a maximal ideal of $R[S]$. Since $R[S]$ is integral over R, the ideal $M \cap R$ is maximal in R — say $M \cap R = M_\lambda$. It follows that $R[S]_M$ is isomorphic to a localization of $R_{M_\lambda}[S]$ at a maximal ideal. Hence if each $R_{M_\lambda}[S]$ is arithmetical, $R[S]_M$ is chained and $R[S]$ is also arithmetical.

THEOREM 19.10. Assume that S is a torsion group and that $\{M_\lambda\}$ is the family of maximal ideals of R. For each prime p, let S_p be the p-primary component of S.

(1) If $R[S]$ is arithmetical, then R is arithmetical. Moreover, for each prime p such that $S_p \neq \{0\}$ and $p = \text{char}(R/M_\lambda)$ for some λ, it follows that R_{M_λ} is a field and S_p is either cyclic or quasicyclic.

(2) The converse of (1) is also valid.

Proof. (1): We have already observed that $R[S]$ arithmetical implies R is arithmetical. Assume that $p = \text{char}(R/M_\lambda)$ and $S_p \neq \{0\}$. Since S_p is a homomorphic image of S, the ring $R[S_p]$ is arithmetical, as is $R_{M_\lambda}[S_p]$. Theorem 19.1 shows, however, that $R_{M_\lambda}[S_p]$ is quasi-local. Therefore $R_{M_\lambda}[S_p]$ is a chained ring, and

Theorem 19.4 shows that R_{M_λ} is a field and that S_p is either cyclic or quasicyclic.

(2) To prove that $R[S]$ is arithemtical under the conditions of (1), it suffices to show that each $R_{M_\lambda}[S]$ is arithmetical. Thus, we assume without loss of generality that R is a chained ring with maximal ideal M. Let $c = \operatorname{char}(R/M)$. The group S is the directed union of its family $\{T_i\}_{i\in I}$ of finite subgroups, and $R[S] = \cup_{i\in I} R[T_i]$. Hence, part (3) of Theorem 19.5 shows that it suffices to prove that each $R[T_i]$ is arithmetical. By a change of notation, we assume that S is finite. If c divides the order of S, then the hypothesis implies that R is a field and S_c is cyclic. Moreover, $S = S_c \oplus T$ for some subgroup T of S, and $R[S] \simeq R[S_c][T]$, where $R[S_c]$ is a chained ring with residue field of characteristic c by Theorem 19.4. Hence, we also assume that c does not divide the order of S. The group S can be expressed as a finite direct sum of cyclic groups $\{U_i\}_{i=1}^m$ of prime-power order. Each maximal ideal of $R[U_1]$ lies over M in R, and hence all maximal ideals of $R[U_1]$ have associated residue field of characteristic c. By induction, it suffices to prove the following result, which we state separately for the sake of clarity.

(19.11) If R is a chained ring with maximal ideal M, if $c = \operatorname{char}(R/M)$, and if S is a finite group of prime power order $m = p^k$, where $c \neq p$, then $R[S]$ is arithmetical.

<u>Proof of</u> (19.11). Since $R[S]$ is integral over R, each maximal ideal of $R[S]$ lies over M in R, so the

maximal ideals of $R[S]$ are in one-to-one correspondence with those of $R[S]/M[S] \simeq F[X]/(X^m - 1)$, where $F = R/M$. Since $c \neq p$, the ring $F[X]/(X^m - 1)$ is reduced and has only finitely many maximal ideals. Let $\{M_i\}_{i=1}^n$ be the set of maximal ideals of $R[S]$; we have $M[S] = \cap_{i=1}^n M_i$. Fix j between 1 and n and let e denote extension of ideals of $R[S]$ to $R[S]_{M_j}$. We have $M[S]^e = \cap_{i=1}^n M_i^e = M_j^e$. Take $f \in M[S]$ and assume that m is a coefficient of f that generates the content ideal of f. Then $f = mf_1$, where $f_1 \in R[S] \setminus M[S]$. Thus $(f)^e = (mf_1)^e = (m)^e$. It follows that $\{(m)^e \mid m \in M\}$ is the set of principal ideals of $R[S]_{M_j}$. Since R is a chained ring, we conclude that the set of principal ideals of $R[S]_{M_j}$ is chained, and hence $R[S]_{M_j}$ is also a chained ring.

The results up to this point are sufficient to show that a group ring $R[G]$ need not be arithmetical if it is a Prüfer ring and that $R[G]$ need not be a Bezout ring if it is arithmetical. For the first example, let R be a field of characteristic $p \neq 0$ and let G be a p-group that is not cyclic. Theorem 19.1 and its proof show that $R[G]$ is a local ring with nilpotent maximal ideal. Hence $R[G]$ is trivially a Prüfer ring, but $R[G]$ is not arithmetical by Theorem 19.4. For the second example, take $D = Z[\sqrt{-5}]$. Then D is a Dedekind domain that is not a PID, and it is easy to show that $p = 11$ generates a prime ideal of D. Let $R = D[1/p]$. The domains R and D have the same class group, so R also fails to be a PID. Let G be the cyclic group of order 11. Theorem 19.10 shows that $R[G]$ is arithmetical, but $R[G]$ is not a Bezout ring since R fails to have this property. Note that $R[G]$ is, in fact, a Noetherian arithmetical ring; the next result shows that such rings are general ZPI–rings.

THEOREM 19.12. The following conditions are equivalent.

(1) R is a general ZPI–ring.

(2) R is a Noetherian arithmetical ring.

Proof. If (1) holds, then $R = R_1 \oplus \ldots \oplus R_n$ is a direct sum of Dedekind domains and SPIR's. Each of the summands is Noetherian and arithmetical, and it is straightforward to show that the direct sum has the same two properties.

(2) $\Longrightarrow$ (1): We note that a local arithmetical ring is a PIR, and hence is either a SPIR or a rank-one discrete valuation domain. This observation has several applications to R, as follows. If P and M are proper prime ideals of R with $P < M$, then R_M is local and arithmetical, hence a Noetherian valuation domain. It follows that P is the only P–primary ideal, P is contained in each M–primary ideal of R, and P is the unique prime ideal of R properly contained in M. Let $(0) = \cap_{i=1}^{n} Q_i$ be a shortest primary representation of (0) in R, where Q_i is P_i–primary for each i. There are no inclusion relations among distinct primes P_i since a relation $P_i < P_j$ would imply that $Q_i < Q_j$. Thus, P_i and P_j are comaximal for $i \neq j$ since a maximal ideal of R properly contains at most one prime ideal. It follows that Q_i and Q_j are also comaximal, and hence $R \simeq \Sigma_1^n \dot{\oplus} (R/Q_i)$. If P_i is maximal in R, then R/Q_i is local, zero-dimensional, and arithmetical, hence a SPIR. If P_i is not maximal, then $P_i = Q_i$ and R/Q_i is a

Dedekind domain. Therefore R is a general ZPI–ring.

In view of Theorem 19.12, equivalent conditions for $R[S]$ to be a general ZPI–ring can be easily obtained from Theorems 19.7 and 19.10.

THEOREM 19.13. Assume that S is nonzero and that S is not torsion–free. Let G be the quotient group of S and let H be the torsion subgroup of G.

(1) If S is not periodic, then $R[S]$ is a general ZPI–ring if and only if R is a finite direct sum of fields and $S \simeq Z \oplus H$ or $S \simeq Z_0 \oplus H$, where H is finite and the order of H is a unit of R. In this case, $R[S]$ a ZPI–ring implies that it is a PIR.

(2) If S is periodic, then $R[S]$ is a general ZPI–ring if and only if R is a general ZPI–ring, $S = G$ is a finite group, and $R[S]$ is arithmetical.

<u>Proof</u>. (1): Assume that $R[S]$ is a general ZPI–ring. Theorem 19.7 shows that $H \subseteq S$, R and $R[H]$ are regular, and S/H is a subgroup of Q or a Prüfer submonoid of Q. Moreover, since $R[S]$ is Noetherian, then R is Noetherian and S, G, H, and S/H are finitely generated. It follows that R is a finite direct sum of fields, H is finite, the order of H is a unit of R, and $S/H \simeq Z$ or Z_0. Finally, the proof of Theorem 19.7 shows that if $S/H \simeq Z$, then $S \simeq H \oplus Z$, and similarly, $S/H \simeq Z_0$ implies that $S \simeq H \oplus Z_0$. Conversely, if R, S, and H are as described in (1), then R is Noetherian and S is finitely generated, so $R[S]$ is Noetherian. Moreover, part (2) of Theorem 19.13 shows that $R[S]$ is a Bezout ring, from which

it follows that R[S] is a PIR .

Statement (2) follows immediately from (19.10), (19.12), and the fact that R[S] is Noetherian if and only if R is Noetherian and S is finitely generated.

The example $R = Z[\sqrt{-5}, 1/11]$, $G = Z_{11}$ shows that a group ring R[G] need not be a PIR if it is a general ZPI-ring. Under what conditions is R[S] a PIR , for S nonzero and not torsion-free? Part (1) of Theorem 19.13 gives equivalent conditions in the case where S is not periodic. We proceed to consider the case where S is periodic. If R[S] is a PIR , then R is a PIR and S is finitely generated, hence finite. It follows that $R = R_1 \oplus \ldots \oplus R_n$ is a finite direct sum of PID's and SPIR's , and each $R_i[S]$ is a PIR . And conversely, if $R = R_1 \oplus \ldots \oplus R_n$ is so decomposed and S is a finite group such that $R_i[S]$ is a PIR for each i , then R[S] is a PIR . Thus, there are three cases to consider for R in determining conditions under which R[S] is a PIR — namely, the cases where (1) R is a field, (2) R is a SPIR that is not a field, and (3) R is a PID that is not a field. The next three results treat these three cases.

THEOREM 19.14. Assume that F is a field of characteristic p and that G is a torsion group. If $p = 0$, then F[G] is a PIR . If $p \neq 0$, then F[G] is a PIR if and only if the p-primary component G_p of G is cyclic.

Proof. If F[G] is a PIR , then part (1) of Theorem 19.10 implies that G_p is cyclic if $p \neq 0$. Conversely, if G satisfies the given conditions, then F[G] is arithmetical by Theorem 19.10, and $F[G]$ is Artinian since it is a finitely generated F-module. Therefore $F[G]$ is a finite direct sum of zero-dimensional local rings. Each of these summands is a local chained ring, hence a PIR, so $F[G]$ is also a PIR.

THEOREM 19.15. Assume that R is a SPIR that is not a field and that G is a finite group of order m. Then $R[G]$ is a PIR if and only if m is a unit of R.

Proof. If m is a unit of R, then part (2) of Theorem 19.10 shows that $R[G]$ is arithmetical. As in Theorem 19.14, $R[G]$ is also Artinian, and the proof that $R[G]$ is a PIR is the same as that given in (19.14) for the corresponding case.

If m is not a unit of R, then m belongs to the maximal ideal M of R and $\text{char}(R/M) = p$ is a divisor of m. Since $G_p \neq \{0\}$ and $R_M = R$ is not a field, part (1) of Theorem 19.10 shows that $R[G]$ is not arithmetical, and hence is not a PIR.

THEOREM 19.16. Assume that D is a PID that is distinct from its quotient field K. Let G be a finite group of order m. The following conditions are equivalent.

(1) $D[G]$ is a PIR.

(2) m is a unit of D, and if k is any divisor of the exponent of G and u is a primitive kth root of unity over K, then $D[u]$ is a PID.

Proof. (1) $\Longrightarrow$ (2): If m is not a unit of D, then some prime divisor p of m is the characteristic of D/M_λ

for some maximal ideal M_λ of D. Since D_{M_λ} is not a field, part (1) of Theorem 19.10 shows that $D[G]$ is not arithmetical in this case. Therefore $D[G]$ a PIR implies m is a unit of D. If k divides the exponent of G, then the fundamental theorem of finite abelian groups implies that there exists a subgroup K of G such that G/K is cyclic of order k. Since $D[G/K]$ is a homomorphic image of $D[G]$, it is a PIR. Thus $D[X]/(X^k - 1)$ is a PIR. Since $D[u]$ is isomorphic to $D[X]/(f(X))$, where $f(X)$ is the minimal polynomial for u over D (which is equal to the minimal polynomial for u over K), it follows that $D[u]$ is a homomorphic image of $D[X]/(X^k - 1)$, and is therefore a PID.

(2) ⟹ (1): Since m is a unit of D, part (1) of Theorem 19.10 implies that $D[G]$ is arithmetical. Hence $D[G]$ is a general ZPI-ring by part (2) of (19.13). Moreover, Theorem 9.17 implies that $D[G]$ is reduced. Hence $D[G]$ is isomorphic to the direct sum of the rings $\{D[G]/P_i\}_{i=1}^n$, where $\{P_i\}_1^n$ is the set of minimal primes of $D[G]$. To show that $D[G]$ is a PIR, we show that each $D[G]/P_i$ is a PID. The domain D is integrally closed, nonzero elements of D are regular in $D[G]$, and $D[G]$ is integral over D; hence the lying under theorem implies that $P_i \cap D = (0)$ for each i. Therefore $D[G]/P_i \simeq D[\{X^g + P_i \mid g \in G\}]$. Consider first the case where G is cyclic, generated by h. Then $u = X^h + P_i$ generates $D[G]/P_i$ as a ring extension of D, and $u^m = 1$. Therefore u is a primitive kth root of unity for some divisor k of m, and the hypothesis of (2) implies that $D[u]$ is a PID

To summarize, if G is cyclic, we've shown that $D[G]$ is isomorphic to a finite direct sum $D[u_1]\oplus\ldots\oplus D[u_n]$, where each u_i is a root of unity in a fixed algebraic closure L of K , and where the (multiplicative) order of u_i divides m . In the general case, express G as a finite direct sum of cyclic groups of prime-power order; say $G = G_1\oplus\ldots\oplus G_r$, where $r > 1$. Let $H = G_1\oplus\ldots\oplus G_{r-1}$. Then $G = H \oplus G_r$, and the exponent of G is the least common multiple of the exponents of H and G_r . By induction we assume that $D[H]$ is a finite direct sum of domains $D[v_i]$, where $v_i \in L$ is a root of unity whose order divides the exponent of H . Thus $D[G]$ is a direct sum of rings $D[v_i][G_r]$, where the hypothesis on D implies that each $D[v_i]$ is a PID . The order of G_r is a unit of D , hence of $D[v_i]$. Thus, the proof in the case where G is cyclic implies that $D[v_i][G_r]$ is isomorphic to a finite direct of rings $D[v_i,w_j]$, where $w_j \in L$ is a root of unity whose order divides the exponent (= order) of G_r . We have $D[v_i,w_j] = D[y_{ij}]$, where $y_{ij} \in L$ is a root of unity whose order is the least common multiple of the orders of v_i and w_j , hence a divisor of the exponent of G . Thus each $D[y_{ij}]$ is a PID , and $D[G] \simeq \Sigma_{i,j}\oplus D[y_{ij}]$ is a PIR . This completes the proof.

Since an SPIR is a local chained ring, Corollary 19.2 and Theorem 19.4 yield the following characterization of monoid rings that are SPIR's .

THEOREM 19.17. If U is a nonzero monoid, the monoid ring $R[U]$ is a SPIR if and only if R is a field of characteristic $p \neq 0$ and U is a cyclic p—group.

Section 19 Remarks

The question of whether each chained ring R is the homomorphic image of a valuation domain has been raised in the literature. If R is Noetherian, an affirmative answer follows easily from Cohen's structure theorem for complete local rings. Ohm and Vicknair show in [113] that the question also has an affirmative answer if R is a monoid ring; their proof is based on Theorem 19.4.

A unitary ring R is said to be <u>semi-quasi-local</u> if R has only finitely many maximal ideals, and R is <u>semilocal</u> if R is Noetherian with only finitely many maximal ideals. It can be shown that a group ring $R[G]$ is semilocal if and only if R is semilocal and G is finite; $R[G]$ is semi-quasi-local if and only if $(R;M_1,\ldots,M_n)$ is semi-quasi-local, G is a torsion group, and either (1) G is finite, or (2) $\mathrm{char}(R/M_1) = \ldots = \mathrm{char}(R/M_n) = p \neq 0$ and G/G_p is finite.

The problems of determining conditions under which $R[S]$ is a Prüfer ring or a Bezout ring seem to be open in the cases not covered in Section 18. Glastad and Hopkins have addressed the problem of determining when $R[S]$ is a PIR in [62]. Specifically, they consider the case where $R[S]$ is a zero-dimensional ring of the form $R[X_1,\ldots,X_n]/(\{X_i^{e_i}(1-X_i^{f_i})\}_{i=1}^n)$, where each e_i and each f_i is a positive integer; each zero-dimensional monoid ring that is a PIR is the homomorphic image of a ring of the type described.

§20. Chain Conditions in Monoid Rings

Two chain conditions — the ascending chain condition and the ascending chain condition for principal ideals — have already been considered in monoid rings, the latter for monoid domains. To wit, Theorem 7.7 shows that a monoid ring $R[S]$ is Noetherian if and only if R is Noetherian and S is finitely generated, and Theorem 14.17 shows that a monoid domain $D[G]$, where G is a group, satisfies a.c.c.p. if and only if D satisfies a.c.c.p. and each nonzero element of G is of type $(0,0,\dots)$. In this section we consider certain other chain conditions in $R[S]$ and $R[G]$. Specifically, the first part of the section (through Theorem 20.8) is concerned with the problem of determining conditions under which $R[S]$ is Artinian or an RM–ring. The remainder of the section treats the property of being locally Noetherian in unitary group rings.

Throughout this section we use the notation R for a unitary ring, S for a monoid, and G for a group. The ring R is Artinian if and only if R is both zero-dimensional and Noetherian. Thus, if $R[S]$ is Artinian, then $R[S]$ satisfies both a.c.c. and d.c.c. on kernel ideals, and hence S satisfies a.c.c. and d.c.c. on congruences. We proceed to show that a monoid with these properties is finite. The proof, like that of Theorem 5.10, is not immediate, but the case where S is a group is an elementary result.

THEOREM 20.1. The group G satisfies both a.c.c. and d.c.c. on subgroups if and only if G is finite.

Proof. We need only show that G is finite if it satisfies both chain conditions. Thus, G is finitely generated since it satisfies a.c.c. , so $G = G_1 \oplus \ldots \oplus G_n$ is a finite direct sum of cyclic groups G_i . Since Z does not satisfy d.c.c. on subgroups, each G_i , and hence G , is finite.

THEOREM 20.2. Assume that T is a cancellative semigroup.

(1) If T satisfies d.c.c. on ideals, then T is a group.

(2) If T satisfies both a.c.c. and d.c.c. on congruences, then T is a finite group.

Proof. (1): Choose $t \in T$ and consider the decreasing sequence $t + T \supseteq 2t + T \supseteq \ldots$ of ideals of T . There exists a positive integer k such that $kt + T = (k + 1)t + T = \ldots$ Thus $(k + 1)t \in kt + T = (k + 2)t + T$ can be written as $(k + 2)t + u$ for some $u \in T$. Since T is cancellative, it follows that $t + u$ is an identity element of T and t has an inverse in T . Therefore T is a group.

(2): Each ideal I of T induces the Rees congruence ρ_I with respect to I on T , and $I < J$ implies $\rho_I < \rho_J$. Therefore T satisfies d.c.c. on ideals, and (1) shows that T is a group. Consequently, T is finite by Theorem 20.1.

Theorems 20.3 and 20.4 are results on congruences that could have been stated in Section 4, but they haven't been needed up to this point. The proof of (20.3) is routine and will be omitted.

THEOREM 20.3. Assume that $\sim$ is a congruence on the semigroup T and for $t \in T$, denote by $[t]$ the class of t in $T/\sim$.

(1) If ρ is a congruence on T such that $p \geq \sim$, then the relation ρ^* on $T/\sim$ defined by $[t]\ \rho^*\ [u]$ if $t\ \rho\ u$ is a congruence on $T/\sim$.

(2) If μ is a congruence on $T/\sim$ and if the relation μ' on T is defined by $a\ \mu'\ b$ if $[a]\ \mu\ [b]$, then μ' is a congruence on T and $\mu' \geq \sim$.

(3) The correspondences in (1) and (2) are inverses of each other and are order-preserving. Hence, $\rho \longrightarrow \rho^*$ is an isomorphism of the lattice of congruences ρ on T such that $\rho \geq \sim$ onto the lattice of congruences on $T/\sim$. Thus, if T satisfies d.c.c. (or a.c.c.) on congruences, then so does $T/\sim$.

Part (3) of Theorem 5.1 and its proof establish the next result.

THEOREM 20.4. Assume that P is a proper prime ideal of the semigroup T and let $U = T \setminus P$. If T satisfies d.c.c. (a.c.c.) on congruences, then so does U.

THEOREM 20.5. If the monoid S satisfies both a.c.c. and d.c.c. on congruences, then S is finite.

Proof. Since S is finitely generated by Theorem 5.10, it suffices to prove that S is periodic. Thus, pick $s \in S$ and let $\sim$ be the cancellative congruence on S. The cancellative monoid $S/\sim$ satisfies both a.c.c. and d.c.c. on congruences by Theorem 20.3, and hence is a finite group by

Theorem 20.2. It follows that there exist distinct positive integers m_1 and n_1 and an element $t_1 \in S$ such that $m_1 s + t_1 = n_1 s + t_1$. Let $I = \{t \in S \mid ms + t = ns + t$ for some $m \neq n\}$. Then I is an ideal of S, and I meets $\langle s \rangle$ if and only if s is periodic. Assume that $I \cap \langle s \rangle = \phi$. Then there exists a prime ideal J of S containing I such that $J \cap \langle s \rangle = \phi$. Let $U = S \backslash J$. Theorem 20.4 shows that U satisfies both a.c.c. and d.c.c. on congruences, so the result above implies that there exists distinct $m,n \in Z^+$ and $u \in U$ such that $ms + u = ns + u$. Thus $u \in I$, and this contradicts the fact that $U \cap I = \phi$. We conclude that S is periodic, and hence finite.

THEOREM 20.6. The following conditions are equivalent.

(1) $R[S]$ is Artinian.

(2) R is Artinian and S satisfies both a.c.c. and d.c.c. on congruences.

(3) R is Artinian and S is finite.

Proof. (1) $\Longrightarrow$ (2): The homomorphic image of an Artinian ring is Artinian, and we have previously observed that S satisfies both chain conditions on congruences if $R[S]$ is Artinian.

That (2) implies (3) is the content of Theorem 20.5. Condition (3) implies that $R[S]$ is a finitely generated module over the Artinian ring R. Thus $R[S]$ is both Noetherian and zero-dimensional, hence Artinian. Therefore (3) implies (1) and the proof is complete.

The proof of Theorem 20.5 is similar to that of Theorem 17.3 and indeed, (17.3) can be used to prove the implication

(1) $\Longrightarrow$ (3) in Theorem 20.6 without appeal to Theorem 20.5. To see this, note that $R[S]$ Artinian implies that R is Artinian, S is finitely generated, and $\dim R = \dim R[S] = 0$. Therefore Theorem 17.3 shows that S is periodic, and hence finite.

Theorem 20.6 is useful in determining the semigroup rings (as opposed to monoid rings) that are Artinian. One part of the characterization extends to the Noetherian condition, and hence is stated as part of Theorem 20.7.

THEOREM 20.7. Let T be a semigroup.

(1) $R[T]$ is Artinian if and only if R is Artinian and T is finite.

(2) If $R[T]$ is Noetherian, then R is Noetherian and T is finitely generated. The converse fails.

Proof. (1): If R is Artinian and T is finite, then $R[T]$ is Artinian as an R-module, and hence Artinian as a ring. Conversely, assume that $R[T]$ is Artinian. Let T^0 be the monoid obtained by adjoining an identity to T. Then $R[T^0] = R + R[T]$. We consider $R[T^0]$ as a module over itself. Then $R[T]$ is a submodule that is Artinian as an $R[T]$-module, hence Artinian as an $R[T^0]$-module. The structure of $R[T^0]/R[T] \simeq R$ as an $R[T^0]$-module is essentially the same as its structure as a ring. But R is Artinian as a ring since it is a homomorphic image of $R[T]$. Consequently, $R[T^0]$ is an Artinian $R[T^0]$-module — that is, an Artinian ring. Theorem 20.6 then shows that T^0 is finite, and hence T is finite.

(2): By mocking the argument just given in (1), it

follows that R is Noetherian and T^0 is finitely generated if $R[T]$ is Noetherian, and hence T is also finitely generated. To show that the converse fails, we show that the ring $R[Z^+] \simeq XR[X]$ is Noetherian if and only if the additive group of R is finitely generated. Thus, if $XR[X]$ is Noetherian, then $(\{rX \mid r \in R\})$ is generated by a finite subset $\{r_iX\}_{i=1}^n$ of $\{rX \mid r \in R\}$. If $r \in R$, it follows that

$$rX = \Sigma_{i=1}^n k_i r_i X + \Sigma_{i=1}^n f_i r_i X ,$$

for integers k_i and elements $f_i \in XR[X]$. Consequently, $r = \Sigma_1^n k_i r_i$, and $\{r_i\}_1^n$ generates the additive group of R. Conversely, if $(R,+)$ is finitely generated by $\{r_i\}_1^n$, then R is Noetherian and is generated by $\{r_i\}_1^n$ as a module over its prime subring Π. Let $U = \Pi + XR[X]$. The ring $R[X]$ is Noetherian and is generated by $\{r_i\}_1^n$ as a module over U. By Eakin's Theorem, U is Noetherian. Since Π is the prime ring of R, the ideals of $XR[X]$ coincide with the ideals of U contained in $XR[X]$. Thus $XR[X]$ is Noetherian since U is Noetherian.

The ring R is said to satisfy the <u>restricted minimum condition</u> (RM-<u>condition</u>), or to be an RM-<u>ring</u>, if R/A satisfies the minimum condition for each nonzero ideal A of R. An RM-ring with proper zero divisors is Artinian, and a unitary integral domain D is an RM-ring if and only if it is Noetherian of dimension at most 1. The next result determines the RM-domains within the class of nontrivial monoid rings.

THEOREM 20.8 Assume that S is nonzero. Then $R[S]$ is an RM–domain if and only if R is a field and S is isomorphic either to Z or to a submonoid of Z_0.

Proof. If R is a field, then $R[Z]$ is a one-dimensional Noetherian domain, and therefore an RM–domain. Moreover, if $S \subseteq Z_0$, then Theorem 2.4 shows that there exists a positive integer d such that $R < R[S] \subseteq R[X^d]$ and $R[X^d]$ is integral over $R[S]$. Therefore $R[S]$ is again Noetherian and one-dimensional, hence an RM–domain.

For the converse, $R[S]$ an integral domain implies that R is a domain and S is torsion–free and cancellative. The augmentation ideal I of $R[S]$ is a nonzero prime ideal, so $R[S]/I \simeq R$ is a field. Let G be the quotient group of S. Then $R[G]$ is an RM–domain, so G is a finite direct sum of copies of Z. Because $R[Z]$ is not a field, it follows that $G \simeq Z$. We assume that $S \subseteq Z$. If S contains both positive and negative integers, then S is isomorphic to Z (Theorem 2.6), and otherwise, S is isomorphic to a submonoid of Z_0.

The ring R is said to be locally Noetherian if each localization R_P of R is Noetherian. In order that R should be locally Noetherian, it is sufficient that R_M is Noetherian for each maximal ideal M of R. Each regular ring and each almost Dedekind domain is locally Noetherian. Hence a locally Noetherian ring need not be Noetherian, even if it is a group ring that is an integral domain. In general a locally Noetherian ring R is Noetherian if and only each ideal of R has only finitely many minimal prime ideals [10], [70]. Theorem 20.9 lists some basic properties of locally

Noetherian rings. In the rest of the section, we use the following notation. If N is a multiplicative system in R and if A is an ideal of R, we write AR_N for the extension of A to R_N, although A need not be naturally imbedded in R_N.

THEOREM 20.9. Each homomorphic image, quotient ring, and finitely generated extension ring of a locally Noetherian ring R is locally Noetherian.

Proof. If A is an ideal of R and P/A is a prime ideal of R/A, then $(R/A)_{P/A} \simeq R_P/AR_P$ is Noetherian since R_P is Noetherian. Hence R/A is locally Noetherian. If N is a multiplicative system in R and if P is a prime ideal of R that misses N, then $(R_N)_{PR_N} \simeq R_P$ is Noetherian, so R_N is locally Noetherian. Finally, let $T = R[a_1,\dots,a_n]$ be a finitely generated extension ring of R. Let M be a proper prime ideal of T and let $P = R \cap M$. To within isomorphism, T_M is a localization of a finitely generated extension ring of R_P, hence of a Noetherian ring. Therefore T_M is Noetherian and T is locally Noetherian.

The problem of determining conditions under which a monoid ring is locally Noetherian is unresolved, but the problem has been solved for group rings. We deal with this case in the rest of the section. Theorem 20.11 gives necessary conditions in order that $R[G]$ should be locally Noetherian, and Theorem 20.14 shows that these conditions are also sufficient. Theorem 20.10 is a special case of Theorem 20.11 that is used in its proof.

THEOREM 20.10. Assume that F is a field and that H is a finitely generated subgroup of the nonfinitely generated group G such that G/H is a p–group. If M is the augmentation ideal of $F[G]$, then $MF[G]_M$ is finitely generated if and only if the characteristic of F is distinct from p.

Proof. Let $\{h_i\}_{i=1}^n$ be a finite set of generators for H. Then $\{1 - X^{h_i}\}_{i=1}^{\infty}$ generates the kernel ideal I of the canonical homomorphism $F[G] \longrightarrow F[G/H]$. If $p \neq \text{char}(F)$, we show that $MF[G]_M = IF[G]_M$. It suffices to show that $1 - X^g \in IF[G]_M$ for each $g \in G$. By assumption, $p^k g \in H$ for some $k \in Z^+$. Let $f = \Sigma_{i=0}^{p^k - 1} X^{ig}$. Then $(1 - X^g)f = 1 - X^{p^k g} \in I$. Moreover, under the augmentation map, f is mapped to $p^k \neq 0$. Therefore $f \notin M$ and $1 - X^g \in IF[G]_M$.

To show that $MF[G]_M$ is not finitely generated in the case where $p = \text{char}(F)$, it suffices to show that no finite subset of $\{1 - X^g \mid g \in G\}$ generates $MF[G]_M$. Assume, to the contrary, that $\{k_i\}_1^m$ is a finite subset of G such that $\{1 - X^{k_i}\}_1^m$ generates $MF[G]_M$. Without loss of generality we assume that $\{h_i\}_1^n \subseteq \{k_i\}_1^m$, so that G/K is a p–group, where K is the subgroup of G generated by $\{k_i\}_1^m$. Since G is not finitely generated, then $G \neq K$ and G/K contains an element $g + K$ of order p. Let J be the ideal of $F[G]$ generated by $\{1 - X^{k_i}\}_1^m$. Since $JF[G]_M = MF[G]_M$, then $J:(1 - X^g)$ contains an element $f \notin M$. Let ϕ be the canonical homomorphism from $F[G]$ onto $F[G/K]$. Then J is the kernel of ϕ. We have $0 = \phi(1 - X^g)\phi(f) = (1 - X^{g+K})\phi(f)$, and hence $\phi(f)$ belongs to $\text{Ann}(1 - X^{g+K})$, which is $(1 - X^{g+K})^{p^k - 1}$ by

Theorem 8.13. Therefore $\phi(f) \in \phi(M)$, which implies that $f \in M$ since $M \supseteq J = \ker \phi$. This contradiction implies that $MF[G]_M$ is not finitely generated, as asserted.

THEOREM 20.11. Let Ω be set of primes p such that p is a nonunit of R. If $R[G]$ is locally Noetherian, then the following three conditions are satisfied.

(1) R is locally Noetherian.

(2) G has finite torsion-free rank.

(3) If F is a free subgroup of G of rank $r_0(G)$, then the p-component $(G/F)_p$ is finite for each $p \in \Omega$.

Proof. Theorem 20.9 shows that (1) is satisfied. To prove (2), let $\{x_\alpha\}_{\alpha \in A}$ be a maximal linearly independent subset of G, and assume that A is infinite. Choose an infinite subset $\{\alpha_i\}_1^\infty$ of A and let F be the subgroup of G generated by $\{x_\alpha\}_{\alpha \in A}$. If $P_j = \{1 - X^{\alpha_i}\}_1^j R[F]$ for each j, then $\{P_j\}_1^\infty$ is an infinite strictly ascending sequence of primes of $R[F]$. Since $R[G]$ is integral over $R[F]$, there exists an infinite strictly ascending sequence $\{Q_j\}_1^\infty$ in $R[G]$ such that $Q_j \cap R[F] = P_j$ for each j. If $Q = \cup_1^\infty Q_j$, then Q is prime in $R[G]$ and $R[G]_Q$ is not Noetherian since $QR[G]_Q$ does not have finite height. Therefore G has finite torsion-free rank.

Let $r_0(G) = n$ and let F be a free subgroup of G of rank n. Assume that $p \in \Omega$ and let M be a maximal ideal of R such that $\operatorname{char}(R/M) = p$. Since $(G/F)_p$ is a homomorphic image of G, the ring $(R/M)[(G/F)_p]$ is a homomorphic image of $R[G]$, and hence is locally Noetherian. Applying Theorem 20.10, we conclude that $(G/F)_p$ is finite.

In order to prove the converse of Theorem 20.11, we use two preliminary results. The first of these belongs to the theory of abelian groups.

THEOREM 20.12. Assume that K is a finite subgroup of the group G and that p is prime.

(1) G_p is finite if and only if $(G/K)_p$ is finite.

(2) If G_p is finite and if G is a torsion group, then $(G/H)_p$ is finite for each subgroup H of G.

(3) Assume that $r_0(G) = n$ is finite and that G contains a free subgroup of rank n such that $(G/F)_p$ is finite. Then G_p is finite and $G = G_p \oplus H$, where H contains a free subgroup F_1 of rank n such that $(H/F_1)_p = \{0\}$.

Proof. (1): Let ϕ be the natural map from G to G/K, and let ϕ^* be the restriction of ϕ to G_p. Then $\phi^*(G_p) \subseteq (G/K)_p$ and $\ker\phi^* = G_p \cap K$ is finite. Hence G_p is finite if $(G/K)_p$ is finite. On the other hand, $\phi^{-1}((G/K)_p)$ is the set of pth roots in G of elements of K. If G_p is finite, then an element $g \in G$ with a pth root in G has exactly $|G_p|$ pth roots, for these roots form a coset of G_p in G. Therefore $(G/K)_p$ is finite if G_p is finite.

(2): Since G is a torsion group, $G = G_p \oplus M$, where $M_p = \{0\}$. Moreover, $H = H_p \oplus (M \cap H)$, and $G/H \simeq (G_p/H_p) \oplus (M/(M \cap H))$, where the p-component of $M/(M \cap H)$ is $\{0\}$. Therefore $(G/H)_p \simeq G_p/H_p$ is finite.

(3): We show first that $(G/F_2)_p$ is finite for each free subgroup F_2 of G of rank n. Thus, let

$F^* = F \cap F_2$. Then $F/F^* \simeq (F_2 + F)/F_2 \subseteq G/F_2$ is a finitely generated torsion group, hence a finite group. Since $(G/F^*)/(F/F^*) \simeq G/F$, it follows from part (1) that $(G/F)_p$ is finite iff $(G/F^*)_p$ is finite iff $(G/F_2)_p$ is finite.

Since $G_p \cap F = \{0\}$ and since $(G/F)_p$ is finite, we conclude that G_p is a finite pure subgroup of G , hence a direct summand — say $G = G_p \oplus H$, where $r_0(H) = n$. By the result of the preceding paragraph, we may assume without loss of generality that $F \subseteq H$. Let T be the torsion subgroup of H . Since H/F is a torsion group with $(H/F)_p$ finite, it follows from (2) that $(H/(F + T))_p$ is also finite. Choose a subgroup K of H containing $F + T$ such that $K/(F + T) = (H/(F + T))_p$. Then K/T is a finitely generated torsion-free group, hence free. Moreover, $r_0(K/T)$ is n since $F \subseteq K$. Choose $x_1, x_2, \ldots, x_n \in K$ such that $\{x_i + T\}_{i=1}^n$ is a free basis for K/T , and let F_2 be the subgroup of K generated by $\{x_i\}_{i=1}^n$. Then F_2 is free of rank n . We show that $(H/F_2)_p = \{0\}$. Since $H_p = \{0\}$, we have $(T + F_2/F_2)_p \simeq T_p = \{0\}$. Thus, the proof of (2) shows that the p-component of H/F_2 is isomorphic to that of $(H/F_2)/((T + F_2)/F_2) \simeq H/(T + F_2) = H/K$. By choice of K , $(H/K)_p = \{0\}$, and hence $(H/F_2)_p = \{0\}$ as well.

THEOREM 20.13. Assume that R is a quasi-local ring with maximal ideal M , that $c = \text{char}(R/M)$, and that F is a subgroup of G such that G/F is a torsion group with no element of order divisible by c . If P is a maximal ideal of $R[G]$ such that $P \cap R = M$, then $P \cap R[F]$ generates $PR[G]_P$.

Proof. Denote by P_F the maximal ideal $P \cap R[F]$ of $R[F]$. Choose $f \in P$. We show that $f \in P_F R[G]_P$. We have $f \in R[E]$ for some finitely generated subgroup E of G containing F, and hence $f \in P_E = P \cap R[E]$. If $f \in P_F R[E]_{P_E}$, then $f \in P_F R[G]_P$. Thus, we assume without loss of generality that $G = E$, so that G/F is a finite group of order not divisible by c. In this case G/F is a finite direct sum of t cyclic groups. To prove the result, we use induction on t. If $G/F = (g + F)$ is cyclic of order k, where c does not divide k, then $R[G] = R[F][X^g]$. The residue class ring $R[G]/P_F[G] = R[F][X^g]/P_F[X^g] \simeq L[u]$, where $L \simeq R[F]/P_F$ has characteristic c and $u^k = v$ is a unit of L. Thus $L[u]$ is a homomorphic image of $L[X]/(X^k - v)$. Since k is a unit of L, it follows that $L[X]/(X^k - v)$ is a finite direct sum of fields, and hence so is $L[u]$. In particular, $P_F[G]$ is the intersection of the family $\{P_i\}_{i=1}^r$ of maximal ideals of $R[G]$. Thus, for $1 \le j \le r$, we have

$$P_F[G]R[G]_{P_j} = (\cap_{i=1}^r P_i)R[G]_{P_j} = \cap_1^r P_i R[G]_{P_j} = P_j R[G]_{P_j} .$$

This proves the result in the case where G/F is cyclic. Assume that $G/F = (g_1 + F)\oplus \ldots \oplus (g_v + F)$, and let K be the subgroup of G generated by $F \cup \{g_i\}_{i=1}^{v-1}$. Then $P^* = P \cap R[K]$ is maximal in $R[K]$ and $P^* \cap R = M$. Hence $\mathrm{char}(R[K]/P^*) = c$ does not divide the order of $C_v = (g_v + F)$. Since $R[G]_P$ is isomorphic to a localization of $(R[K]_{P^*})[C_v]$, the case where G/F is cyclic shows that $PR[G]_P$ is generated by $P^*R[K]_{P^*}$. By induction it follows that $P^*R[K]_{P^*}$ is generated by P_F, and hence P_F also generates $PR[G]_P$ in $R[G]_P$.

Assume that R is a quasi-local ring with finitely generated maximal ideal M . A result of Nagata [109, § 31] shows that R is Noetherian if and only if each finitely generated ideal of R is closed in the M-adic topology on R . This is the main tool we use to show that conditions (1)–(3) of Theorem 20.11 imply that $R[G]$ is locally Noetherian.

THEOREM 20.14. Let Ω be the set of primes p such that p is a nonunit of R . Assume that R is locally Noetherian, that G has finite torsion-free rank n , and that there exists a free subgroup F of G such that $(G/F)_p$ if finite for each $p \in \Omega$. Then $R[G]$ is locally Noetherian.

<u>Proof</u>. Let P be a maximal ideal of $R[G]$, let $M = P \cap R$, and let $c = \mathrm{char}(R/M)$. If $c \neq 0$, then G_c is finite and Theorem 20.12 shows that $G = G_c \oplus H$, where H has a free subgroup F_1 of rank n such that $(H/F_1)_c = \{0\}$. If $c = 0$, we let $G_c = \{0\}$. Then $R[G] = R[G_c][H]$, where $R[G_c]$ is locally Noetherian by Theorem 20.9. Without loss of generality we assume that $R = R[G_c]$ and $G = H$. Since $R[G]_P$ is isomorphic to a localization of $R_M[G]$ at a maximal ideal, we also assume that R is a local ring with maximal ideal M . Theorem 20.13 then shows that $P \cap R[F]$ generates $PR[G]_P$. But $R[F]$ is a Noetherian ring, so $P \cap R[F]$ and $PR[G]_P$ are finitely generated. To complete the proof, it suffices to show that $IR[G]_P$ is closed in the natural topology on $R[G]_P$ for each finitely generated ideal I of $R[G]$ contained in P . Thus, take

$f/h \in \cap_{i=1}^{\infty}(I + P^i)R[G]_P$, the closure of $IR[G]_P$ in $R[G]_P$, where $f \in R[G]$ and $h \in R[G]\backslash P$. Choose a finitely generated subgroup E of G containing F such that f,h, and each element of a finite set of generators of I belongs to R[E] . Let $I_E = I \cap R[E]$ and let $P_E = P \cap R[E]$. Then $I = I_E R[G]$ by choice of E and $PR[G]_P = P_E R[G]_P$ by Theorem 20.13. Hence

$$f/g \in [\cap_{i=1}^{\infty}(I_E + P_E^i)R[G]_P] \cap R[E]_{P_E} .$$

Since $R[G]_P$ is faithfully flat as an $R[E]_{P_E}$-module, each ideal of $R[E]_{P_E}$ is contracted from $R[G]_P$. Therefore

$$f/g \in \cap_1^{\infty}(I_E + P_E^i)R[E]_{P_E} ,$$

the closure of $I_E R[E]_{P_E}$ in the natural topology. Since $R[E]_{P_E}$ is a local ring, $I_E R[E]_{P_E}$ is closed, and hence $f/g \in I_E R[E]_{P_E} \subseteq IR[G]_P$. This shows that $IR[G]_P$ is closed in the natural topology on $R[G]_P$, thereby completing the proof of Theorem 20.14.

As an application of Theorem 20.14, we determine those group rings that are almost Dedekind domains. Recall that an almost Dedekind domain may be defined as a locally Noetherian Prüfer domain (see the remarks at the end of Section 13). Thus, the following result is a consequence of Theorems 13.6 and 20.14.

COROLLARY 20.15. The group ring R[G] is an almost Dedekind domain if and only if the following conditions are satisfied.

(1) R is a field,

(2) G is a subgroup of Q , and

(3) either char(R) = 0 or char R = p ≠ 0 and for each (some) nonzero element $g \in G$, there exists a positive integer $k(g)$ such that $g/p^{k(g)} \notin G$.

We note in particular that Corollary 20.15 is strong enough to show that R[G] need not be Noetherian if it is an almost Dedekind domain. For example, Q[Q] is such a domain.

Section 20 Remarks

Additional material on Artinian rings can be found in [136, Section 2 of Chapter IV]. See [33] and Section 40 of [51] concerning RM–rings. The papers [10] and [70] treat locally Noetherian rings in general, but the material on locally Noetherian group rings is more closely related to the papers [90], [23], and [52]. In particular, Lantz in [90] also determines conditions under which R[G] is locally Cohen–Macaulay, locally Gorenstein, or locally regular (not in the sense of von Neumann); he also considers briefly the problem of determining when R[G] satisfies the chain condition or the second chain condition on prime ideals (see also [97]).

The proof of Theorem 20.12 uses the fact that G_p is a direct summand of G if G_p is finite. This is a consequence of a more general result due to Kulikov — see Theorem 27.5 of [47].

CHAPTER V

DIMENSION THEORY AND THE ISOMORPHISM PROBLEMS

Chapter V contains material on two topics with little overlap. The first of these topics is that of the (Krull) dimension of $R[S]$, which is treated in Section 21. Theorem 21.4 states that $\dim R[S] = \dim R[G]$ for a cancellative monoid S with quotient group G, and $\dim R[G]$ is determined in terms of an appropriate polynomial ring over R by Theorem 17.1. The other sections of Chapter V concern the isomorphism problems for semigroup rings. More specifically, Sections 23 and 24 deal with the question of whether isomorphism of $R_1[S]$ and $R_2[S]$ implies isomorphism of R_1 and R_2, and Section 22 considers R-automorphisms of $R[S]$, a basic tool in the development of Section 24. A survey of some results concerning the other isomorphism problem — whether isomorphism of S_1 and S_2 follows from that of $R[S_1]$ and $R[S_2]$ — is given in Section 25, but no proofs are included.

§21. Dimension Theory of Monoid Rings

In this section we seek to determine the (Krull) dimension of a monoid ring $R[S]$. While some results are known for general monoids S (for example, see Section 17), the theory is complete only in the case where S is cancellative, and that is the case we treat. For S cancellative, we show that $\dim R[S] = \dim R[G]$, where G is the quotient group of S. By Theorem 17.1, this reduces the problem to the study of the dimension of a polynomial ring $R[\{X_\lambda\}_{\lambda\in\Lambda}]$ over R, a topic which has been the subject of intensive investigation. It turns out that even the proof of the equality $\dim R[S] = \dim R[G]$ depends heavily on the known prime ideal structure of $R[X_1,\ldots,X_n]$. Thus, to begin the section, we review some of the notation and results from this theory. Except for the statement that we label as part (4) of (21.3), all the results cited can be found in [51, §§30, 31] or in [137, Chapter VII, §7]. First we introduce some notation and terminology. For a unitary ring R, we write $R^{(m)}$ for the polynomial $R[X_1,\ldots,X_m]$ in m indeterminates over R, and for an ideal A of R, $A^{(m)} = A\cdot R[X_1,\ldots,X_m] = A[X_1,\ldots,X_m]$. A chain $P_0 \subsetneq P_1 \subsetneq\cdots\subsetneq P_k$ of prime ideals of $R^{(m)}$ is said to be <u>special</u> if $(P_i \cap R)^{(m)}$ is a member of the chain for each i between 0 and k. Part (2) of the following statement (21.1) is usually known as Jaffard's Special Chain Theorem; the labeling of the parts is significant in the sense that (2) follows from (1) and (3) follows from (2).

(21.1) Assume that $R^{(m)}$ is finite-dimensional and that

Q is a proper prime ideal of $R^{(m)}$. Let $P = Q \cap R$ and let $\{M_\lambda\}_{\lambda \in \Lambda}$ be the set of maximal ideals of R.

(1) $ht(Q) = ht(P^{(m)}) + ht(Q/P^{(m)})$; in particular, $ht(Q)$ can be realized as the length of a special chain of primes of $R^{(m)}$ with terminal element Q.

(2) $\dim R^{(m)}$ can be realized as the length of a special chain $P_0 < P_1 < \ldots < P_n$ of proper primes of $R^{(m)}$. Moreover, $P_n \cap R$ is maximal in R for each such special chain.

(3) $\dim R^{(m)} = \sup \{ht(M_\lambda^{(m)}) + m\}_{\lambda \in \Lambda}$.

We subsequently show that (21.1) extends to each monoid ring $R[S]$ such that S is torsion-free, cancellative, and finitely generated.

A finite-dimensional unitary ring R is said to satisfy the <u>saturated chain condition</u> if any two saturated chains $P < P_1 < \ldots < P_r = Q$ and $P < Q_1 < \ldots < Q_s = Q$ between two fixed primes have the same length. Affine domains over a field satisfy this condition.

(21.2) Let $D = k[x_1, \ldots, x_n]$ be an affine domain over the field k. Let $d = \text{tr.d.}(D/k)$ and let P be a nonzero proper prime ideal of D.

(1) $ht(P) = d - \text{tr.d.}[(D/P)/k]$.

(2) $\dim D = d$.

(3) D satisfies the saturated chain condition.

Unlike the parts of (21.1) and (21.2), the first three parts of (21.3) are not closely related to the fourth; we state the four parts together to avoid repetition of hypotheses.

(21.3) Assume that D is a unitary domain with quotient field K and let $\{t_i\}_{i=1}^m$ be a finite subset of K.

(1) If $\phi: D^{(m)} \longrightarrow D[t_1,\ldots,t_m]$ is the homomorphism defined by $\phi(f(X_1,\ldots,X_m)) = f(t_1,\ldots,t_m)$, then $\ker(\phi)$ has height m.

(2) If D admits an overring of dimension $r \geq 1$, then there exist $u_1,\ldots,u_{r-1} \in K$ such that $\dim D[u_1,\ldots,u_{r-1}] \geq r$.

(3) If D is an n-dimensional Noetherian domain, then each overring of D has dimension at most n.

(4) $\dim D^{(m)} = m + \sup\{D[u_1,\ldots,u_m] \mid \{u_i\}_{i=1}^m \subseteq K\}$.

A proof of part (4) of (21.3) can be found in [41]. In the rest of the section, we assume that R is a unitary ring, S is a cancellative monoid, and G is the quotient group of S. Although its proof will be delayed, we state the main result of the seciton at once.

THEOREM 21.4. $\dim R[G] = \dim R[S]$.

To prove (21.4), we establish a series of results that have the effect of showing that it is sufficient to prove the result in cases where various restrictions are imposed on R and S. The first such reduction is immediate: since $R[G]$ is a quotient ring of $R[S]$, then $\dim R[G] \leq \dim R[S]$, and hence (21.4) holds if $R[G]$ is infinite-dimensional — that is, if $\dim R = \infty$ or if $r_0(G) = \infty$. Thus, in proving (21.4), we need only consider the case where R is finite-dimensional and G has finite torsion-free rank. Theorem 21.5 provides another reduction.

THEOREM 21.5. If the equality in (21.4) holds for a finite-dimensional unitary integral domain and a finitely generated torsion-free monoid, then it holds in general.

Proof. To prove the equality of (21.4) for general R and S, we may assume that R is finite-dimensional and that G has finite torsion-free rank k. Let F be a free subgroup of G of rank k. Let $S' = S \cap F$ and let F' be the quotient group of S'. We observe that $R[S]$ is integral over $R[S']$ and that $R[G]$ is integral over $R[F']$. Thus, if $s \in S$, then $ns \in S' = F \cap S$ for some $n \in Z^+$. Moreover, if $g \in G$, then $g = a - b$ for some $a, b \in S$. We choose $n \in Z^+$ such that $na, nb \in S'$. Then $ng = na - nb \in F'$, and this establishes the two statements concerning integrality. We have $\dim R[S] = \dim R[S']$ and $\dim R[G] = \dim R[F']$, so by replacing S with S', it suffices to prove the equality in the case where S is torsion-free and G is finitely generated.

Suppose that $\dim R[G] < \dim R[S]$, where S is torsion-free. Let $P_0 < P_1 <\ldots< P_t$ be a chain of proper prime ideals of $R[S]$ of length $t > \dim R[G]$. If $P = P_0 \cap R$, then $P[S]$ is prime in $R[S]$ by Corollary 8.2, and the same result shows that $P[G]$ is prime in $R[G]$. Moreover,

$$\dim (R/P)[S] = \dim R[S]/P[S] \geq t > \dim R[G] \geq \dim (R/P)[G]$$

and $D = R/P$ is a finite-dimensional unitary domain. Let $Q_i = P_i/P[S]$ for each i, and choose $f_i \in Q_i - Q_{i-1}$ for $1 \leq i \leq t$. Let $\{g_i\}_{i=1}^m$ be a finite set of generators for G and write $g_i = s_i - t_i$, with $s_i, t_i \in S$ for each i.

Let T be the submonoid of S generated by $\{s_i\}_1^m \cup \{t_i\}_1^m \cup \mathrm{Supp}(f_1) \cup \ldots \cup \mathrm{Supp}(f_t)$. Then T is finitely generated, G is the quotient group of T , and $\dim D[T] \geq t > \dim D[G]$, for $f_i \in (Q_i \cap D[T]) \setminus (Q_{i-1} \cap D[T])$ for each i . This establishes Theorem 21.5.

In the case where S is finitely generated and G is torsion-free of rank n , we can reinterpret the equality $\dim R[G] = \dim R[S]$ as follows. If N is the set of non-zero integers, then G_N is an n-dimensional vector space over Q . Since S generates G_N as a Q-vector space, S contains a basis $\{s_i\}_1^n$ of G_N . If $\{t_j\}_{j=1}^k$ is a finite set of generators for S , then each t_j is uniquely expressible in the form $\Sigma_{i=1}^n a_{ji} s_i$, where $a_{ji} \in Q$. It follows that there exists $M \in Z^+$ such that $Mt_j \in Zs_1 + \ldots + Zs_n = Zs_1 \oplus \ldots \oplus Zs_n$ for each j . Then $R[S]$ is integral over $R[T]$, where T is the monoid generated by $\{s_i\}_1^n \cup \{Mt_j\}_1^k$, and hence $\dim R[S] = \dim R[T]$; moreover, T has quotient group $Zs_1 \oplus \ldots \oplus Zs_n \simeq G$. Thus, what we're dealing with in $R[T]$, to within isomorphism, is a subring of $R[X_1^{\pm 1}, \ldots, X_n^{\pm 1}]$ containing $R[X_1, \ldots, X_n]$ that is finitely generated over $R[X_1, \ldots, X_n]$ by a set $\{h_i\}_{i=1}^k$ of pure monomials $h_i = X_1^{a_1} \ldots X_n^{a_n}$, where each a_j is an integer (not necessarily nonnegative); we note that this use of the term "pure monomial" is a extension of that used in Section 15, where the exponents a_j were assumed to be non-negative. In this notation, (21.4) is equivalent to the assertion that $\dim R[X_1^{\pm 1}, \ldots X_n^{\pm 1}] = \dim R[X_1, \ldots, X_n, h_1, \ldots, h_k]$. Since $\dim R[Z^{(n)}] = \dim R^{(n)}$ by Theorem 17.1, we have established the following result.

THEOREM 21.6. The equality in Theorem 21.4 is equivalent to the following statement. If D is a finite-dimensional unitary domain and if $\{h_j\}_{j=1}^k$ is a finite set of pure monomials in the indeterminates $\{X_i\}_1^n$ over D, then $\dim D[X_1,\ldots,X_n] = \dim D[X_1,\ldots,X_n,\{h_j\}_1^k]$.

We turn to the proof of the statement in (21.6). The first step in the proof is a result about polynomial rings.

THEOREM 21.7. Let D be a unitary integral domain, let P be a proper prime ideal of D, and let m and n be integers such that $1 \le m \le n$. If $\operatorname{ht} P^{(n)} \ge m + 1$, then $\operatorname{ht} P^{(k)} \ge m + 1$ for $m \le k \le n$.

Proof. Clearly it suffices to prove that $\operatorname{ht} P^{(m)} \ge m + 1$. By passage to D_P , we assume that D is quasi-local with maximal ideal P . If $\operatorname{ht} P \ge m + 1$, then certainly $\operatorname{ht} P^{(m)} \ge m + 1$. Thus, assume that $m \ge \operatorname{ht} P = \dim D$. Part (3) of (21.1) implies that $\dim D^{(n)} = n + \operatorname{ht} P^{(n)} \ge n + m + 1$. But it is also the case that $\dim D^{(n)} = n + \sup\{\dim D[t_1,\ldots,t_n] \mid \{t_i\}_1^n \subseteq K\}$, where K is the quotient field of D, by part (4) of (21.3). Therefore, there exists an overrring of D of dimension at least $m + 1$, so part (2) of (21.3) implies that $\dim D[s_1,\ldots,s_m] \ge m + 1$ for some subset $\{s_i\}_{i=1}^m$ of K . Then $\dim D^{(m)} = m + \sup\{\dim D[t_1,\ldots,t_m] \mid \{t_i\}_1^m \subseteq K\} \ge 2m + 1$. Since $\dim D^{(m)} = m + \operatorname{ht} P^{(m)}$, it follows that $\operatorname{ht} P^{(m)} \ge m + 1$.

If D is a unitary integral domain, the set B of polymomials $f \in D^{(m)}$ of unit content forms a multiplicative system in $D^{(m)}$, and the quotient ring $(D^{(m)})_B$ of $D^{(m)}$ is denoted by $D(X_1,\ldots,X_m)$. Assume that J is a subring of $D(X_1,\ldots,X_m)$ containing $D^{(m)}$. Thus $J_S = (D^{(m)})_B$. If Q is a prime ideal of $D^{(m)}$ that misses B , there exists a unique prime $\overline{Q}$ of J lying over Q , namely $\overline{Q} = QD(X_1,\ldots,X_m) \cap J$. Therefore $\text{ht}\, Q = \text{ht}\, \overline{Q}$. In particular, if P is a proper prime ideal of D , then there exists a unique prime $\overline{P^{(m)}}$ of J such that $\overline{P^{(m)}} \cap D^{(m)} = P^{(m)}$. Domains of the form $T = D^{(m)}[\{h_j\}_1^k]$, where each h_j is a pure monomial in $\{X_i\}_1^m$, satisfy the condition that $D^{(m)} \subseteq T \subseteq D[X_1^{\pm 1},\ldots,X_n^{\pm 1}] \subseteq D(X_1,\ldots,X_m)$. Moreover, PT is prime in T for each prime ideal P of D , so the following result applies to the domains $D^{(m)}[\{h_j\}]$ of Theorem 21.6. Theorem 21.8 is a generalization of part (1) of (21.1).

THEOREM 21.8. Let D and B be as in the preceding paragraph. Assume that $J = D^{(m)}[t_1,\ldots,t_n]$ is a finitely generated extension of $D^{(m)}$ in $D(X_1,\ldots,X_m)$ and that each prime of D extends to a prime ideal of J . If Q is a proper prime ideal of J and if $P = Q \cap D$, then $\text{ht}\, Q = \text{ht}(Q/\overline{P^{(m)}}) + \text{ht}(\overline{P^{(m)}}) \le m + \text{ht}(\overline{P^{(m)}})$.

Proof. Observe that since $PJ = P^{(m)}J$ is prime in J , then $\overline{P^{(m)}} = P^{(m)}J_B \cap J = P^{(m)}J$ for each prime P of D . We establish first the inequality $\text{ht}(Q/P^{(m)}J) \le m$. Thus, to within isomorphism,

$$(D/P)^{(m)} = D^{(m)}/P^{(m)} \subseteq J/P^{(m)}J \subseteq (D^{(m)})_B/P^{(m)}(D^{(m)})_B ,$$

and under this identification, $(Q/P^{(m)}J) \cap (D/P) = (0)$. If N is the set of nonzero elements of D/P , it follows that

$k^{(m)} \subseteq (J/P^{(m)}J)_N \subseteq (k^{(m)})_{B*}$, where k is the quotient field of D/P and $B*$ is a multiplicative system in $k^{(m)}$. Since each overring of $k^{(m)}$ has dimension at most m by part (3) of (21.3), we conclude that $ht(Q/P^{(m)}J)_N = ht(Q/P^{(m)}J) \leq m$.

The inequality $ht\ Q \geq ht(P^{(m)}J) + ht(Q/P^{(m)}J)$ clearly holds; we establish the reverse inequality by induction on $ht\ Q$. If $ht\ Q = 1$, then either $Q = P^{(m)}J$ or $P = (0)$. In either case, the result holds. Assume that $ht\ Q = k > 1$ and let $(0) < Q_1 < \ldots < Q_{k-1} < Q$ be a chain of prime ideals of J . Set $P_1 = Q_{k-1} \cap D$. If $P = P_1$, then the induction hypothesis implies that $ht\ Q = k = ht\ Q_{k-1} + 1 = ht\ P^{(m)}J + ht(Q_{k-1}/P^{(m)}J) + 1 \leq ht\ P^{(m)}J + ht(Q/P^{(m)}J)$. Thus, assume that $P_1 < P$ and set $r = ht(Q/P^{(m)}J)$, $s = ht(Q_{k-1}/P_1^{(m)}J)$. By assumption we have $k - 1 = s + ht\ P_1^{(m)}J$. Therefore $s + ht\ P_1^{(m)}J + 1 = k = ht\ Q \geq r + ht\ P^{(m)}J \geq r + ht(P^{(m)}J/P_1{}^{(m)}J) + ht\ P_1^{(m)}J$. Clearly then, we need only show that $r + ht(P^{(m)}J/P_1^{(m)}J) \geq s + 1$; this inequality is patent if $r \geq s$, and hence we assume that $s > r$.

Let $\phi: D^{(m)}[Y_1,\ldots,Y_m] \longrightarrow J$ be the canonical $D^{(m)}$-homomorphism determined by $\phi(Y_i) = t_i$ for each i , and denote by U the kernel of ϕ . The next step in the proof is to show that $ht(H^{(m+n)} + U)/H^{(m+n)} = n$ for each proper prime ideal H of D . Consider the following commutative diagram.

$$\begin{array}{ccc} D[X_1,\ldots,X_m,Y_1,\ldots,Y_n] & \xrightarrow{\tau} & (D/H)[X_1,\ldots,X_m,Y_1,\ldots,Y_n] \\ \downarrow \phi & & \downarrow \alpha \\ J = D[X_1,\ldots,X_m,t_1,\ldots t_n] & \xrightarrow{\psi} & (D/H)[X_1,\ldots,X_m,\tau^*(t_1),\ldots,\tau^*(t_n)] \\ \downarrow i & & \downarrow i \\ (D^{(m)})_B & \xrightarrow{\tau^*} & [(D/H)^{(m)}]_{\tau(B)} \end{array}$$

Here τ is the canonical map. We note that $\tau(D^{(m)}) = (D/H)^{(m)}$ and $0 \notin \tau(B)$ since $H^{(m)} \cap B = \phi$. Hence τ induces an epimorphism τ^* of $(D^{(m)})_B = D(X_1,\ldots,X_m)$ onto $[(D/H)^{(m)}]_{\tau(B)} = (D/H)(X_1,\ldots,X_m)$, and the restriction ψ of τ^* to J is an epimorphism of J onto $(D/H)[X_1,\ldots,X_m,\tau^*(t_1),\ldots,\tau^*(t_n)]$. Finally, α is the $(D/H)^{(m)}$-homomorphism of $(D/H)^{(m+n)}$ determined by the relations $\alpha(Y_i) = \tau^*(t_i)$; we note that part (1) of (21.3) shows that $\ker \alpha$ has height n. The kernel of ψ is $H^{(m)}(D^{(m)})_B \cap J = H^{(m)}J$. Therefore $\psi \circ \phi$ has kernel $\phi^{-1}(H^{(m)}J) = H^{(m+n)} + U$. Considering $\psi\circ\phi$ as $\alpha\circ\tau$, it follows that $\tau(H^{(m+n)} + U)$ is the kernel of α. Therefore $\tau(H^{(m+n)} + U)$ has height n. Finally, $\tau(H^{(m+n)} + U)$ corresponds to $(H^{(m+n)} + U)/H^{(m+n)}$ under the canonical isomorphism between $(D/H)^{(m+n)}$ and $D^{(m+n)}/H^{(m+n)}$, from which we obtain the desired equality $\operatorname{ht}(H^{(m+n)} + U)/H^{(m+n)} = n$.

We return to the proof that $r + \operatorname{ht}(P^{(m)}J/P_1{}^{(m)}J) \geq s + 1$. Consider a chain $P^{(m+n)} < U_1 < \ldots < U_n = P^{(m+n)} + U$ of prime ideals of $D^{(m+n)}$ of length n, and let $M = \phi^{-1}(Q)$. Then $P^{(m+n)} < U_1 < \ldots < U_n \subseteq M$ is a chain of primes of $D^{(m+n)}$, each meeting D in P. By passage first to $(D^{(m+n)}/P^{(m+n)}) \simeq (D/P)^{(m+n)}$ and then to $k^{(m+n)}$, where k is the quotient field of D/P, it follows from (21.2) that this chain can be refined to a chain

$$P^{(m+n)} < U_1 < \ldots < U_n = P^{(m+n)} + U < U_{n+1} < \ldots < U_{n+w} = M$$

of length $n + w = ht(M/P^{(m+n)}) = n + ht(M/P^{(m+n)} + U)$. Since $P^{(m+n)} + U$ is the preimage of $P^{(m)}J$ under ϕ, it it follows that $ht(M/P^{(m+n)}) = n + ht(Q/P^{(m)}J) =$ $n + r \leq n + m$. Similarly, if $M_{k-1} = \phi^{-1}(Q_{k-1})$, then $M_{k-1} \cap D = P_1$, $ht(M_{k-1}/(P_1^{(m+n)} + U)) = ht(Q_{k-1}/P_1^{(m)}J) = s$, and $ht(M_{k-1}/P_1^{(m+n)}) = n + s \leq n + m$; in particular, $s \leq m$. Applying (1) of (21.1) to the polynomial ring $(D/P_1)[X_1,\ldots,X_m,Y_1,\ldots,Y_n]$, we conclude that $ht(M/P_1^{(m+n)}) = ht(P^{(m+n)}/P_1^{(m+n)}) + ht(M/P^{(m+n)})$; hence $n + r + ht(P^{(m+n)}/P_1^{(m+n)}) = ht(M/P_1^{(m+n)}) \geq 1 +$ $ht(M_{k-1}/P_1^{(m+n)}) = 1 + s + n$. Thus, we have $ht(P^{(m+n)}/P_1^{(m+n)}) \geq 1 + s - r$. Since $s > r$, Theorem 21.7 shows that $ht(P^{(t)}/P_1^{(t)}) \geq 1 + s - r$ for $s - r \leq t \leq n + m$. Now $s - r \leq s \leq m$, so we have $ht(P^{(m)}/P_1^{(m)}) \geq s - r + 1$. But then $ht(P^{(m)}J/P_1J) \geq s - r + 1$ also, and this is equivalent to the desired inequality $r + ht(P^{(m)}J/P_1J) \geq s + 1$. This completes the proof of Theorem 21.8.

<u>Proof of Theorem</u> 21.4. We prove the equivalent statement in Theorem 21.6 concerning $D^{(n)}$ and $J = D^{(n)}[\{h_j\}_1^k]$. If Q is a proper prime ideal of J and if

$P = Q \cap D$, then Theorem 21.8 implies that $\operatorname{ht} Q = \operatorname{ht}(P^{(n)}J) + \operatorname{ht}(Q/P^{(n)}J) \le \operatorname{ht}(P^{(n)}J) + n \le \dim R^{(n)}$. This inequality is sufficient to imply the inequality $\dim J \le \dim R^{(n)}$, and the reverse inequality has already been noted.

The prime ideal structure of $R^{(k)}$ has clearly been used in a strong way in the proof of Theorem 21.8, and hence in the proof of (21.4). On the other hand, the part of the theory pertaining strictly to $R^{(k)}$ in the proof of (21.4) is confined to (21.1) and to Theorem 21.7. Theorem 21.8 extends part (1) of (21.1) to monoid rings $D[T]$, where T is a finitely generated monoid lying between $Z_0 s_1 \oplus \ldots \oplus Z s_n$ and $Z s_1 \oplus \ldots \oplus Z s_n$. We proceed to extend each part of (21.1) to monoid rings $R[S]$, where S is finitely generated and torsion-free. To accomplish the extension, we define a <u>special chain</u> in $R[S]$ as a chain $P_0 \subseteq P_1 \subseteq \ldots \subseteq P_k$ of prime ideals of $R[S]$ with the property that $(P_i \cap R)[S]$ is a member of the chain for each i between 0 and k .

THEOREM 21.9. Assume that R is finite-dimensional and that S is torsion-free and finitely generated. Let Q be a proper prime ideal of $R[S]$ and let $P = Q \cap R$. Then $\operatorname{ht} Q = \operatorname{ht}(P[S]) + \operatorname{ht}(Q/P[S])$. Moreover, $\operatorname{ht} Q$ can be realized as the length of a special chain of primes of $R[S]$ with terminal element Q .

<u>Proof</u>. As in the paragraph following the proof of Theorem 21.5, there exists a finitely generated submonoid T of S such that S is integral over T , T contains a basis for its quotient group G_1 , and $G_1 \simeq G \simeq Z^n$ for

some $n \in Z^+$. Let $Q_1 = Q \cap R[T]$. We show that $ht(P[S]) = ht(P[T])$, that $ht(Q/P[S]) = ht(Q_1/P[T])$, and that $ht(Q) = ht(Q_1)$. Since $P[S]$ does not meet the multiplicative system $\{X^s \mid s \in S\}$, we have $ht(P[S]) = ht(P[G]) = ht(P^{(n)})$; by the same argument, $ht(P[T]) = ht(P^{(n)})$, and hence $ht(P[S]) = ht(P[T])$. Let $Q_1^* = Q_1/P[T]$. By passage to $(R[S]/P[S])_{[(R/P)\setminus\{0\}]}$ and $(R[T]/P[T])_{[(R/P)\setminus\{0\}]}$, it suffices to establish the equality $ht\ Q^* = ht\ Q_1^*$ in the case where R is a field. In that case $R[S]$ and $R[T]$ are affine domains over R and we have $ht\ Q^* = n - k_1$, where k_1 is the transcendence degree of $R[S]/Q^*$ over R, and $ht\ Q_1^* = n - k_2$, where $k_2 = \text{tr.d.}\ [R[T]/Q_1^*]/R$. Since $R[S]/Q^*$ is integral over $R[T]/Q_1^*$, it follows that $k_1 = k_2$, and hence $ht\ Q^* = ht\ Q_1^*$. We next prove that $ht\ Q_1 = ht(P[T]) + ht\ Q_1^*$. Theorem 21.8 establishes this result in the case where R is an integral domain. In the general case, let $W_0 < W_1 < \ldots < W_{v-1} < Q_1$ be a chain of primes in $R[T]$ of length $v = ht(Q_1)$. Let $P_0 = W_0 \cap R$. Since $P_0[T]$ is prime in $R[T]$, we necessarily have $W_0 = P_0[T]$. From the domain case, it follows that $v = ht(Q_1/P_0[T]) = ht(P[T]/P_0[T]) + ht(Q_1/P[T]) \le ht(P[T]) + ht(Q_1/P[T]) \le ht\ Q_1 = v$. Consequently, $ht\ Q_1 = ht(P[T]) + ht\ Q_1^* = ht(P[S]) + ht\ Q^* \le ht\ Q$. But since $R[S]$ is integral over $R[T]$, the inequality $ht\ Q \le ht\ Q_1$ also holds. This proves the assertion in (21.9) that $ht\ Q = ht(P[S]) + ht(Q/P[S])$.

To prove the statement concerning special chains, we use induction on $ht\ Q$. If $ht\ Q = 0$, then $Q = (Q \cap R)[S]$ and the statement holds. In the case where $ht\ Q = v > 0$, the equality $v = ht(P[S]) + ht(Q/P[S])$ implies that there exists a chain $Q_0 < Q_1 < \ldots < Q_{v-1} < Q_v = Q$ of primes of length v such that $Q_u = P[S]$ for some $u \leq v$. If $P = Q_i \cap R$ for each i, then the given chain is special. In the contrary case it follows that $(Q_{u-1} \cap R) < P$. The ideal Q_{u-1} has height $u - 1 < v$, and hence the induction hypothesis implies that there exists a special chain $Q_0^* < \ldots < Q_{u-1}^* = Q_{u-1}$. Since the chain $Q_0^* < \ldots < Q_{u-1}^* < Q_u < \ldots < Q_v = Q$ is also special, this completes the proof.

THEOREM 21.10. Assume that R is finite-dimensional and that S is finitely generated and torsion-free. Then $n = \dim R[S]$ can be realized as the length of a special chain $P_0 < P_1 < \ldots < P_n$ of proper primes of $R[S]$. Moreover, $P_n \cap R$ is maximal in R for each such special chain.

Proof. We have $n = ht\ M$ for some maximal ideal M of $R[S]$, and hence the statement concerning the existence of a special chain of length n follows from Theorem 21.9. To prove the statement concerning maximality of $P_n \cap R$, let $P_i = (P_n \cap R)[S]$. By passage to $(R/P_n \cap R)[S]$ and the special chain $(P_n \cap R)[S]/(P_n \cap R)[S] < \ldots < P_n/(P_n \cap R)[S]$ in $(R/(P_n \cap R))[S]$, we can assume that R is an integral domain with quotient field F and that $P_n \cap R = (0)$. Our object is to prove that $R = F$. Since the chain extends to a chain of proper primes of $F[S]$, it follows that $n \leq \dim F[S] = m$, where $m = r_0(G)$. On the other hand, $\dim R[S] = \dim R^{(m)} \geq m + \dim R$, so $\dim R = 0$ and $R = F$.

THEOREM 21.11. Assume that R is finite-dimensional,

that S is finitely generated and torsion-free, and that $r_0(G) = m$. Let $\{M_\lambda\}_{\lambda\in\Lambda}$ be the set of maximal ideals of R .

(1) If Q is a maximal ideal of $R[S]$ and if $P = Q \cap R$, then $\text{ht } Q = \text{ht}(P[S]) + m$.

(2) $\dim R[S] = \sup\{\dim R_{M_\lambda}[S]\}_{\lambda\in\Lambda} = m + \sup\{\text{ht}(M_\lambda[S])\}_{\lambda\in\Lambda}$.

<u>Proof</u>. (1): Set $R_1 = R_P$, $P_1 = PR_P$, and $Q_1 = QR_P[S]$. Then Q_1 is maximal in $R_1[S]$, P_1 is maximal in R_1 , and $P_1 = Q_1 \cap R_1$. Since R_1/P_1 is a field, (21.2) shows that $\text{ht}(Q_1/P_1[S]) = m = \dim(R_1/P_1)[S]$. Then Theorem 21.9 shows that $\text{ht } Q = \text{ht } Q_1 = \text{ht}(P_1[S]) + m = \text{ht}(P[S]) + m$.

(2): Assume first that R is quasi-local with maximal ideal M . Choose a special chain $P_0 < P_1 < \ldots < P_n$ in $R[S]$ of length $n = \dim R[S]$. Theorem 21.10 shows that $P_n \cap R = M$, and then part (1) implies that $\dim R[S] = n = \text{ht } P_n = \text{ht}(M[S]) + m$. It follows that $\sup_\lambda\{\dim R_{M_\lambda}[S]\} = m + \sup_\lambda\{\text{ht}(M_\lambda[S])\}$, and it is clear that $\dim R[S] \geq \dim R_{M_\lambda}[S]$ for each λ . On the other hand, Theorem 21.10 implies that $\dim R[S] \leq \dim R_{M_\lambda}[S]$ for some λ , and this completes the proof.

Section 21 Remarks

The treatment in this section follows closely that of Arnold and Gilmer in [12]. At the same time, we have omitted proofs of some cases of (21.4) that are derived in [12] without using (21.7) or (21.8); these include the cases where R is a Noetherian ring, a ring of nonzero characteristic, or a Prüfer domain. The proof in the case of a ring of nonzero characteristic uses the Noether Normalization Lemma in an affine domain $k[X_1,\ldots,X_n,\{h_j\}_1^k]$ over a finite prime field k. Arnold and Gilmer recognized that an analogous proof in the case of characteristic 0 could be obtained if the normalization lemma could be extended to $Z[X_1,\ldots,X_n,\{h_j\}_1^k]$ Using a theorem of Shimura, Moh [105] subsequently proved the required normalization lemma over Z. An extra dividend of the proof using (21.8) is that it essentially yields as corollaries the extra information about the prime ideal structure of $R[S]$ contained in Theorems 21.9, 21.10, and 21.11.

While the equality $\dim R[S] = \dim R[G]$ is valid for an arbitrary cancellative monoid, an example in [12] shows that the hypothesis that S is finitely generated is necessary in (21.9), (21.10), and (21.11), even if G is free of finite rank and R is a field.

§22. R–automorphisms of R[S]

Let R be a unitary ring and let S be a monoid. An automorphism ϕ of R[S] is called an R–<u>automorphism</u> if $\phi(r) = r$ for each $r \in R$. The set of R–automorphisms of R[S] is a group under composition; it is denoted by $\mathrm{Aut}_R R[S]$. Not much is known about $\mathrm{Aut}_R R[S]$ for general S. For example, $\mathrm{Aut}_R R[Z_0] \simeq \mathrm{Aut}_R R[X]$ is known, but $\mathrm{Aut}_R R[X,Y]$ is not known, even for a field R. Group rings are more tractable in this regard, due to the fact that a ring automorphism induces an automorphism of the unit group of the ring. Thus, in Theorem 22.4 we determine $\mathrm{Aut}_R R[Z^n]$ We begin considering the polynomial ring R[X].

If μ is an R–endomorphism of R[X], then for $g(X) = \Sigma_{i=0}^{n} g_i X^i \in R[X]$, we have $\mu(g(X)) = \Sigma_0^n g_i[\mu(X)]^i = g(\mu(X))$. Thus μ is the substitution determined by $X \longrightarrow \mu(X)$. Conversely, if $f(X) \in R[X]$, then the substitution map $g(X) \longrightarrow g(f(X))$ is an R–endomorphism, which we denote by μ_f. Therefore the problem of determining $\mathrm{Aut}_R R[X]$ is that of determining conditions on f in order that μ_f is a bijection. Theorem 22.1 resolves this question.

THEOREM 22.1. Assume that R is a unitary ring and that $f = \Sigma_{i=0}^{n} f_i X^i \in R[X]$. Let μ_f be the substitution map $g \longrightarrow g(f)$. The following conditions are equivalent.

(1) μ_f is an automorphism.

(2) μ_f is surjective (that is, $R[X] = R[f]$).

(3) f_1 is a unit of R and f_i is nilpotent for $i \geq 2$.

<u>Proof</u>. The implication (1) $\Longrightarrow$ (2) is clear. Assume

that (2) is satisfied. If P is a proper prime of R and if α is the canonical map of $R[X]$ onto $(R/P)[X]$, then $(R/P)[X] = (R/P)[\alpha(f)]$. Since R/P is an integral domain, the degree of any nonzero element of $(R/P)[\alpha(f)]$ is a multiple of $\deg \alpha(f)$. Since $X \in (R/P)[\alpha(f)]$, it follows that $\alpha(f)$ has degree 1. Consequently, $a_1 \notin P$ while $a_i \in P$ for each $i \geq 2$, so that a_1 is a unit of R and a_i is nilpotent for $i \geq 2$.

(3) $\Longrightarrow$ (1): We show that the conditions given on f imply that μ_f is both injective and bijective. Thus, assume that $g(f) = 0$. To show that $g = 0$, it suffices to show that $g(X - f_0) = 0$, for the substitution mapping μ_{X-f_0} is clearly an automorphism of $R[X]$. Hence, we assume without loss of generality that $f_0 = 0$. Let $g(X) = \Sigma_{i=0}^{m} g_i X^i$. The constant term of $g(f)$ is $g_0 = 0$. Assume that $g_0 = \ldots = g_{i=1} = 0$, where $i \leq m$. Then $0 = g(f) = f^i(g_i + g_{i+1}f + \ldots + g_m f^{m-i})$. Corollary 8.6 shows, however, that f is not a zero divisor in $R[X]$, and hence $g_i + g_{i+1}f + \ldots + g_m f^{m-i} = 0$ and $g_i = 0$. By induction it follows that $g = 0$, and μ_f is injective. Since $R[f] = R[f_1^{-1}(f - f_0)]$, in proving surjectivity of μ_f, we can assume without loss of generality that $f_0 = 0$ and $f_1 = 1$. We then establish the inclusion $R[X] \subseteq R[f]$ by induction on k, the order of nilpotence of the ideal $B_f = (f_2, \ldots, f_n)$ of R. If $k = 1$, then $f = X$ and the inclusion is clear. We assume that the inclusion $A[X] \subseteq A[f]$ holds for each element $f \in R[X]$ of the desired form for which $k < r$, where $r \geq 2$. Consider $h = X + h_2X^2 + \ldots + h_v X^v \in R[X]$, where B_h has order of nilpotence r. For $i \geq 2$, note

that $h_i h^i$ has the form $h_i X^i + \Sigma_{j=i+1}^{n_i} v_{ij} X^j$, where each $v_{ij} \in (B_h)^2$. Thus, if $g = h - h_2 h^2 - \ldots - h_v h^v$, then g has the form $X + g_2 X^2 + \ldots + g_m X^m$, where each $g_i \in (B_h)^2$. It follows that $B_g^{r-1} \subseteq B_h^{2(r-1)} = (0)$, so the induction hypothesis implies that $R[X] \subseteq R[g]$. Since the inclusion $R[g] \subseteq R[h]$ is clear, this completes the proof.

Some of the ideas in the proof of Theorem 22.1 are useful in determining $\mathrm{Aut}_R R[Z^n]$ as well. We consider the case $n = 1$ separately, not because it is necessary to do so, but primarily because the notation is much simpler in this case. Thus, we regard $R[Z]$ as $R[X,X^{-1}]$. An R—automorphism ϕ of $R[Z]$ maps X to a unit u of $R[X,X^{-1}]$, and is completely determined by u; to wit, if $\phi(X) = u$, then $\phi(\Sigma a_i X^i) = \Sigma a_i u^i$. Conversely, if $u \in R[X,X^{-1}]$ is a unit, then the substitution map $f(X) \longrightarrow f(u)$ of $R[X]$ onto $R[u]$ has a canonical extension to a homomorphism of $R[X]_{\{X^i\}_1^\infty} = R[X,X^{-1}]$ onto $R[u]_{\{u^i\}_1^\infty} = R[u,u^{-1}]$, and the extension maps $f(X) = \Sigma a_i X^i$ to $\Sigma a_i u^i$ for arbitrary $f(X) \in R[X,X^{-1}]$. For a unit u of $R[X,X^{-1}]$, we denote by ϕ_u the R—endomorphism of $R[X,X^{-1}]$ that maps X to u. Theorem 22.3 is the analogue of (22.1). The proof of Theorems 22.3 and 22.4 use the following auxillary result.

THEOREM 22.2. Assume that R is a unitary ring with prime subring Π. Let u be a unit and n be a nilpotent of the ring R. Then $u - n$ is a unit of the ring $\Pi[u,u^{-1},n]$.

Proof. Assume that $n^k = 0$. Then $u^k = u^k - n^k = (u - n)(\Sigma_{i=0}^{k-1} u^i n^{k-1-i})$, so $(u - n)^{-1} = \Sigma_{i=0}^{k-1} u^{i-k} n^{k-i-1}$, which is in $\Pi[u^{-1},n]$.

THEOREM 22.3. Assume that R is a unitary ring and that $u = \Sigma_{i=a}^{b} u_i X^i$ is a unit of $R[X,X^{-1}]$. Let ϕ_u be the R—endomorphism of $R[X,X^{-1}]$ that maps X to u . The following conditions are equivalent.

(1) ϕ_u is an automorphism.

(2) ϕ_u is surjective (that is, $R[X,X^{-1}] = R[u,u^{-1}]$) .

(3) u_i is nilpotent for $i \notin \{-1,1\}$.

Proof. (1) <==> (2): We need only prove that (1) follows from (2) . Assume that $\Sigma_{i=m}^{w} r_i u^i = 0$, where $r_i \in R$. We write $X = \Sigma a_i u^i$, $X^{-1} = \Sigma b_i u^i$, $u^{-1} = \Sigma c_i X^i$. Let R_1 be the unitary subring of R generated by 1 and the coefficients r_i , a_i , b_i , c_i , and u_i . Then R_1 is Noetherian and $R_1[X,X^{-1}] = R_1[u,u^{-1}]$. The restriction of ϕ_u to the Noetherian ring $R_1[X,X^{-1}]$ is surjective, hence injective. Since $r_i \in R_1$ for each i , it follows that each $r_i = 0$, and hence ϕ_u is also surjective.

(2) ==> (3): Let P be a proper prime ideal of R and let α be the canonical map of $R[X,X^{-1}]$ onto $D[X,X^{-1}]$, where $D = R/P$. Then $D[X,X^{-1}] = D[\alpha(u),\alpha(u)^{-1}]$. Since Z is torsion—free and cancellative, Theorem 11.1 shows that $\alpha(u)$ is a trivial unit of $D[X,X^{-1}]$ —— say $\alpha(u) = dX^k$, where d is a unit of D . Thus $D[X,X^{-1}] = D[X^k,X^{-k}]$, and this implies that $k = \pm 1$. It follows that $u_i \in P$ for

$i \notin \{-1,1\}$, and hence u_i is nilpotent.

(3) ==> (2): We assume that (3) is satisfied. Let $u^{-1} = \Sigma_{i=c}^{d} v_i X^i$. Theorem 11.3 shows that u has unit content and that $R = (u_{-1}^q) \oplus (u_1^q)$ for all sufficiently large integers q . We first establish the inclusion $R[X,X^{-1}] \subseteq R[u,u^{-1}]$ under the additional hypothesis that u_1 is a unit of R and u_{-1} is nilpotent. Let B_u be the ideal of R generated by the cofficients u_i of u , where $i \neq 1$. Then B_u is nilpotent, and to prove the result, we use induction on the order of nilpotence k of B_u . If $k = 1$, then $u = u_1 X$ and $R[X,X^{-1}] = R[u,u^{-1}]$. Consider the case where $B = B_u$ has order of nilpotence $k \geq 2$. If α is the canonical map of $R[X,X^{-1}]$ to $(R/B)[X,X^{-1}]$, then $\alpha(u) = u_1^* X$, where $u_1^* = u_1 + B$, and hence $\alpha(u^{-1}) = (u_1^*)^{-1} X^{-1}$. It follows that $u^n \equiv (u_1 X)^n \pmod{B[X,X^{-1}]}$ for each integer n . Thus $u^n = r_n X^n + t_n$, where r_n is a unit of R and $t_n \in B[X,X^{-1}]$ for each n . Let $s_n = r_n^{-1}$. For $a \leq i \leq b$, $i \neq 1$, we have $u_i s_i u^i \equiv u_i X^i$ $\pmod{B^2[X,X^{-1}]}$. Hence if $v = u - \Sigma_{i \neq 1} u_i s_i u^i$, taken over over the range $a \leq i \leq b$, then $v \equiv u_1 X \pmod{B^2[X,X^{-1}]}$ and the induction hypothesis implies that $R[X,X^{-1}] \subseteq R[v,v^{-1}]$ To complete the proof for this case (u_1 a unit, u_{-1} nilpotent), we need to show that v is a unit of $R[u,u^{-1}]$; this follows immediately from Theorem 22.2.

We observe that the case just considered also handles the case where u_{-1} is a unit and u_1 is nilpotent, for the proof shows that under these conditions the cofficient of X in u^{-1} is a unit and all other coefficients are nilpotent. The remaining case is that where neither u_1 nor u_{-1} is nilpotent. Let e_1 and e_2 be idempotents corresponding to a decomposition $R = (u_1^q) \oplus (u_{-1}^q)$ of R. Then $u = e_1u + e_2u$, and $R[u,u^{-1}] = Re_1[e_1u,e_1u^{-1}] \oplus Re_2[e_2u,e_2u^{-1}]$, where e_1u^{-1} is the inverse of e_1u in $Re_1[e_1u,e_1u^{-1}]$, and similarly for e_2u^{-1}. As an element of $Re_1[e_1X,e_1X^{-1}]$, the coefficient of e_1X in e_1u is e_1u_1, a unit of $Re_1 = (u_1^q)$, and the coefficient of e_1X^{-1} is e_1u_{-1}, a nilpotent element. Since all other coefficients of e_1u_1 are nilpotent as well, it follows from the case considered above that $Re_1[e_1X,e_1X^{-1}] = Re_1[e_1u,e_1u^{-1}]$. By a similar argument, $Re_2[e_2X,e_2X^{-1}] = Re_2[e_2u,e_2u^{-1}]$. Therefore $R[X,X^{-1}] = Re_1[e_1X,e_1X^{-1}] \oplus Re_2[e_2X,e_2X^{-1}] = R[u,u^{-1}]$, and this completes the proof of Theorem 22.3.

In order to pass from $\mathrm{Aut}_R R[Z]$ to $\mathrm{Aut}_R R[Z^n]$, recall from Section 11 that a set $E = \{e_1,\ldots,e_k\}$ of nonzero orthogonal idempotents of R with $1 = \Sigma_1^k e_i$ is said to <u>split</u> the unit $u \in R[S]$ if there exist a unit $v \in R$, invertible elements $g_1,\ldots,g_k \in S$, and a nilpotent element $t \in R[S]$ such that $u = v(\Sigma_1^k e_i X^{g_i}) + t$. Let $\{g_i\}_{i=1}^n$ be a free basis for Z^n. We regard $R[Z^n]$ as $R[X^{\pm g_1},\ldots,X^{\pm g_n}]$, where $R[X^{g_1},\ldots,X^{g_n}]$ is the polynomial ring in n variables over R. Under an R–endomorphism of $R[Z^n]$, each X^{g_i} maps to a unit u_i of $R[Z^n]$. And conversely, if $u = (u_1,u_2,\ldots,u_n)$ is an n–tuple of units of $R[Z^n]$, there exists a unique R–endomorphism ϕ_u

of $R[Z^n]$ such that $\phi_u(X^{g_i}) = u_i$ for each i ; ϕ_u is the canonical extension to $R[Z^n]$ of the substitution map on $R[X^{g_1},\ldots,X^{g_n}]$ that maps X^{g_i} to u_i for each i . As in (22.1) and (22.3), we are therefore left with the problem of determining conditions on the n-tuple $u = (u_1,\ldots,u_n)$ in order that ϕ_u should be an automorphism. This is where the notion of splitting comes in; Theorem 11.16 shows that there exists a set E of idempotents of R that splits each u_i .

THEOREM 22.4. Assume that R is a unitary ring and that $u = (u_1,\ldots,u_n)$ and ϕ_u are as described in the preceding paragraph. The following conditions are equivalent.

(1) ϕ_u is an automorphism.

(2) ϕ_u is surjective (that is, $R[Z^n] = R[u_1^{\pm 1},\ldots,u_n^{\pm 1}]$) .

(3) Let $E = \{e_i\}_{i=1}^m$ be a set of nonzero orthogonal idempotents of R such that $1 = \Sigma_1^m e_i$ and such that E splits each u_i . For $1 \le i \le n$, write $u_i = t_i + \Sigma_{j=1}^m u_{ij}X^{b_{ij}}$, where $t_i \in R[S]$ is nilpotent and where u_{ij} is a unit of Re_j . Then $\{b_{ij}\}_{i=1}^n$ is a basis of Z^n for $1 \le j \le m$.

(4) There exists a set E as in (3) such that if each u_i is as represented there, then $\{b_{ij}\}_{i=1}^n$ is a basis of Z^n for $1 \le j \le m$.

<u>Proof</u>. (1) <==> (2): We show that (2) implies (1) . Thus assume that $\Sigma r_{i_1 \ldots i_n} u_1^{i_1}\ldots u_n^{i_n} = 0$. As in the proof of Theorem 22.3, there exists a Noetherian unitary subring R_1

of R containing each $r_{i_1 \ldots i_n}$ such that $R_1[X^{\pm g_1},\ldots,X^{\pm g_n}] = R_1[u_1^{\pm 1},\ldots,u_n^{\pm 1}]$. Therefore the restriction of ϕ_u to the Noetherian ring $R_1[Z^n]$ is surjective, hence injective. It follows that each $r_{i_1 \ldots i_n} = 0$, so ϕ_u is also injective.

(2) ⟹ (3): Let N be the nilradical of R , and for $1 \leq j \leq m$, let C_j be the ideal of R generated by $N \cup \{e_i\}_{i \neq j}$. If $v_{ij} = u_{ij} + C_j$ for each i , then v_{ij} is a unit of $R_j = R/C_j$, and under the canonical homomorphism of $R[Z^n]$ to $R_j[Z^n]$, we obtain the relation

$$R_j[Z^n] = R_j[(v_{1j}X^{b_{1j}})^{\pm 1},\ldots,(v_{nj}X^{b_{nj}})^{\pm 1}] =$$

$$R_j[X^{\pm b_{1j}},\ldots,X^{\pm b_{nj}}] .$$

(We have tacitly used the fact that $N[Z^n]$ is the nilradical of $R[Z^n]$; see Theorem 9.9.) It follows that $\{b_{ij}\}_1^n$ generates Z^n , and since Z^n has torsion-free rank n , $\{b_{ij}\}_{i-1}^n$ is a basis for Z^n .

That (3) implies (4) follows from Theorem 11.16.

(4) ⟹ (2): Let the notation be as in (3) . Since $R[X^{\pm g_1},\ldots,X^{\pm g_n}] = \Sigma_{i=1}^m \oplus Re_i[e_iX^{\pm g_1},\ldots,e_iX^{\pm g_n}]$ and since $R[u_1^{\pm 1},\ldots,u_n^{\pm 1}] = \Sigma_{i=1}^m \oplus Re_i[e_iu_1^{\pm 1},\ldots,e_iu_n^{\pm 1}]$, it suffices to consider the case where $m = 1$. Thus, we assume without loss of generality that $u_i = t_i + v_iX^{b_i}$, where $t_i \in R[Z^n]$ is nilpotent and v_i is a unit of R . Let B_u be the ideal of R generated by the coefficients of $t_1,t_2,\ldots,t_n$.

Then B_u is a nilpotent ideal of R, and to establish the inclusion $R[X^{\pm g_1},\ldots,X^{\pm g_n}] \subseteq R[u_1^{\pm 1},\ldots,u_n^{\pm 1}]$, we use induction on the order of nilpotence k of B_u. If $k = 1$, then $R[u_1^{\pm 1},\ldots,u_n^{\pm 1}] = R[X^{\pm b_1},\ldots,X^{\pm b_n}] = R[X^{\pm g_1},\ldots,X^{\pm g_n}]$. Consider the case where $B = B_u$ has order of nilpotence $k \geq 2$. If α is the canonical map from $R[Z^n]$ to $(R/B)[Z^n]$, then $\alpha(u_i) = v_i^* X^{b_i}$, where $v_i^* = v_i + B$, and hence $\alpha(u_i^{-1}) = (v_i^*)^{-1} X^{-b_i}$. It follows that for arbitrary integers $a_1,\ldots,a_n$ that

$$u_1^{a_1}\cdots u_n^{a_n} \equiv v_1^{a_1}\cdots v_n^{a_n} X^{a_1b_1+\cdots+a_nb_n} \pmod{B[Z^n]}.$$

Therefore

$$(*) \qquad X^{a_1b_1+\cdots+a_nb_n} = c_{a_1a_2\cdots a_n} u_1^{a_1}\cdots u_n^{a_n} + f_{a_1a_2\cdots a_n},$$

where $c_{a_1a_2\cdots a_n}$ is a unit of R and $f_{a_1a_2\cdots a_n} \in B[Z^n]$.

Next we modify $u_1,\ldots,u_n$ to obtain elements $h_1,h_2,\ldots,h_n$, as follows. For $1 \leq i \leq n$, t_i is a sum of monomials $y_{ij}X^{w_{ij}}$, where $y_{ij} \in B$ and where $w_{ij} \in Zb_1 \oplus \cdots \oplus Zb_n$. Thus, $(*)$ implies that $y_{ij}X^{w_{ij}}$ can be expressed in the form $z_{ij} + P_{ij}$, where $z_{ij} \in R[u_1^{\pm 1},\ldots,u_n^{\pm n}]$ is nilpotent, and where $p_{ij} \in B^2[Z^n]$. Let $h_i = u_i - \Sigma z_{ij}$, the sum being taken over all monomials $y_{ij}X^{w_{ij}}$ in t_i. By choice of the elements z_{ij}, $h_i \in R[u_1^{\pm 1},\ldots,u_n^{\pm 1}]$ and h_i has the form $v_iX^{b_i} + t_i^*$, where $t_i^* \in B^2[Z^n]$. The induction hypothesis implies that

$R[Z^n] \subseteq R[h_1^{\pm 1},\ldots,h_n^{\pm 1}]$, and Theorem 22.2 shows that $R[h_1^{\pm 1},\ldots,h_n^{\pm 1}] \subseteq R[u_1^{\pm 1},\ldots u_n^{\pm 1}]$. By the principle of mathematical induction, the equality $R[Z^n] = R[u_1^{\pm 1},\ldots,u_n^{\pm 1}]$ follows.

The automorphism group of $R[G]$, where G is torsion—free but not finitely generated, is not known in general, but in the case where R is reduced and indecomposable, Theorem 11.6 shows that $R[G]$ has only trivial units. For any group ring with this property, Theorem 22.5 indicates how the R—automorphisms of $R[G]$ arise.

THEOREM 22.5. Let R be a unitary ring with group of units $U(R)$. Assume that the group ring $R[G]$ admits only trivial units.

(1) If $\phi \in \mathrm{Aut}_R R[G]$, then for each $g \in G$, $\phi(X^g) = u_g X^{h_g}$ for some unit $u_g \in R$ and some $h_g \in G$. The mapping $\phi:G \longrightarrow U(R)$ defined by $\alpha(g) = u_g$ is a homomorphism; similarly, the mapping $\psi:G \longrightarrow G$ defined by $\psi(g) = h_g$ is an automorphism.

(2) Conversely, if $\alpha:G \longrightarrow U(R)$ is a homomorphism and $\psi \in \mathrm{Aut}(G)$, then there exists a unique element $\phi \in \mathrm{Aut}_R R[G]$ such that $\phi(X^g) = \alpha(g)X^{\psi(g)}$ for each $g \in G$.

Proof. (1): By definition, $\alpha = \mu\circ\phi$, where μ is the augmentation map on $R[G]$. Hence α is a homomorphism. If $g_1,g_2 \in G$, then $\phi(X^{g_1+g_2}) = u_{g_1}u_{g_2}X^{h_{g_1}+h_{g_2}}$, so $\psi(g_1 + g_2) = h_{g_1} + h_{g_2} = \psi(g_1) + \psi(g_2)$. Moreover, if $\psi(g) = 0$, then $\phi(X^g) \in R$, so $X^g \in R$ since ϕ is an

R-automorphism. Consequently, $g = 0$ and ψ is injective. To see that ψ is surjective, take $s \in G$. Then $\phi^{-1}(X^s) = uX^g$ for some $u \in U(R)$, $g \in G$. It follows that $\phi(X^g) = u^{-1}X^s$, and hence $\psi(g) = s$.

(2): Since $\{X^g \mid g \in G\}$ is a free basis for $R[G]$ as an R-module, there exists at most one $\phi \in \mathrm{Aut}_R R[G]$ with the property described — namely, the map ϕ defined by $\phi(\Sigma a_g X^g) = \Sigma a_g \alpha(g) X^{\psi(g)}$. To verify that ϕ so defined is a homomorphism, we need only show that it preserves multiplication of basis elements. Thus,

$$\phi(X^{g_1}X^{g_2}) = \alpha(g_1 + g_2)X^{\psi(g_1+g_2)} = \alpha(g_1)\alpha(g_2)X^{\psi(g_1)+\psi(g_2)}$$
$$= \phi(X^{g_1})\phi(X^{g_2}) .$$

Since ψ is an injection, it follows that an equality $\Sigma a_g \alpha(g) X^{\psi(g)} = 0$ implies that $a_g \alpha(g) = 0$, and hence $a_g = 0$, for each g. Therefore ϕ is injective. Finally, it is clear that R and $\{X^{\psi(g)} \mid g \in G\} = \{X^g \mid g \in G\}$ are contained in the image of ϕ so that ϕ is also surjective. This establishes (2).

If the notation and hypothesis are as in the statement of Theorem 22.5, we denote by $\mathrm{Hom}(G,U(R))$ the set of homomorphisms from G into $U(R)$. The set $\mathrm{Hom}(G,U(R))$ is an abelian group under pointwise multiplication — that is, $(\alpha_1\alpha_2)(g) = \alpha_1(g)\alpha_2(g)$; the identiy element of $\mathrm{Hom}(G,U(R))$ is the trivial homomorphism of G into $U(R)$. If $\alpha_1, \alpha_2 \in \mathrm{Hom}(G,U(R))$, if $\psi_1, \psi_2 \in \mathrm{Aut}(G)$, and if $\phi_1, \phi_2 \in \mathrm{Aut}_R R[G]$ are defined as in (2) of Theorem 22.5, then

$$(\phi_1 \circ \phi_2)(X^g) = \phi_1(\alpha_2(g)X^{\psi_2(g)}) = \alpha_2(g)\phi_1(X^{\psi_2(g)})$$

$$= \alpha_2(g)\alpha_1(\psi_2(g))X^{\psi_1(\psi_2(g))} .$$

Thus, if ϕ_i is denoted by (ψ_i,α_i) , then the multiplication in $\mathrm{Aut}_R R[G]$ is given by

$$(\#) \qquad (\psi_1,\alpha_1)(\psi_2,\alpha_2) = (\psi_1\psi_2,(\alpha_1\circ\psi_2)\alpha_2) .$$

It follows that the maps $\psi \longrightarrow (\psi,1)$ and $\alpha \longrightarrow (1,\alpha)$ are isomorphisms of Aut(G) and Hom(G,U(R)) , respectively, into $\mathrm{Aut}_R R[G]$. If we identify Aut(G) and Hom(G,U(R)) with their images in $\mathrm{Aut}_R R[G]$, then $\mathrm{Aut}_R R[G]$ is the product of Aut(G) and Hom(G,U(R)) , and part (1) of (22.5) shows that $\mathrm{Aut}(G) \cap \mathrm{Hom}(G,U(R)) = \{1\}$. Finally, (#) shows that Hom(G,U(R)) is a normal subgroup of $\mathrm{Aut}_R R[G]$, and hence $\mathrm{Aut}_R R[G]$ is a split extension Hom(G,U(R)) by Aut(G) ; we state the result formally in Theorem 22.6.

THEOREM 22.6. Assume that R is a unitary ring and that the group ring R[G] has only trivial units. Let U(R) be the unit group of R , let Hom(G,U(R)) be the group of homomorphisms of G into U(R) under pointwise multiplication, and let Aut(G) be the automorphism group of G . Then $\mathrm{Aut}_R R[G] \simeq [\mathrm{Hom}(G,U(R))]\mathrm{Aut}(G)$, a split extension of Hom(G,U(R)) by Aut(G) .

Section 22 Remarks

An R–automorphism ϕ of $R[X_1,\ldots,X_n]$ is determined by the n–tuple $(f_1,\ldots,f_n)$, where $f_i = \phi(X_i)$. In the case of a field R , work on the problem of determining $\mathrm{Aut}_R R[X_1,\ldots,X_n]$ has been focused on what is known as the

Jacobian Conjecture. To wit, if $X_i \longrightarrow f_i$ determines an automorphism of $R[X_1,\ldots,X_n]$, then it is straightforward to show that the determinant of the Jacobian matrix $J(f_1,\ldots,f_n) = [\partial f_i/\partial X_j]$ is a nonzero element of R. The converse fails for $\text{char } R \neq 0$, even for $n = 2$; the Jacobian Conjecture states that for $\text{char } R = 0$, the condition that $\det(J(f_1,\ldots,f_n)) \in R\backslash\{0\}$ implies that $X_i \longrightarrow f_i$ determines an automorphism of $R[X_1,\ldots,X_n]$. For a nice survey of this topic, see [17]. We note that a non-surjective R—endomorphism of $R[X_1,\ldots,X_n]$ that maps each X_i to a unit of $R[Z^n]$ may induce an automorphism of $R[Z^n]$. For example, the automorphism of $R[X,Y]$ determined by $X \longrightarrow XY$ and $Y \longrightarrow X^2Y$ has this property.

The method used in the proofs of (22.3) and (22.4) can be used to show that, in general, a surjective R—endomorphism of an affine ring $T = R[t_1,\ldots,t_n]$ over a unitary ring R is injective. In particular, $R[S]$ has this property for each finitely generated monoid S. Without the hypothesis that S is finitely generated, the conclusion need not hold; for example, $X_1 \longrightarrow 0$, $X_i \longrightarrow X_{i-1}$ for $i \geq 2$ determines a surjective R—endomorphism of the polynomial ring $R[\{X_i\}_1^\infty]$ that is not injective. The power series ring $R[[X_1,\ldots,X_n]]$ also has the property that each of its surjective R—endomorphisms is injective (see [42], [57]). For a general discussion of conditions under which injectivity follows from surjectivity, see [115].

As indicated in the remarks at the end of Section 11, the class of group rings $R[G]$ with only trivial units has been investigated in the literature, but it seems unreasonable to expect a strong result characterizing such rings for arbitrary R and G . Besides the references on this topic listed in §11, see also [37–38].

If $R = R_1 \oplus \ldots \oplus R_n$ is such that $R_i[G]$ has only trivial units for each i , then $\mathrm{Aut}_R R[G]$ is isomorphic to the direct sum of the groups $\mathrm{Aut}_{R_i} R_i[G]$, and the latter groups are known from Theorem 22.6. Lantz in [89] has also given a description of $\mathrm{Aut}_R R[G]$ in the case where R is reduced and G is torsion-free.

Further results concerning $\mathrm{Aut}_R R[G]$, including a generalization of Theorem 22.6, can be found in [102].

§23. Coefficient Rings in Isomorphic Monoid Rings. I.

Assume that the semigroup rings $R_1[S]$ and $R_2[S]$ are isomorphic. Are R_1 and R_2 isomorphic? This question has been considered extensively in the case $S = Z_0^n$, and hence $R_i[S]$ is the polynomial ring in n indeterminates over R_i Even in the case of Z_0, it is known that R_1 and R_2 need not be isomorphic. In fact, an example due to Hochster [77] exhibits nonisomorphic Noetherian domains R_1 and R_2 such that $R_1[X] \simeq R_2[X]$. Work on the isomorphism question for polynomial rings is too extensive for inclusion here, but references to some work on the question are included in Section 23 Remarks. Instead, we concentrate in this section on the case where R_1 and R_2 are fields, and in Section 24 on the case where S is a torsion-free group.

This is the only section of the text where results are proved for semigroups that need not be commutative. Theorem 23.1, Corollary 23.2, and the summarizing result of the section — Theorem 23.17 — do not require that the semigroups in question are commutative. To distinguish these cases, we use the notation RS for the semigroup ring of S over R in the case where S need not be commutative, we write the semigroup operation in S as multiplication, and the elements of RS are written in the form $\Sigma_{i=1}^n r_i s_i$. The semigroups considered in results (23.3) through (23.16) are commutative, and the more familiar notation $R[S]$ and $\Sigma r_i X^{s_i}$ is used in that part of the section. Theorem 23.17 shows that if F and K are fields, S and T are semigroups, S contains a periodic element, and $FS \simeq KT$, then $F \simeq K$. The proof proceeds by reducing the question for general S and T to the case where S and T are periodic commutative semigroups. The first result provides a reduction to the case of commutative semigroups. We continue to assume that all rings considered are commutative.

THEOREM 23.1. Assume R is a ring with $R = R^2$ and S is a semigroup. There exists a smallest congruence $\sim$ on S such that the factor semigroup $S/\sim$ is commutative. Let I be the kernel of the canonical homomorphism from RS onto $R[S/\sim]$. Then I is the unique minimal ideal of RS such that the residue class ring RS/I is commutative.

Proof. Let $A = \{(xy,yx) \mid x,y \in S\} \subseteq S \times S$. Then A generates a congruence $\sim$ on S, and it is straightforward to show that $\sim$ is the least congruence on S with commutative associated factor semigroup.

We show that I is generated by

$$B = \{rx - ry \mid r \in R, x,y \in S, \text{ and } x \sim y\}.$$

Let J be the ideal of RS generated by B. The inclusion $J \subseteq I$ is clear. To prove the reverse inclusion, take $f = \Sigma_{i=1}^{n} r_i s_i \in I\backslash(0)$, and use induction on $n = |\mathrm{Supp}(f)|$. The case $n = 1$ is impossible, and if $n = 2$, then $r_1[s_1] + r_2[s_2] = 0$, implying $[s_1] = [s_2]$ and $r_1 = -r_2$. Therefore $f = r_1 s_1 - r_1 s_2$ with $s_1 \sim s_2$ and $f \in J$. For $n > 2$, the relation $\Sigma_1^n r_i[s_i] = 0$ implies $s_i \sim s_j$ for some $i \neq j$. Then $r_i s_i - r_i s_j \in J$ and $f_1 = (f - (r_i s_i - r_i s_j)) \in I$ has fewer than n elements in its support. By the induction hypothesis, $f_1 \in J$, so $f \in J$ and $I = J$.

Clearly RS/I is commutative. Let C be any ideal of

RS such that RS/C is commutative. Define a relation μ on S by $x \mu y$ if and only if $rx - ry \in C$ for all $r \in R$. It is routine to show that μ is an equivalence relation on S. We verify that μ is a congruence on S by showing that it is compatible with the semigroup operation. Thus, assume $x \mu y$, $z \in S$, and $r \in R$. We must show that $rxz - ryz$ and $rzx - rzy$ are in C. Write $r = \Sigma_{j=1}^{k} a_j b_j$ where $a_j, b_j \in R$; this is possible since $R = R^2$. For each j we have

$$a_j b_j xz - a_j b_j yz = (a_j x - a_j y) b_j z \in C \text{, and}$$

$$a_j b_j zx - a_j b_j zy = a_j z (b_j x - b_j y) \in C .$$

Hence $rxz - ryz = \Sigma_1^k (a_j b_j xz - a_j b_j yz) \in C$, and similarly, $rzx - rzy \in C$. We show that S/μ is commutative. If $s, t \in S$ and if $r = \Sigma_1^k a_j b_j \in R$, then $a_j b_j st - a_j b_j ts = a_j s \cdot b_j t - b_j t \cdot a_j s \in C$ since RS/C is commutative. As above this implies that $rst - rts \in C$, so $st \mu ts$ and S/μ is commutative. Therefore $\mu \geq \sim$. Let $x, y \in S$ be such that $x \sim y$. Then $x \mu y$ so $rx - ry \in C$ for each $r \in R$. Since $B = \{rx - ry \mid r \in R, x \sim y\}$ generates I, it follows that $I \subseteq C$, and this complets the proof of Proposition 23.1.

COROLLARY 23.2. Assume that $R_1 S \approx R_2 T$, where R_1 and R_2 are idempotent. Let $S' = S/\sim$ and let $T' = T/\mu$, where $\sim$ and μ are minimal congruences on S and T, respectively, such that the factor semigroups are commutative. Then $R_1 S' \approx R_2 T'$.

If S and T are commutative and if ϕ is an

isomorphism of $R_1[S]$ onto $R_2[T]$, then ϕ induces an isomorphism ϕ^* of $R_1[S]/N(R_1[S])$ onto $R_2[T]/N(R_2[T])$, where $N(R_1[S])$ and $N(R_2[T])$ denote the nilradicals of $R_1[S]$ and $R_2[T]$, respectively. If R_1 and R_2 are reduced rings of prime characteristic p , Theorem 9.4 shows that $N(R_1[S])$ is the kernel ideal of the congruence $\sim_p$ of p—equivalence on S , and a similar statement holds for $R_2[T]$. On the other hand, if R_1 is a domain of characteristic 0 , then Corollary 9.12 implies that $R_1[S]/N(R_1[S]) \simeq R_1[S/\sim]$, where $\sim$ is the congruence of asumptotic equivalence on S . The next result follows immediately from these observations.

THEOREM 23.3. Assume that S and T are commutative semigroups and $R_1[S] \simeq R_2[T]$.

(1) If R_1 and R_2 are reduced rings of prime characteristic p , then $R_1[S/\sim_p] \simeq R_2[T/\sim_p]$, and each of these rings is reduced.

(2) If R_1 and R_2 are domains of characteristic 0 , then $R_1[S/\sim] \simeq R_2[T/\sim]$, and each of these rings is reduced.

Our next reduction of the isomorphism problem is to the case of periodic semigroups. We denote by S^* the set of periodic semigroups. We denote by S^* the set of periodic elements of the commutative semigroup S . The hypotheses of results through Theorem 23.16 either assume or imply that $S^* \neq \phi$. In particular, $S^* \neq \phi$ if S is a monoid.

THEOREM 23.4. Assume R is a reduced ring and G is an abelian torsion—free group. If $f \in R[G]$ is such that the ideals (f) and $(f - f^2)$ are idempotent, then $f \in R$.

Proof. Write $f = uf^2$ and $f - f^2 = v(f - f^2)^2$ with $u,v \in R[G]$. Let $f = \Sigma a_g X^g$. We show that $a_g = 0$ for $g \neq 0$ by showing that a_g belongs to each proper prime ideal P of R.

Let ϕ be the canonical homomorphism from $R[G]$ to the integral domain $(R/P)[G]$. Then

$$\phi(f) = \phi(u)\phi(f)^2 \text{ and}$$

$$\phi(f - f^2) = \phi(v)\phi(f - f^2)^2 .$$

If $\phi(f) = 0$, then $a_g \in P$. If $\phi(f) \neq 0$, then the first of the equations above implies that $(R/P)[G]$ is unitary and $\phi(f)$ is a unit of $(R/P)[G]$. If $\phi(f) = \phi(f^2) = \phi(f)^2$, then $\phi(f)$ is the identity element of $(R/P)[G]$; hence $\mathrm{Supp}(\phi(f)) = \{0\}$ and $a_g \in P$. Finally, if $\phi(f)$ and $\phi(f - f^2)$ are nonzero, then each is a unit. But the units of $(R/P)[G]$ are trivial by Theorem 11.1, so that if $\mathrm{Supp}(\phi(f)) = \{s\}$, then $\mathrm{Supp}(\phi(f) - \phi(f)^2) = \{s,2s\}$, which must be a singleton set. Thus $s = 0$, $a_g \in P$, and $f \in R$.

THEOREM 23.5. Assume that G is an abelian group such that $R[G]$ is reduced. If $f \in R[G]$ is such that the ideals (f) and $(f - f^2)$ are idempotent, then $f \in R[G^*]$, where G^* is the torsion subgroup of G.

Proof. Write $f = uf^2$, $f - f^2 = v(f - f^2)^2$ as in the proof of Theorem 23.4, and let H be the subgroup of G generated by $\mathrm{Supp}(f) \cup \mathrm{Supp}(u) \cup \mathrm{Supp}(v)$. Since H is finitely generated, the torsion subgroup H^* of H is a direct summand of H, say $H = H^* \oplus W$ with W torsion-free. We have $f, u, v \in R[H] \simeq RH^*[W]$. Since $R[H^*]$ is reduced,

$f \in R[H^*]$ by Theorem 23.4. The result then follows since $R[H^*] \subseteq R[G^*]$.

THEOREM 23.6. Assume that S is a commutative cancellative semigroup such that $R[S]$ is reduced. If $f \in R[S] \setminus (0)$ is such that the ideals (f) and $(f - f^2)$ are idempotent, then the set S^* of periodic elements of S is nonempty, S is a monoid, and $f \in R[S^*]$.

Proof. The nonzero ideal (f) is generated by an idempotent element e, and Theorem 10.6 shows that each element of $\mathrm{Supp}(e)$ is periodic. Hence $S^* \neq \phi$ and S^* contains an idempotent w. Because S is cancellative, w is an identity element for S and S is a monoid.

Let G be the quotient group of S. The group ring $R[G]$ is a quotient ring of $R[S]$ and is therefore reduced. From Theorem 23.5 we conclude that $f \in R[G^*]$, where G^* is the torsion subgroup of G. Thus $f \in R[S] \cap R[G^*] = R[S \cap G^*]$. To complete the proof, we check that $S \cap G^* = S^*$. If $s \in S \cap G^*$, then $ns = 0$ for some $n \in Z^+$, and thus $(n + 1)s = s$, which says $s \in S^*$. If $t \in S^*$, there exists $n, m \in Z^+$ with $n > m$ such that $nt = mt$, implying $(n - m)t = 0$ since S is cancellative. Thus $t \in S \cap G^*$ and $S \cap G^* = S^*$.

Theorem 23.10 is the last of the series of four propositions beginning with Theorem 23.4 that have basically the same form. The hypothesis in (23.10) differs from that in (23.6) in that we no longer assume that S is cancellative. The proof of Theorem 23.10 uses the Archimedean decomposition $S = \cup S_a$ of S introduced in Section 17. We recall from

Theorem 17.10 that each S_a is a cancellative subsemigroup of S if S is free of asymptotic torsion. Two preliminary results are needed in the proof of (23.10); the first of these concerns the Archimedean decomposition of S.

THEOREM 23.7. Assume that S is a commutative semigroup that is free of asymptotic torsion. For $s \in S$, define $U_s = \{t \in S \mid s + S \subseteq \mathrm{rad}(t + S)\}$.

(1) U_s is a subsemigroup of S containing the Archimedean component S_s of s as an ideal. Moreover, $U_s = U_t$ for each $t \in S_s$.

(2) If $a, b \in S_s$ and $c \in U_s$ are such that $a + c = b + c$, then $a = b$.

<u>Proof</u>. The assertions in (1) that $S_s \subseteq U_s$ and $U_s = U_t$ for $t \in S_s$ follow immediately from the definition of S_s. To see that U_s is closed under addition, take $t_1, t_2 \in U_s$. Choose $k_1, k_2 \in Z^+$ and $s_1, s_2 \in S$ such that $k_1 s = t_1 + s_1$ and $k_2 s = t_2 + s_2$. Then $(k_1 + k_2)s = t_1 + t_2 + s_1 + s_2 \in t_1 + t_2 + S$. Therefore $s + S \subseteq \mathrm{rad}(t_1 + t_2 + S)$ and $t_1 + t_2 \in U_s$. To establish the inclusion $U_s + S_s \subseteq S_s$, take $u \in U_s$, $v \in S_s$. We show that $\mathrm{rad}(u + v + S) = \mathrm{rad}(v + S)$; the inclusion $\mathrm{rad}(u + v + S) \subseteq \mathrm{rad}(v + S)$ follows since $u + v + S \subseteq v + S$. Write $kv = u + w$, where $k \in Z^+$ and $w \in S$. Then $(k + 1)v = u + v + w$ implies $v \in \mathrm{rad}(u + v + S)$, and hence $\mathrm{rad}(v + S) \subseteq \mathrm{rad}(u + v + S)$. This completes the proof of (1).

(2): Since $a + c = b + c$, then $a + (c + a) = b + (c + a)$. (1) shows that $c + a \in S_s$, and S_s is a cancellative semigroup since S is free of asymptotic torsion. Therefore $a = b$.

THEOREM 23.8. Assume that S is a commutative semigroup and that $p \in Z^+$ is prime. Let $\sim$ be the cancellative congruence on S.

(1) If S is free of asymptotic torsion, then so is $S/\sim$.

(2) If S is p-torsion-free, then so is $S/\sim$.

Proof. We recall that $\sim$ is defined on S by $a \sim b$ if and only if $a + x = b + x$ for some $x \in S$. We prove (1); the proof of (2) is similar, but easier. Assume that $[a], [b] \in S/\sim$ are asymptotically equivalent. Choose relatively prime integers n, m such that $[na] = [nb]$ and $[ma] = [mb]$. There exists $x, y \in S$ such that $na + x = nb + x$ and $ma + y = mb + y$. Therefore $m(a + x + y) = m(b + x + y)$, $n(a + x + y) = n(b + x + y)$, and $a + x + y$ and $b + x + y$ are asymptotically equivalent. Hence $a + x + y = b + x + y$ since S is free of asymptotic torsion. Consequently, $a \sim b$ and $[a] = [b]$.

Assume that S is a commutative semigroup. Theorem 9.17 shows that $R[S]$ is reduced if and only if (i) R is reduced, (ii) S is free of asymptotic torsion, and (iii) p is regular on R for each prime p such that S is not p-torsion-free. Theorem 23.8 shows that properties (ii) and (iii) transfer from R and S to R and $S/\sim$, where $\sim$ is the cancellative congruence on S. Thus, we obtain the following corollary.

COROLLARY 23.9. Let $\sim$ be the cancellative congruence on the commutative semigroup S. If $R[S]$ is reduced, then so is $R[S/\sim]$.

Corollary 23.9 is the final preliminary result needed in the proof of Theorem 23.10, which will prove to be a key result in the ultimate proof of Theorem 23.17.

THEOREM 23.10. Assume that S is a commutative semigroup such that $R[S]$ is reduced. If $f \in R[S]\setminus(0)$ is such that the ideals (f) and $(f - f^2)$ are idempotent, then $S^* \neq \phi$ and $f \in R[S^*]$.

Proof. The proof that $S^* \neq \phi$ is the same as that given in the proof of Theorem 23.6. Let $S = \cup S_a$ be the decomposition of S into its Archimedean components, and let η be the congruence on S that gives rise to this decomposition — that is, $a \, \eta \, b$ if and only if $\text{rad}(a + S) = \text{rad}(b + S)$. Since $R[S]$ is reduced, S is free of asymptotic torsion, and hence each S_a is cancellative. To prove the result, we use induction on the cardinality m of the equivalence classes under η represented by elements of $\text{Supp}(f)$.

If $m = 1$, then $\text{Supp}(f) \subseteq S_a$ for some $a \in S$ and $f \in R[S_a]$. To verify the theorem in this case, it is sufficient, by Theorem 23.6, to show that f and $f - f^2$ generate idempotent ideals in $R[S_a]$, for then $f \in R[S_a^*] \subseteq R[S^*]$. Toward this end, we write $f = uf^3$ and $f - f^2 = v(f - f^2)^3$ for some $u,v \in R[S]$. The supports of f, f^3, $f - f^2$, and $(f - f^2)^3$ are all contained in S_a. Write $u = u_1+\dots+u_r$, where $\text{Supp}(u_i) \subseteq S_{a_i}$ and where

$S_{a_i} \neq S_{a_j}$ for $i \neq j$. Then $f = \Sigma_1^r u_i f^3$, where $\mathrm{Supp}(u_i f^3) \subseteq S_{a_i+a}$. It follows that $f = (\Sigma u_j)f^3$, the sum being taken over those j such that $S_{a_j+a} = S_a$. Summing over the same indices j, we note that $\Sigma u_j f \in R[S_a]$ and $f = (\Sigma u_j f)f^2 \in R[S_a]f^2$. Therefore f generates an idempotent ideal of $R[S_a]$. By the same argument, $f - f^2$ also generates an idempotent ideal of $R[S_a]$, and this establishes the case $m = 1$.

We assume that the result is true for $m < n$, where $n > 1$, and consider the case where the elements of $\mathrm{Supp}(f)$ determine n classes under η. Assume that $\mathrm{Supp}(f)$ contains an aperiodic element b. We first show that $b + S \subseteq \mathrm{rad}(s + S)$ for each $s \in \mathrm{Supp}(f)$. If not, then there exists $s \in \mathrm{Supp}(f)$ and a prime ideal P of S containing s but not b. Consider the decomposition $R[S] = R[S\backslash P] + R[P]$ of $R[S]$; here $S\backslash P$ is a subsemigroup of S and $R[P]$ is an ideal of $R[S]$. Let $f = f_1 + f_2$, where $f_1 \in R[S\backslash P]$ and $f_2 \in R[P]$. Then $f_1 - f_1^2$ is the $R[S\backslash P]$—component of $f - f^2$, so f_1 and $f_1 - f_1^2$ generate idempotent ideals of $R[S\backslash P]$ and of $R[S]$. The number of Archimedean components represented by elements of $\mathrm{Supp}(f_1)$ is less than n since $S_s \subseteq P$. The induction hypothesis implies that $f_1 \in R[S^*]$, and this is a contradiction, for $b \in \mathrm{Supp}(f_1)\backslash S^*$. We conclude that $b + S \subseteq \mathrm{rad}(s + S)$, as asserted. It follows that S_b contains each aperiodic element of $\mathrm{Supp}(f)$ and that $\mathrm{Supp}(f) \subseteq U_b = \{t \in S \mid b + S \subseteq \mathrm{rad}(t + S)\}$. Define

$$I = \{x \in S \mid t_1 + x = t_2 + x \text{ for some } t_1, t_2 \in S_b, t_1 \neq t_2\}.$$

We show that the assumption that I is empty leads to a contradiction. Write $f = f_1 + f_2$, where $f_2 \in R[S_b]$ and where $\mathrm{Supp}(f_1) \subseteq (S \setminus S_b) \cap S^*$. Let $\sim$ be the cancellative congruence on S and let ϕ be the natural homomorphism of S onto $S/\sim$. Theorem 23.8 shows that $R[S/\sim]$ is reduced. Moreover, $\phi(f)$ and $\phi(f - f^2) = \phi(f) - \phi(f)^2$ generate idempotent ideals of $R[S/\sim]$. Consider $[b] \in S/\sim$. Since $I = \phi$, then $[b] \neq [c]$ for $c \in \mathrm{Supp}(f_2)$, $c \neq b$. Therefore either $[b] \in \mathrm{Supp}(\phi(f))$ and $[b] \in (S/\sim)^*$ by Theorem 23.6, or else $[b] \notin \mathrm{Supp}(\phi(f))$ and $[b] = [s]$ for some $s \in \mathrm{Supp}(f_1)$. In the latter case, $[b] = [s] \in (S/\sim)^*$ since $s \in S^*$. Thus, in either case there exist $m,n \in Z^+$, $m \neq n$, and $x \in S$ such that $mb + x = nb + x$. But this contradicts the assumption that $I = \phi$ since mb and nb are distinct elements of S_b. We conclude that $I \neq \phi$. Thus I is an ideal of S, and part (2) of Theorem 23.7 shows that $I \cap U_b = \phi$. Choose a prime ideal P of S containing I such that $P \cap U_b = \phi$, and let $T = S \setminus P$. As above, $R[T]$ is reduced and $f \in R[T]$ is such that f and $f - f^2$ generate idempotent ideals of $R[T]$. We note, moreover, that the definitions of I and T imply that $J = \phi$, where

$$J = \{x \in T \mid t_1 + x = t_2 + x \quad \text{for some} \quad t_1, t_2 \in S_b,\ t_1 \neq t_2\}.$$

By considering the homomorphic image of f in $R[T/\sim]$, where $\sim$ is the cancellative congruence on T, we obtain, as above, a contradiction to the assumption that b is aperiodic. Hence $b \in S^*$, and this completes the proof of Theorem 23.10.

Theorem 23.10 is related to the isomorphism problem because it can be used to show that if R is a (von Neumann) regular ring and S is a commutative semigroup such that $S^* \neq \phi$, then $R[S^*]$ is the maximal regular subring of $R[S]$. This result is the content of Theorem 23.12; the proof of that result uses Theorem 23.11.

THEOREM 23.11. Assume that R is a regular ring.

(1) If A is an ideal of R, then A, considered as a ring, is regular.

(2) If S is a periodic commutative semigroup and if $R[S]$ is reduced, then $R[S]$ is a regular ring.

Proof. (1): If $a \in A$, then $a = ra^3 = ra \cdot a^2$ for some $r \in R$. Since $ra \in A$, it follows that A is regular as a ring.

To prove (2), we consider first the case where R is unitary. Let $S^0 = S \cup \{0\}$ be the monoid obtained by adjoining an identity element to S. Then S^0 is periodic, and hence $R[S^0]$ is integral over R. It follows that $\dim R[S^0] = \dim R = 0$. We have $R[S^0] = R + R[S]$, where $R[S]$ is an ideal of $R[S^0]$ and where $R \cap R[S] = (0)$. If $g \in R[S^0]$ is nilpotent, we write $g = r + t$, where $r \in R$ and $t \in R[S]$. The element r is nilpotent, and hence $r = 0$. Because $R[S]$ is reduced, we then have $t = 0$, so $R[S^0]$ is a zero-dimensional reduced ring, hence regular. Part (1) then shows that $R[S]$ is also regular in the case where R is unitary. In the general case, take $f = \Sigma_1^n r_i X^{s_i} \in R[S]\setminus(0)$. We wish to show that $(f) = (f^2)$. Let I be the ideal of R generated by $\{r_i\}_1^n$. Since R is regular, I is principal, generated by an idempotent e. As a ring, I is unitary and regular; moreover, $f \in I[S]$. By the case just considered, it follows that f generates an idempotent ideal of $I[S]$, and hence of $R[S]$. Therefore $R[S]$ is regular.

THEOREM 23.12. Assume R is a regular ring, S is a commutative semigroup with $S^* \neq \phi$, and $R[S]$ is reduced. Let $W = \{f \in R[S] \mid f$ and $f - f^2$ generate idempotent ideals of $R[S]\}$. Then $W = R[S^*]$ and W is the unique maximal regular subring of $R[S]$.

<u>Proof</u>. Theorem 23.10 shows that $W \subseteq R[S^*]$. It is clear that W contains each regular subring of $R[S]$, and hence $W \supseteq R[S^*]$ by Theorem 23.11. Therefore $W = R[S^*]$ is the unique maximal regular subring of $R[S]$.

COROLLARY 23.13. Assume R_1 and R_2 are regular rings, S and T are commutative semigroups with $S^* \neq \phi$, $R_1[S]$ is reduced, and $R_1[S] \simeq R_2[T]$. Then $T^* \neq \phi$ and $R_1[S^*] \simeq R_2[T^*]$.

<u>Proof</u>. The ring $R_2[T]$ is reduced since $R_1[S]$ is. Theorem 23.12 shows that $R[S]$ contains a nonzero element f such that (f) and $(f - f^2)$ are idempotent, and hence $R_2[T]$ also has such an element. Theorem 23.10 then shows that $T^* \neq \phi$. By Theorem 23.12, it follows that the isomorphism $R_1[S] \simeq R_2[T]$ induces an isomorphism $R_1[S^*] \simeq R_2[T^*]$.

We show presently (Theorem 23.16) that isomorphism of $F[S]$ and $K[T]$ implies isomorphism of F and K for periodic commutative semigroups S and T. To handle the case where S is not a monoid, we need one preliminary ring-theoretic result.

THEOREM 23.14. Let A be an ideal of the ring R and let P be a proper prime ideal of A, considered as a ring.

(1) P is an ideal of R and is of the form $Q \cap A$ for some prime ideal Q of R.

(2) If $P_1 < \ldots < P_n$ is a chain of proper primes of A, then there exists a chain $Q_1 < \ldots < Q_n$ of proper primes of R such that $P_i = Q_i \cap A$ for each i. Hence $\dim R \geq \dim A$.

Proof. (1): Choose $r \in R$ and $x \in A \backslash P$. Then $rx \in A$ implies $rxP \subseteq P$. Moreover, $x \in A$ and $rP \subseteq A$. Since P is prime in A, it follows that $rP \subseteq P$, and hence P is an ideal of R. There exists an ideal Q of R containing P such that Q is maximal with respect to failure to meet the multiplicative system $A \backslash P$ in R. The ideal Q is prime in R and $Q \cap A = P$.

(2): Assume that there exists a chain $Q_1 < \ldots < Q_m$ of primes of R with $m < n$ and $Q_i \cap A = P_i$ for $1 \leq i \leq m$. Then $A/P_m \simeq (A + Q_m)/Q_m$ is an ideal of R/Q_m and P_{m+1}/P_m is a proper prime ideal of A/P_m. By (1) we conclude that $P_{m+1} = Q_{m+1} \cap A$ for some prime Q_{m+1} of R properly containing Q_m. By induction, (2) then follows.

THEOREM 23.15. Assume that R is unitary and M is a maximal ideal of R. Let $\{M_\alpha\}$ be the set of maximal ideals of R distinct from M, and consider M as a ring.

(1) $\{M_\alpha \cap M\}$ is a set of (distinct) maximal prime ideals of M; moreover, $M/(M \cap M_\alpha) \simeq R/M_\alpha$ for each α.

(2) If $\dim R = 0$, then $\{M_\alpha \cap M\}$ is the set of

maximal prime ideals of M .

Proof. (1): We have $M/(M \cap M_\alpha) \simeq (M + M_\alpha)/M_\alpha \simeq R/M_\alpha$, so each $M \cap M_\alpha$ is a maximal prime ideal of M . These ideals are distinct since $M_\alpha \cap M \not\subseteq M_\beta$ for $\alpha \neq \beta$.

(2): If P is a maximal prime of M , then Theorem 23.14 shows that P is contracted from a prime ideal of R , and hence $P = M \cap M_\alpha$ for some α since R is zero-dimensional.

THEOREM 23.16. Assume that F and K are fields and that S and T are commutative semigroups. If $F[S] \simeq K[T]$ and if S is periodic, then T is also periodic and $F \simeq K$.

Proof. Let $S^0 = S \cup \{0\}$ be the monoid obtained by adjoining an identity element 0 to S ; this construction is possible, even if S is already a monoid. Then $F[S^0] = F + F[S]$, so $F[S]$ is a maximal ideal of $F[S^0]$. Since S^0 is periodic, $\dim F[S^0] = 0$, and hence $\dim F[S] \leq 0$ by part (2) of Theorem 23.14. Therefore $\dim F[S] = 0$ since F is a homomorphic image of $F[S]$. Theorem 23.15 shows that the residue fields of $F[S]$ are the same as the residue fields of $F[S^0]$, with the possible exception of $F[S^0]/F[S] \simeq F$. But F is a residue field of $F[S]$, and hence the residue fields of $F[S]$ are the same as the residue fields of $F[S^0]$.

We wish to show that the residue fields of $K[T]$ are the same as those of $K[T^0]$. To do so, it suffices to show that T is periodic. Because $K[T] \simeq F[S]$, it follows that $\dim K[T] = 0$ and that $K[T]/P$ is a field for each maximal prime P of $K[T]$. We show that this implies $\dim K[T^0] = 0$.

Suppose not, and let $Q_1 < Q_2$ be a chain of proper primes of $K[T^0]$. Since Q_1 is not maximal in $K[T^0]$, then $Q_1 \not\supseteq K[T]$, and hence $Q_1 \cap K[T] < Q_2 \cap K[T]$. Therefore $Q_2 \cap K[T] = K[T]$, which implies $Q_2 = K[T]$. It follows that Q_1 is a maximal prime of $K[T]$, and this is a contradiction, for the nonzero proper ideal $K[T]/Q_1$ of the domain $K[T^0]/Q_1$ does not have an identity element as a ring. We conclude that $\dim K[T^0] = 0$, which implies that T^0 is periodic by Theorem 17.3.

We examine the residue fields of $F[S^0]$. If M_α is a maximal ideal of $F[S^0]$, then $F \cap M_\alpha = \phi$, so F is naturally imbedded in $\Delta_\alpha = F[S^0]/M_\alpha$. Moreover, since S^0 is periodic, Δ_α is generated over F by elements u satisfying an equation of the form $u^n - u^m = 0$, where $n,m \in Z^+$ and $n > m$. Then $u^m(u^{n-m} - 1) = 0$, so either $u = 0$ or u is a root of unity. Therefore Δ_α is generated over F by roots of unity.

Let ϕ be an isomorphism of $F[S]$ onto $K[T]$. Then ϕ induces a bijection ϕ^* of the residue fields $\{\Delta_\alpha\}_{\alpha\in A}$ of $F[S]$ onto the residue fields of $K[T]$ in such a way that Δ_α and $\phi^*(\Delta_\alpha)$ are isomorphic. Since $F \simeq \Delta_\alpha$ for some α and since K is imbeddable in each residue field of $K[T]$, it follows that K is imbeddable in F. Similarly, F is imbeddable in K. Moreover, these imbeddings may be accomplished by isomorphisms $\sigma: F \longrightarrow K$ and $\psi: K \longrightarrow F$ in such a way that K is generated over $\sigma(F)$ by a set $\{u_i\}$ of roots of unity. Assume that the (multiplicative) order of u_i is n_i. Since $\psi(u_i) \in F$ has order n_i, it follows that $\sigma(F)$ contains the n_ith roots of unity.

Consequently, $u_i \in \sigma(F)$ for each i and $\sigma(F) = K$. Therefore $F \simeq K$.

Unlike the results in this section since Theorem 23.3, the semigroups in the statement of Theorem 23.17 are not assumed to be commutative.

THEOREM 23.17. Assume F and K are fields, S and T are semigroups, S contains a periodic element, and $FS \simeq KT$. Then $F \simeq K$.

<u>Proof</u>. Corollary 23.2 and Theorem 23.3 show that the isomorphism $FS \simeq KT$ induces an isomorphism $F[S_1] \simeq K[T_1]$, where S_1 and T_1 are commutative, $F[S_1]$ is reduced, and S_1 is a homomorphic image of S. Hence $S_1^* \neq \phi$. Corollary 23.13 then shows that $T_1^* \neq \phi$ and $F[S_1^*] \simeq K[T_1^*]$. Applying Theorem 23.16, we conclude that $F \simeq K$.

Beyond Theorem 23.3, the material in this section appears to have slight application to the isomorphism question in the case of a commutative semigroup S such that $S^* = \phi$. The next result resolves what is probably the most natural subcase of this problem.

THEOREM 23.18. Assume F and K are fields and S and T are subsemigroups of Z^+ such that $F[S] \simeq K[T]$. Then $F \simeq K$.

<u>Proof</u>. Let m and n be the smallest elements of S and T, respectively. Define

$$U = \{f \in F[S] \mid \text{for each } k \in Z^+, f^k \text{ is not a product of more than } k \text{ elements of } F[S]\}.$$

We show first that U is the set of elements of $F[S]$ of order m. If $f \in F[S]$ has order m, it is clear that $f \in U$. To prove the converse, let d be the greatest common divisor of the elements of S and choose $M \in Z^+$ such that $rd \in S$ for each $r \geq M$. If $g \in F[S]$ has order $w > m$, then there exists $k \in Z^+$ such that $kw - (k+1)m \geq Md$. For this k, we have $g^k = (X^m)^{k+1}h$, where the degree of each term in h is of the form rd for some $r \geq M$. Hence $h \in F[S]$, and $g \notin U$. Similarly, the set of elements of $K[T]$ of order n is characterized as the set V, where

$$V = \{f \in K[T] \mid \text{for each } k \in Z^+, f^k \text{ is not a product of more than } k \text{ elements of } K[T]\}.$$

Let ϕ be an isomorphism of $F[S]$ onto $K[T]$. Then the above shows that $\phi(U) = V$. Let $a \in F$, $a \neq 0$. Then $\phi(aX^m)$ has the form $bX^n + q$ for some $b \in K\backslash\{0\}$ and some $q \in K[T]$ of order $> n$. We define $\sigma: F \longrightarrow K$ as follows: $\phi(a)$ is the coefficient of X^n in $\phi(aX^m)$. If $\phi(1) = u$, we show that the mapping $\tau: F \longrightarrow K$ defined by $\tau(a) = u^{-1}\phi(a)$ is an isomorphism of F onto K. Suppose $a_1, a_2 \in F$ and

$$\phi(a_1X^m) = b_1X^n + q_1, \ \phi(a_2X^m) = b_2X^n + q_2,$$

$$\phi(X^m) = uX^n + q$$

Then $\phi((a_1 + a_2)X^m) = (b_1 + b_2)X^n + q_1 + q_2$, so $\sigma(a_1 + a_2) = \sigma(a_1) + \sigma(a_2)$ and $\tau(a_1 + a_2) = \tau(a_1) + \tau(a_2)$. Moreover,

$$\phi(a_1a_2X^{3m}) = \phi(a_1a_2X^m)\phi(X^m)^2 = (\sigma(a_1a_2)X^n + q^*)(uX^n + q)^2$$
$$= u^2\sigma(a_1a_2)X^{3n} + \text{(terms of higher degree)}.$$

Also,

$$\phi(a_1a_2X^{3m}) = \phi(a_1X^m)\phi(a_2X^m)\phi(X^m) = (b_1X^n + q_1)(b_2X^n + q_2)(uX^n + q)$$
$$= b_1b_2uX^{3n} + \text{(higher degree terms)}.$$

Therefore $u^2\sigma(a_1a_2) = b_1b_2u = u\sigma(a_1)\sigma(a_2)$ and $\tau(a_1a_2) = u^{-1}\sigma(a_1a_2) = u^{-2}\sigma(a_1)\sigma(a_2) = \tau(a_1)\tau(a_2)$. Therefore τ is an injective homomorphism from F into K. To prove that τ is surjective, take $b \in K\backslash(0)$ and let $\phi^{-1}(ubX^n) = aX^m + p$, where $\text{ord}(p) > m$. Then $\phi(aX^m) = cX^n + q$, where $\text{ord}(q) > m$. If $c \neq ub$, then $(cX^n + q) - ubX^n = (c - ub)X^n + q$ has order n, whereas $\phi^{-1}(cX^n + q) - \phi^{-1}(ubX^n) = aX^m - (aX^m + p) = -p$ has order greater than m, a contradiction. Therefore $c = ub$, $b = u^{-1}c$, and $b = \tau(a)$. Thus F and K are isomorphic under the mapping τ.

We remark that not only are the fields F and K in the statement of Theorem 23.18 isomorphic, but the semigroups S and T are also isomorphic in that case.

Section 23 Remarks

The unitary ring R is said to be <u>invariant</u> if $R[X] \simeq S[Y]$ implies $R \simeq S$, and R is n-<u>invariant</u> if isomorphism of the polynomial rings $R^{(n)}$ and $S^{(n)}$ implies invariance of R and S. Coleman and Enochs initiated the study of invariant rings in [35]; see [1] and [24] for other results. The class of invariant rings includes the classes of Artinian rings, von Neumann regular rings, and quasi-local domains; the class is closed under finite direct sum, and it contains the ring of all algebraic integers in an arbitrary finite algebraic number field. The analogous problem for power series rings is considered in [114], [84], and [68], where the term power-invariant is used to describe a ring R such that $R[[X]] \simeq S[[Y]]$ implies that $R \simeq S$. In particular, [68] contains an example of a Noetherian ring R that is not power-invariant.

The case of Theorem 23.17 where S and T are groups was proved in [2] by Adjaero and Spiegel. Numerous other cases of the isomorphism $R_1[G] \simeq R_2[G]$ are considered in [2] besides the case where R_1 and R_2 are fields.

I am unaware of an example that shows for fields F and K that isomorphism of the semigroup rings FS and KT does not imply isomorphism of F and K. Even if such an example exists, it's possible that isomorphism of FS and KS implies $F \simeq K$. Examples mentioned in Section 25 show that an isomorphism $F[G_1] \simeq F[G_2]$ for groups G_1 and G_2 and a field F need not imply isomorphism of G_1 and G_2.

§24. Coefficient Rings in Isomorphic Monoid Rings. II.

In considering the isomorphism $R_1[S] \simeq R_2[S]$ in Section 23, we imposed conditions on the coefficient rings R_1 and R_2. In this section, conditions are imposed on the monoid S instead. Most frequently S is taken to be a finitely generated torsion-free group. Corollary 24.9 shows that $R_1[Z^n] \simeq R_2[Z^n]$ for some n if and only if $R_1[Z] \simeq R_2[Z]$, and hence the primary emphasis is on the case where $S = Z$. Another reason for the stress on this case is that Sehgal poses as Problem 27 of [125, p. 230] the question of whether isomorphism of $R_1[Z]$ and $R_2[Z]$ implies isomorphism of R_1 and R_2 for Noetherian unitary rings R_1 and R_2. In Example 24.7 we present an example due to Krempa [86] which shows that the question has a negative answer. All rings considered in this section are assumed to be unitary. Before beginning, a word of caution concerning notation is in order. Both polynomial rings and group rings over R are considered in the section, and brackets are used to denote each. But in combination, the group under consideration is always Z, whereas X and Y are consistently used to denote indeterminates. For example, $R[X][Z]$ denotes the group ring of Z over the polynomial ring in X over R, while $R[Z][X,Y]$ means the polynomial ring in X and Y over the group ring of Z over R.

THEOREM 24.1. Let S be either Z_0^n or Z^n. If ϕ is an isomorphism of $R[S]$ onto $T[S]$ such that $\phi(R) \subseteq T$, then $R \simeq T$.

<u>Proof</u>. Let $\{g_i\}_{i=1}^n$ be a free basis for S and let

$u_i = \phi(X^{g_i})$ for each i. In the case of Z_0^n, we have $T[S] = T[\{X^{g_i}\}_1^n] = \phi(R[S]) \subseteq T[u_1,\ldots,u_n] \subseteq T[S]$, and hence $T[S] = T[u_1,\ldots,u_n]$. If $S = Z^n$, then each u_i is a unit of $T[S]$ and as in the preceding sentence, $T[X^{\pm g_1},\ldots,X^{\pm g_n}] = T[u_1^{\pm 1},\ldots,u_n^{\pm 1}]$. By Theorem 22.4 or by the remarks at the end of Section 22, it follows in either case that there exists a T-automorphism σ of $T[S]$ such that $\sigma(u_i) = X^{g_i}$ for each i. Thus $\sigma\phi\colon R[S] \twoheadrightarrow T[S]$ is an isomorphism that maps the augmentation ideal $I = (\{1 - X^{g_i}\}_1^n)$ of $R[S]$ onto the augmentation ideal $J = (\{1 - X^{g_i}\}_1^n)$ of $T[S]$. Consequently, $R \simeq R[S]/I \simeq T[S]/J \simeq T$.

If G and H are groups and if ϕ is an isomorphism of $R[G]$ onto $T[H]$, then ϕ is said to be elementary if, for each $g \in G$, $\phi(X^g)$ is of the form $uX^h + t$, where u is a unit of T and $t \in T[H]$ is nilpotent. Theorem 24.2 permits a frequent reduction of questions concerning isomorphisms ϕ of $R[G]$ onto $T[H]$ to the case where ϕ is elementary.

THEOREM 24.2. Assume that G and H are torsion-free groups, that G is finitely generated, and that ϕ is an isomorphism of $R[G]$ onto $T[H]$. Then there exist decompositions $R = R_1 \oplus \ldots \oplus R_n$ and $T = T_1 \oplus \ldots \oplus T_n$ into nonzero subrings R_i, T_i such that, for each i, ϕ restricts to an elementary isomorphism of $R_i[G]$ onto $T_i[H]$.

Proof. Since $\{\phi(X^g) \mid g \in G\}$ is a finitely generated subgroup of the unit group of $T[H]$, Corollary 11.18 implies that there exists a decomposition $1 = \Sigma_1^n f_i$ of $1 \epsilon T$ into nonzero orthogonal idempotents such that $F = \{f_1, f_2, \ldots f_n\}$ splits each $\phi(X^g)$. Let $e_i = \phi^{-1}(f_i)$ for each i. Corollary 10.8 shows that each e_i is in R. Let $R_i = Re_i$ and let $T_i = Tf_i$. Then $\phi(R_i[G]) = \phi(e_i R[G]) = f_i T[H] = T_i[H]$ for each i, and by definition of the statement that F splits each $\phi(X^g)$, it follows that the restriction of ϕ to $R_i[G]$ is elementary.

We regard $R[Z]$ and $T[Z]$ as $R[X,X^{-1}]$ and $T[X,X^{-1}]$, respectively. If ϕ is an elementary isomorphism of $R[Z]$ onto $T[Z]$ and if $\phi(X) = uX^r + t$, where u is a unit of T and $t \epsilon T[Z]$ is nilpotent, then $|r|$ is called the <u>degree</u> of ϕ and is denoted by $\deg \phi$.

THEOREM 24.3. Assume that $\phi : R[Z] \twoheadrightarrow T[Z]$ is an elementary isomorphism of degree r.

(1) If $r = 1$, then $R \simeq T$.

(2) There exists an R–automorphism σ of $R[Z]$ such that $\phi\sigma(X) = vX^r$ for some unit v of R.

(3) If $r = 0$, then $R \simeq T$.

<u>Proof</u>. (1): If $r = 1$, then Theorem 22.3 implies that there exists a T–automorphism σ of $T[Z]$ such that $\sigma\phi(X) = X$. Considering the isomorphism $\sigma\phi$ of $R[Z]$ onto $T[Z]$, it follows as in the proof of Theorem 24.1 that $R \simeq T$.

(2): Let $\phi(X) = uX^{\varepsilon r} + t$, where $\varepsilon = \pm 1$. Then $\phi^{-1}(t) \epsilon R[Z]$ is nilpotent. Let σ_1 be the R–automorphism of $R[Z]$ such that $\sigma_1(X) = X - \phi^{-1}(t)$ and let $\sigma_2 \epsilon \mathrm{Aut}_R R[Z]$ be such that $\sigma_2(X) = X^{\varepsilon}$. Then $\phi\sigma_1(X) = \phi(X - \phi^{-1}(t)) = \phi(X) - t = uX^{\varepsilon r}$. Therefore

$\phi\sigma_1\sigma_2(X) = \phi\sigma_1(X^{\varepsilon}) = u^{\varepsilon}X^r$. Letting $v = u^{\varepsilon}$ and $\sigma = \sigma_1\sigma_2$, this establishes (2).

(3): By part (2), we can assume that $\phi(X) = v$ is a unit of T. Theorem 24.2 shows that there exist decompositions $R = \Sigma_1^n \oplus Re_i$ and $T = \Sigma_1^n \oplus Tf_i$ such that $\phi(e_i) = f_i$ for each i and such that ϕ^{-1} restricts to an elementary isomorphism of $Tf_i[Z]$ onto $Re_i[Z]$ for each i. Moreover, the restriction ϕ_i of ϕ to $Re_i[Z]$ maps e_iX to f_iv, a unit of Tf_i for each i, so $\deg \phi_i = 0$. If we show that $Re_i \simeq Tf_i$ for each i, it follows that $R \simeq T$. Thus, we assume without loss of generality that ϕ^{-1} is elementary. Write $\phi^{-1}(X) = uX^r + p$, where u is a unit of R and $p \in R[Z]$ is nilpotent. Let $\sigma \in \mathrm{Aut}_R R[Z]$ be such that $\sigma(X) = uX$. Then $\phi(\sigma(X)) = \phi(uX) = \phi((uX^r + p)X^{1-r} - pX^{1-r}) = u^{1-r}X - \phi(pX^{1-r})$, where u^{1-r} is a unit of T and $\phi(pX^{1-r}) \in T[Z]$ is nilpotent. Therefore $\phi\sigma$ has degree 1 and part (1) implies that $R \simeq T$.

Recall that rings R_1 and R_2 are <u>subismorphic</u> if each can be imbedded in the other. While isomorphism of $R[Z]$ and $T[Z]$ need not imply isomorphism of R and T, it does imply that R and T are subisomorphic. A proof of this statement uses the following result.

THEOREM 24.4. Let X and Y be indeterminates over the ring R, let r be a positive integer, and let u be a unit of R.

(1) $R[X,X^{-1},Y]/(Y^r - X) \simeq R[Z]$.

(2) $R[X,X^{-1},Y]/(Y^r - u) \simeq R[y][Z]$, where $y^r = u$ and where $\{1,y,\ldots,y^{r-1}\}$ is a free module basis for $R[y]$ over R.

Proof. (1): Let $I = (Y^r - X)$, $y = Y + I$, and let $x = X + I$. We show that $R[X,X^{-1},Y]/I = R[y,y^{-1}]$, where $\{y^i\}_{-\infty}^{\infty}$ is free over R. We have $y^r = x$, a unit of $R[X,X^{-1},Y]/I = R[x,x^{-1},y]$. Moreover, $x^{-1} = (y^{-1})^r$, so $x,x^{-1} \in R[y,y^{-1}]$. To show that $\{y^i\}_{-\infty}^{\infty}$ is free over R, it suffices to show that $r_0 + r_1Y+\ldots+r_kY^k$ belongs to I only if each $r_i = 0$. Thus, write

$$\Sigma_0^k r_i Y^i = (Y^r - X)f(X,Y)/X^m ,$$

where $f(X,Y) \in R[X,Y]$ and $m \geq 0$. Substituting $X = Y^r$ yields $\Sigma_0^k r_i Y^i = 0$, and hence each $r_i = 0$. This implies that $R[y,y^{-1}] \simeq R[Z]$, as asserted.

(2): As in (1), let $J = (Y^r - u)$, $y = Y + J$, and let $x = X + J$. It is clear that $R[X,X^{-1},Y]/J = R[y,x,x^{-1}]$ and to prove (2), it suffices to prove that $\{x^i y^j \mid i \in Z$ and $0 \leq j \leq r - 1\}$ is free over R. In fact, it is enough to show that $\{x^i y^j \mid i \in Z_0$ and $0 \leq j \leq r - 1\}$ is free over R. Thus, assume that $f_0(Y),\ldots,f_s(Y) \in R[Y]$ are polynomials of degree at most $r - 1$ such that

$$f_0(Y) + f_1(Y)X+\ldots+f_s(Y)X^s \in (Y^r - u), \text{ say}$$

$$\Sigma_0^s f_i(Y)X^i = (Y^r - u)h(X,Y)/X^k , \text{ so that}$$

$$\Sigma_0^s f_i(Y)X^{k+i} = (Y^r - u)h(X,Y) .$$

Considering $R[X,Y]$ as $R[Y][X]$, it follows that $Y^r - u$ divides each $f_i(Y)$ in $R[Y]$. Since $Y^r - u$ is monic, each nonzero multiple of $Y^r - u$ has degree at least r. By choice of the polynomials $f_i(Y)$, it then follows that $f_i(Y) = 0$ for each i. This completes the proof.

THEOREM 24.5. Assume that $\phi: R[Z] \twoheadrightarrow T[Z]$ is an elementary isomorphism of degree r. There exists an extension ring $T[v]$ of T such that v^r is a unit of $T, \{1, v, \ldots, v^{r-1}\}$ is a free basis for $T[v]$ over T, and $R \simeq T[v]$.

Proof. Without loss of generality we assume that $\phi(X) = uX^r$, where u is a unit of T. Let $\phi^*: R[Z][Y] \longrightarrow T[Z][Y]$ be the isomorphism of the polynomial ring induced by ϕ — that is, $\phi^*(Y) = Y$ and $\phi^*|_{R[Z]} = \phi$. Let I be the ideal of $R[Z][Y]$ generated by $Y^r - X$, and let $J = \phi^*(I) = (Y^r - uX^r) = ((X^{-1}Y)^r - u)$. The residue class rings $R[Z][Y]/I$ and $T[Z][Y]/J$ are isomorphic under an isomorphism σ induced by ϕ^*. Moreover, the proof of Theorem 24.4 shows that $R[Z][Y]/I = R[y, y^{-1}] \simeq R[Z]$, where $y = Y + I$. The proof of (24.4) also shows that $T[Z][Y]/J = T[v][x, x^{-1}] \simeq T[v][Z]$, where $v = X^{-1}Y + J$, $x = X + J$, $v^r = u + J$ is a unit of T, and $\{1, v, \ldots, v^{r-1}\}$ is a free basis for $T[v]$ over T. We have $\sigma(y) = \phi^*(Y) + J = Y + J = XX^{-1}Y + J = vx$, where v is a unit of $T[v]$. Consequently, σ is elementary of degree 1, and part (1) of Theorem 24.3 shows that $R \simeq T[v]$.

COROLLARY 24.6. If $R[Z]$ and $T[Z]$ are isomorphic, then R and T are subisomorphic. In fact, there exists a subring R_1 of T and an injective endomorphism σ of T such that $\sigma(T) \subseteq R_1 \subseteq T$, where $R_1 \simeq R$ and where R_1 and T are finitely generated free modules over $\sigma(T)$ and R_1, respectively.

Proof. It suffices to show that T can be imbedded in

R by means of an isomorphism τ such that R is a finitely generated free module over $\tau(T)$. Let $\phi: R[Z] \twoheadrightarrow T[Z]$ be an isomorphism, and choose decompositions $R = R_1 \oplus \ldots \oplus R_n$ and $T = T_1 \oplus \ldots \oplus T_n$ of R and T such that $\phi(R_i[Z]) = T_i[Z]$ for each i and such that ϕ_i, the restriction of ϕ to $R_i[Z]$, is elementary. Theorem 24.5 shows that there exists an extension ring $T_i[v_i]$ of T_i such that $R_i \simeq T_i[v_i]$, $v_i^{r_i} = u_i$ is a unit of R_i, and $\{1, v_i, \ldots, v_i^{r_i - 1}\}$ is a free basis for $T_i[v_i]$ over T_i. Let $W = T_1[v_1] \dot{\oplus} \ldots \dot{\oplus} T_n[v_n]$, let $v = (v_1, \ldots, v_n)$, and let $r = \sup\{r_i\}_1^n$. Under the natural imbedding μ of T in W, we have $W = \mu(T)[v]$, where $\{1, v, \ldots, v^{r-1}\}$ is a free basis for W over $\mu(T)$. If $\alpha: W \twoheadrightarrow R$ is the natural isomorphism, then $\tau = \alpha\mu: T \longrightarrow R$ has the desired properties.

We next present an example of non–isomorphic two–dimensional Noetherian domains R and T such that $R[Z]$ and $T[Z]$ are isomorphic; the example is taken from [86].

EXAMPLE 24.7. Let $a = (1 + \sqrt{-47})/2$, let $K = Q(a)$, and let $D = Z[a]$ be the integral closure of Z in K. Then D is a Dedekind domain with class number 5. If ϕ is the nontrivial automorphism of K, then ϕ induces the unique nontrivial automorphism of D and it also induces an automorphism of the group of fractional ideals of D. Let I be the maximal ideal $(2, a)$. It is straightforward to show that I is not principal, that $I \cap \phi(I) = 2D$, and that $I^5 = bD$, where $b = 4 + a$. Thus, in the class group of D, we have $\phi(I) \equiv I^4 \equiv I^{-1}$ and $\phi(I^{-1}) \equiv I$. Therefore I^2 is not equivalent to $\phi^i(I^j)$, where $i = 1, 2$ and

$j = \pm 1$. Let $R = \Sigma_{-\infty}^{\infty} \oplus I^{2n}X^n$ and $T = \Sigma_{-\infty}^{\infty} \oplus I^n X^n$ be the Rees subrings of the group ring $D[X,X^{-1}]$ corresponding to the sequences $\{I^{2n}\}_{n=-\infty}^{\infty}$ and $\{I^n\}_{n=-\infty}^{\infty}$ of fractional ideals of D . Since I , I^{-1} , I^2 , and I^{-2} are finitely generated D–modules, it follows that R and T are affine domains over D , hence Noetherian. As overrings of the two–dimensional Noetherian domain $D[X]$, the domains R and T have dimension at most 2 . On the other hand, $D[X,X^{-1}]$ is a two–dimensional quotient ring of both R and T , so $\dim R = \dim T = 2$.

Assume that R and T are isomorphic under the map σ Under localization at $Z\backslash\{0\}$, σ induces an automorphism of $K[X,X^{-1}]$. Therefore $\sigma(K) = K$, $\sigma|_K = \phi^i$ for $i = 1$ or 2 , and $\sigma(X) = cX^j$, where c is a nonzero element of K and $j = 1$ or -1 . Clearly σ maps I^2X onto I^jX^j , since σ is surjective and since it maps monomials of degree n onto monomials of degree nj . On the other hand, $\sigma(I^2X) = \sigma(I^2)\cdot cX^j = \phi^i(I^2)cX^j$. Therefore $\phi^i(I^2) = I^jX^j$, which implies that $I^2 \equiv \phi^i(I^j)$, a contradiction. Hence R and T are not isomorphic. We show that the group rings $R[Y,Y^{-1}]$ and $T[Y,Y^{-1}]$ are isomorphic. Let τ be the K–automorphism of $K[X^{\pm 1},Y^{\pm 1}]$ determined by $\tau(X) = X^2Y$, $\tau(Y) = bX^5Y^2$. Then τ^{-1} is determined by $\tau^{-1}(X) = b^{-1}X^{-2}Y, \tau^{-1}(Y) = b^2X^5Y^{-2}$. We observe that $\tau(R[Y,Y^{-1}]) \subseteq T[Y,Y^{-1}]$ and that $\tau^{-1}(T[Y,Y^{-1}]) \subseteq R[Y,Y^{-1}]$; this will imply that τ is an isomorphism of $R[Y,Y^{-1}]$ onto $T[Y,Y^{-1}]$. Since $R[Y,Y^{-1}]$ is generated by $\{I^{2n}X^nY^j \mid n,j \in Z\}$ and since $T[Y,Y^{-1}]$ is generated by $\{I^nX^nY^j \mid n,j \in Z\}$, it suffices to show that

$\tau(I^{2n}X^nY^j) \subseteq T[Y,Y^{-1}]$ and that $\tau^{-1}(I^nX^nY^j) \subseteq R[Y,Y^{-1}]$. Since $b \in I^5$, we have

$$\tau(I^{2n}X^nY^j) = I^{2n}(X^{2n}Y^nb^jX^{5j}Y^{2j}) =$$

$$I^{2n}b^jX^{2n+5j}Y^{n+2j} \subseteq I^{2n+5j}X^{2n+5j}Y^{n+2j} ,$$

which is in $T[Y,Y^{-1}]$. Similarly,

$$\tau^{-1}(I^nX^nY^j) = I^nb^{-n}X^{-2n}Y^nb^{2j}X^{5j}Y^{2j} =$$

$$I^nb^{2j-n}X^{5j-2n}Y^{2j+n} \subseteq I^{n+10j-5n}X^{5j-2n}Y^{2j+n} =$$

$$I^{10j-4n}X^{5j-2n}Y^{2j+n} \subseteq R[Y,Y^{-1}] .$$

This completes the presentation of the example.

Given a semigroup S , we say that the ring R is S-<u>invariant</u> if $R[S] \simeq T[S]$ implies $R \simeq T$ for any ring T We investigate the class $\mathcal{S}$ of all Z-invariant rings; we distinguish a subclass $\mathcal{S}^*$ of $\mathcal{S}$ as follows: the ring R is in $\mathcal{S}^*$ if each isomorphism $\sigma:R[Z] \twoheadrightarrow T[Z]$, for any ring T , is elementary of degree 1 . Part (1) of Theorem 24.3 shows that, indeed, $\mathcal{S}^*$ is a subclass of $\mathcal{S}$.

THEOREM 24.8. The ring $R[Z]$ is Z-invariant for each unitary ring R .

<u>Proof</u>. Let $\phi:R[Z^2] \twoheadrightarrow T[Z]$ be an isomorphism. By Theorem 24.2, it suffices to show that $R[Z] \simeq T$ under the assumption that ϕ is elementary. Thus, for $g \in Z^2$ write $\phi(X^g) = u(g)X^{h(g)} + t(g)$, where $u(g)$ is a unit of T and $t(g) \in T[Z]$ is nilpotent. The mapping $g \longrightarrow h(g)$ is easily seen to be a homomorphism of Z^2 into Z . If

$h(g) = 0$ for each g, then viewing $R[Z^2]$ as $R[Z][Z]$, ϕ has degree 0 and Theorem 24.3 implies that $R[Z] \simeq T$. Otherwise, $Z^2 = (g_1) \oplus (g_2)$, where $(g_2) = \ker h$. Considering $R[Z^2]$ as $R[(g_1)][(g_2)]$, ϕ has degree 0 as an isomorphism of $R[Z^2]$ to $T[Z]$, and again $T \simeq R[(g_1)] \simeq R[Z]$.

COROLLARY 24.9. The following conditions are equivalent.

(1) $R[Z] \simeq T[Z]$.

(2) $R[Z^n] \simeq T[Z^n]$ for each $n \in Z^+$.

(3) $R[Z^n] \simeq T[Z^n]$ for some $n \in Z^+$.

Proof. The implications (1) $\Longrightarrow$ (2) and (2) $\Longrightarrow$ (3) are clear. To show that (3) $\Longrightarrow$ (1), it suffices to show that $R[Z^n] \simeq T[Z^n]$, for $n > 1$, implies $R[Z^{n-1}] \simeq T[Z^{n-1}]$. Thus $R[Z^n] \simeq R[Z^{n-2}][Z^2] \simeq T[Z^n] \simeq T[Z^{n-1}][Z]$ implies, by Theorem 24.7, that $R[Z^{n-2}][Z] \simeq R[Z^{n-1}] \simeq T[Z^{n-1}]$.

COROLLARY 24.10. The integral group ring $Z[G]$ is Z–invariant for each abelian group G.

Proof. Assume that $\phi: Z[G \oplus Z] \twoheadrightarrow R[Z]$ is an isomorphism. Theorem 9.17 shows that $Z[G \oplus Z]$ is reduced, and part (1) of Corollary 10.17 shows that $Z[G \oplus Z]$ is also indecomposable. Hence R is also a reduced indecomposable ring, so $R[Z]$ has only trivial units by Theorem 11.6. For each $g \in G$, we write $\phi(X^g) = u(g)X^{h(g)}$, where $u(g)$ is a unit of R. The mapping $g \longrightarrow h(g)$ is a homomorphism of G into Z. If it is the zero mapping, then $\phi(Z[G]) \subseteq R$ and Theorem 24.1 shows that $Z[G] \simeq R$. If $h \neq 0$, then $\mathrm{Im}(h) = H \simeq Z$ and H is a direct summand of G — say $G = H \oplus K$. Then $Z[G] \simeq Z[K][Z]$ is Z–invariant by

Theorem 24.8, and hence $Z[G] \simeq R$.

THEOREM 24.11. Let R be a unitary ring with Jacobson radical $J(R)$ and nilradical $N(R)$. Let S and S^* be as in the paragraph preceding the statement of Theorem 24.8.

(1) If R is a field, then $R \in S^*$.

(2) If R is an integral domain and if $J(R) \neq (0)$, then $R \in S^*$.

(3) If R is indecomposable and if there exists a minimal prime P of R such that $R/P \in S^*$, then $R \in S^*$.

(4) If R is indecomposable and if $J(R) \neq N(R)$, then $R \in S^*$.

(5) If R is quasi-local, then $R \in S^*$.

(6) If $\dim R = 0$, then $R \in S^*$.

(7) Each of the classes S^* and S is closed under taking finite direct sums.

Proof. We assume throughout the proof that $\phi: R[Z] \twoheadrightarrow T[Z]$ is an isomorphism. The ring R is indecomposable in each of cases (1)–(5) , and Corollary 10.17 implies that the ring T is also indecomposable. Thus ϕ is elementary in each of these cases — say $\phi(X) = uX^r + t$, and our object is to show that $|r| = 1$. We begin to consider the individual cases.

(1): If R is a field, then $R[Z]$ and $T[Z]$ are integral domains, and hence T is also a domain. Therefore $T[Z]$ has only trivial units. We show that $\phi(R) \subseteq T$. This is clear if $R = GF(2)$. If $|R| > 2$, take $a \in R \setminus \{0\}$. Then $\phi(a) = vX^m$ for some unit $v \in T$. Since $a + 1$ or $a - 1$ is a unit of R , either $vX^m + 1$ or $vX^m - 1$ is a unit of $T[Z]$, hence trivial. In either case we conclude that $m = 0$, so $\phi(a) \in T$. Thus $T[Z] = \phi(R[Z]) \subseteq T[uX^r, u^{-1}X^{-r}] = T[X^r, X^{-r}]$, and this implies that $|r| = 1$.

(2): As in (1) , $T[Z]$ has only trivial units. We first show that $\phi(J(R)) \subseteq T$. Take $r \in J(R)$. Then $1 + r$ is a unit of R , so $1 + \phi(r) = uX^m$ for some unit $u \in T$ and some integer $m \in Z$. Moreover, $\phi(1 + r + r^2) = uX^m + (uX^m - 1)^2 = uX^{2m} - uX^m + 1$ is also a unit of $T[Z]$, and this implies that $m = 0$. Consequently, $\phi(r) \in T$ and $\phi(J(R)) \subseteq T$. Take any element $y \in R$. Then $yr \in J(R)$, so $\phi(y)\phi(r) \in T$ with $\phi(r) \in T\setminus\{0\}$. Since $\phi(y) \in T[Z]$ and T is an integral domain, it then follows that $\phi(y) \in T$ and $\phi(R) \subseteq T$. The proof that $|r| = 1$ is then completed as in (1) .

(3): Consider the minimal prime $P[Z]$ of $R[Z]$. Then $\phi(P[Z])$ is minimal in $T[Z]$, and hence $\phi(P[Z]) = Q[Z]$ for some minimal prime Q of T . The isomorphism ϕ induces an isomorphism ϕ^* of $(R/P)[Z]$ onto $(T/Q)[Z]$ and $|\phi| = |\phi^*|$ since ϕ is elementary. But $|\phi^*| = 1$ since $R/P \in S^*$.

(4): Since $J(R) \neq N(R)$, there exists a minimal prime P of R that is not expressible as an intersection of maximal ideals of R . Hence R/P is a domain with nonzero Jacobson radical, so $R/P \in S^*$ by (2) . Whence $R \in S^*$ by (3) .

(5): Let M be the maximal ideal of R . If $M \neq N(R)$, the result follows from (4) . In the contrary case, M is a minimal prime of R and the result follows from (1) and (3).

(6): In this case ϕ need not be elementary, but there

exist decompositions $R = R_1 \oplus \ldots \oplus R_n$ and $T = T_1 \oplus \ldots \oplus T_n$ such that $\phi_i = \phi|_{R_i[Z]}$ is an elementary isomorphism of $R_i[Z]$ onto $T_i[Z]$ for each i. Since $\dim R_i = 0$ for each i and since S^* is closed under taking finite direct sums by (7), we assume without loss of generality that ϕ is elementary. Thus, if M is a proper prime of R, then M is minimal in R and ϕ induces an isomorphism $\phi^*: (R/M)[Z] \twoheadrightarrow (T/P)[Z]$ for some minimal prime P of T; moreover, $|\phi| = |\phi^*|$ since ϕ is elementary. Because $R/M \in S^*$, we have $1 = |\phi^*| = |\phi|$.

(7): We treat the cases of S and S^* simultaneously. Thus, assume that $R = R_1 \oplus \ldots \oplus R_n$, where each R_i belongs to a specified class. Let e_i be the identity element of R_i. Since the idempotents of $T[Z]$ are in T by Corollary 10.8, each $\phi(e_i)$ is in T and $T = T_1 \oplus \ldots \oplus T_n$, where $T_i = T\phi(e_i)$ for each i. It follows that ϕ induces an isomorphism ϕ_i of $R_i[Z]$ onto $T_i[Z]$ for each i. If $R_i \in S$ for each i, we conclude that $R_i \simeq T_i$ for each i and that $R \simeq T$. On the other hand, if each R_i is in S^*, then $\phi_i(e_i X) = u_i X + t_i$, where u_i is a unit of T_i and $t_i \in T_i[Z]$ is nilpotent for each i. Thus $\phi(X) = (\Sigma_1^n u_i)X + \Sigma_1^n t_i$, where $\Sigma_1^n u_i$ is a unit of T and $\Sigma_1^n t_i \in T[Z]$ is nilpotent. Consequently, $R \in S^*$ if each $R_i \in S^*$, and this completes the proof of Theorem 24.11.

Section 24 Remarks

Much of the material in this section stems from the paper [87] of Krempa. K. Yoshida has also considered consequences of isomorphism of $R[Z]$ and $S[Z]$ in [134]. In particular, Yoshida also gives an example of non-isomorphic two-dimensional affine domains R and S over a field such that R[Z] and S[Z] are isomorphic. Vleduc in [130] investigates the class of S-invariant rings, where S is a torsion-free cancellative monoid with quotient group of finite torsion-free rank.

§25. Monoids in Isomorphic Monoid Rings — a Survey

The dual of the question considered in Sections 23 and 24 is that of determining relations between monoids S_1 and S_2 if the monoid rings $R[S_1]$ and $R[S_2]$ over a fixed unitary ring R are isomorphic. In particular, are S_1 and S_2 isomorphic? The question has a negative answer in general — for example, $R[\{X_i\}_1^\infty][Z_0^k]$ and $R[\{X_i\}_1^\infty][Z_0^m]$ are isomorphic for arbitrary positive integers k and m, but Z_0^k and Z_0^m are isomorphic only if $m = k$. The same type of example is available for group rings: if G, H, and K are abelian groups such that $G \oplus H \simeq G \oplus K$, then $R[G][H] \simeq R[G][K]$, but H and K need not be isomorphic. (Theorem 24.8 provides an indirect proof of the fact that no example of this type is possible for $G = Z$.) An example of a different type is obtained as follows. Assume that G is a finite abelian group of order n and that K is a field such that n is a unit of K and K contains a primitive n^{th} root of unity. Using the fact that $K[G]$ is zero-dimensional and reduced, it follows easily that $K[G] \simeq K^n$, the direct sum of n copies of K. If H is another abelian group of order n, then $K[H] \simeq K^n \simeq K[G]$, but H and G need not be isomorphic.

While this isomorphism problem has hardly been considered in the literature for monoids that are not groups, the situation for group rings is quite different. In fact, the isomorphism question and the closely related problem of determining the structure of the unit group of $R[G]$ were two of the earliest topics concerning group rings considered. Because our treatment of units of $R[S]$ in Section 11 was abbreviated, many of the known results concerning the isomorphism problem are not accessible from results of this text. While some results are accessible, this material is already available in the books of Passman [121, Chapter 14] and Schgal [125, Chapter 3], and it seemed pointless to include proofs of inferior results here. Hence, this section merely surveys some of the known results in the area, with a few references to the literature.

We have already alluded to the paper [76] of Higman in which he determines, among other things, all finite abelian groups G such that the integral group ring $Z[G]$ has only trivial units. This result can be used to show that isomorphism of $Z[G]$ and $Z[H]$, for finite abelian groups G and H, implies isomorphism of G and H. More generally, if R is an integral domain of characteristic 0, if G and H are abelian, and if G_R (resp., H_R) is the sum of the p-components of G (resp. H) such that p is invertible in R, then isomorphism of $R[G]$ and $R[H]$ implies isomorphism of G/G_R and H/H_R; the same conclusion holds for a finitely generated indecomposable ring of characteristic 0 (see [102–103]).

If F is a field and G is a finite group whose order is a unit of F, then the group ring $F[G]$ is a finite direct sum of cyclotomic extensions of F. Using an explicit description of this decomposition in the case of the rational field Q, it can be shown that $Q[G] \simeq Q[H]$ implies $G \simeq H$ for finite abelian groups G and H. For a more general field F, isomorphism of $F[G]$ and $F[H]$ implies $G \simeq H$ if G is abelian and $F[G]$ has only trivial units; if $F[G]$

and $F[H]$ are F-isomorphic, then $G/T(G)$ and $H/T(H)$ are isomorphic, where $T(G)$ and $T(H)$ are the torsion subgroups of G and H, respectively. Finally, if G is a countable abelian p-group and F is a field of characteristic p, then $F[G] \simeq F[H]$ implies $G \simeq H$. Each of the results mentioned in this paragraph can be found in [121] or in [125]. In particular, much of the work on the isomorphism question for infinite abelian p-groups over fields is due to Berman [18–19].

SELECTED BIBLIOGRAPHY

1. S. Abhyankar, P. Eakin, and W. Heinzer, On the uniqueness of the coefficient ring in a polynomial ring, J. Algebra 23(1972), 310-342.

2. I. Adjaero and E. Spiegel, On the uniqueness of the coefficient ring in a group ring, Canad. J. Math. (to appear).

3. T. Akiba, Integral—closedness of polynomial rings, Japan. J. Math. 6(1980), 67-75.

4. D. D. Anderson, Multiplication ideals, multiplication rings, and the ring R(X), Canad. J. Math. 28(1976), 760-768.

5. D. D. Anderson and D. F. Anderson, Divisibility properties of graded domains, Canad. J. Math. 34(1982), 196-215.

6. D. F. Anderson, Subrings of k[X,Y] generated by monomials, Canad. J. Math. 30(1978), 215-224.

7. _____, Graded Krull domains, Commun. Algebra 7(1979), 79-106.

8. _____, The divisor class group of a semigroup ring, Commun. Algebra 8(1980), 467-476.

9. D. F. Anderson and J. Ohm, Valuations and semivaluations of graded domains, Math. Ann. 256(1981), 145-156.

10. J. T. Arnold and J. W. Brewer, Commutative rings which are locally Noetherian, J. Math. Kyoto Univ. 11(1971), 45-49.

11. J. T. Arnold and R. Gilmer, Dimension theory of commutative rings without identity, J. Pure Appl. Algebra 5(1974), 209-231.

12. _____, The dimension theory of commutative semigroup rings, Houston J. Math. 2(1976), 299-313.

13. K. Asano, Uber kommutative Ringe, in denen jedes Ideal als Produkt von Primidealen darstellbar ist, J. Math. Soc. Japan 3(1951), 82-90.

14. K. E. Aubert, On the ideal theory of commutative semigroups, Math. Scand. 1(1953), 39-54.

15. C. Ayoub, Restricted chain conditions on groups and rings, Houston J. Math. 7(1981), 303-316.

16. R. G. Ayoub and C. Ayoub, On the group ring of a finite abelian group, Bull. Austral. Math. Soc. 1(1969), 245-261.

17. H. Bass, E. H. Connell, and D. Wright, The Jacobian conjecture: Reduction of degree and formal expansion of the inverse, Bull. Amer. Math. Soc. 7(1982), 287-330.

18. S. D. Berman, Group algebras of countable abelian p-groups, Soviet Math. Dokl. 8(1967), 871-873.

19. _____, Group algebras of countable abelian p-groups (in Russian), Publ. Math. Debrecen 14(1967), 365-405.

20. N. Bourbaki, Elements of Mathematics. Commutative Algebra, Addision-Wesley, Reading, Mass., 1972.

21. A Bouvier, Anneaux de Krull gradues (preprint).

22. H. Bresinsky, Symmetric semigroups of integers generated by 4 elements, manuscr. math. 17(1975), 205-219.

23. J. W. Brewer, D. L. Costa, and E. L. Lady, Prime ideals and localization in commutative group rings, J. Algebra 34(1975), 300-308.

24. J. W. Brewer and E. A. Rutter, Isomorphic polynomial rings, Arch. Math. 23(1972), 484-488.

25. B. Brown and N. H. McCoy, Rings with unit element which contain a given ring, Duke Math. J. 13(1946), 9-20.

26. L. Budach, Struktur Noetherscher kommutativer Halbgruppen, Monatsb. Deutsch. Akad. Wiss. Berlin 6(1964), 81-85.

27. H. S. Butts and W. W. Smith, Prüfer rings, Math. Z. 95(1967), 196-211.

28. L. G. Chouinard II, Krull semigroups and divisor class groups, Canad. J. Math. 33(1981), 1459-1468.

29. _____, Projective modules over Krull semigroup rings, Mich. Math. J. 29(1982), 143-148.

30. L. G. Chouinard II, B. R. Hardy, and T. S. Shores, Arithmetical and semihereditary semigroup rings, Commun. Algebra 8(1980), 1593-1652.

31. A. H. Clifford, Arithmetic and ideal theory of commutative semigroups, Annals Math. 39(1938), 594-610.

32. A. H. Clifford and G. B. Preston, The Algebraic Theory of Semigroups, Vol. I, Amer. Math. Soc., Providence, R. I., 1961.

33. I. S. Cohen, Commutative rings with restricted minimum condition, Duke Math. J. 17(1950), 27-42.

34. J. A. Cohn and D. Livingstone, On the structure of group algebras I, Canad. J. Math. 17(1965), 583-593.

35. D. B. Coleman and E. A. Enochs, Polynomial invariance of rings, Proc. Amer. Math. Soc. 25(1970), 559-562.

36. I. G. Connell, On the group ring, Canad. J. Math. 15(1963), 650-685.

37. R. K. Dennis, Units of group rings, J. Algebra 43(1976), 655-664.

38. _____, The structure of the unit group of group rings, Ring Theory II, Marcel Dekker, New York, 1977, pp. 103-130.

39. K. Drbohlav, On finitely generated commutative semigroups, Comment. Math. Univ. Carolinae 4(1963), 87-92.

40. _____, Zur Theorie der Kongruenzrelationen auf kommutativen Halbgruppen, Math. Nachr. 26(1963/64), 233-245.

41. P. Eakin, On Arnold's formula for the dimension of a polynomial ring, Proc. Amer. Math. Soc. 54(1976), 11-15.

42. P. Eakin and A. Sathaye, R—endomorphisms of R[[X]] are essentially continuous, Pacific J. Math. 66(1976), 83-87.

43. E. Formanek, Idempotents in Noetherian group rings, Canad. J. Math. 25(1973), 366-369.

44. R. Fossum, The Divisor Class Group of a Krull Domain, Springer—Verlag, New York, 1973.

45. L. Fuchs, Über die Ideale arithmetischer Ringe, Comment. Math. Helv. 23(1949), 43-47.

46. _____, Partially Ordered Algebraic Systems, Pergamon, London, 1963.

47. _____, Infinite Abelian Groups, Vol. I, Academic Press, New York, 1970.

48. _____, Infinite Abelian Groups, Vol. II, Academic Press, New York, 1973.

49. A. Garcia, Cohen—Macaulayness of the associated graded ring of a semigroup ring, Commun. Algebra 10(1982), 393-415.

50. R. Gilmer, A note on semigroup rings, Amer. Math. Monthly, 75(1969), 36-37.

51. _____, Multiplicative Ideal Theory, Marcel Dekker, New York, 1972.

52. _____, A two—dimensional non—Noetherian factorial ring, Proc. Amer. Math. Soc. 44(1974), 25-30.

53. R. Gilmer and A. Grams, The equality $(A \cap B)^n = A^n \cap B^n$ for ideals, Canad. J. Math. 24(1972), 792-798.

54. R. Gilmer, A. Grams, and T. Parker, Zero divisors in power series rings, J. Reine Angew. Math. 278(1975), 145-164.

55. R. Gilmer and R. Heitmann, The group of units of a commutative semigroup ring, Pacific J. Math. 85(1979), 49-64.

56. R. Gilmer and J. L. Mott, Multiplication rings as rings in which ideals with prime radical are primary, Trans. Amer. Math. Soc. 114(1965), 40-52.

57. R. Gilmer and M. O'Malley, R–endomorphisms of $R[[X_1,\ldots,X_n]]$, J. Algebra 48(1977), 30-45.

58. R. Gilmer and T. Parker, Divisibility properties in semigroup rings, Mich. Math. J. 21(1974), 65-86.

59. _____, Semigroup rings as Prüfer rings, Duke Math. J. 41(1974), 219-230.

60. R. Gilmer and M. L. Teply, Units of semigroup rings, Commun. Algebra 5(1977), 1275-1303.

61. _____, Idempotents of commutative semigroup rings, Houston J. Math. 3(1977), 369-385.

62. B. Glastad and G. Hopkins, Commutative semigroup rings which are principal ideal rings, Comment. Math. Univ. Carolinae 21(1980), 371-377.

63. S. Goto, N. Suzuki, and K. Watanabe, On affine semigroup rings, Japan J. Math. 2(1976), 1-12.

64. A. Grams, Atomic rings and the ascending chain condition for principal ideals, Proc. Cambridge Philos. Soc. 75(1974), 321-329.

65. M. Griffin, Prüfer rings with zero divisors, J. Reine Angew. Math. 239(1969), 55-67.

66. _____, Multiplication rings via their total quotient rings, Canad. J. Math. 26(1974), 430-449.

67. T. Gulliksen, P. Ribenboim, and T. M. Viswanathan, An elementary note on group rings, J. Reine Angew. Math. 242(1970), 148-162.

68. E. Hamann, On the R–invariance of $R[x]$, J. Algebra 35(1975), 1-16.

69. B. R. Hardy and T. S. Shores, Arithmetical semigroup rings, Canad. J. Math. 32(1980), 1361-1371.

70. W. Heinzer and J. Ohm, Locally Noetherian commutative rings, Trans. Amer. Math. Soc. 158(1971), 273-284.

71. J. Herzog, Generators and relations of abelian semigroups and semigroup rings, manuscr. math. 3(1970), 175-193.

72. J. Herzog and E. Kunz, Die Werthalbgruppe eines lokalen Rings der Dimension 1, S. B. Heidelberg. Akad. Wiss. (1971), 27-67.

73. E. Hewitt and H. S. Zuckerman, Finite dimensional convolution algebras, Acta Math. 93(1955), 67-119.

74. _____, The ℓ_1–algebra of a commutative semigroup, Trans. Amer. Math. Soc. 83(1956), 70-97.

75. J. C. Higgins, Representing N–semigroups, Bull. Austral. Math. Soc. 1(1969), 115-125.

76. G. Higman, The units of group rings, Proc. London Math. Soc. 46(1940), 231-248.

77. M. Hochster, Non–uniqueness of coefficient rings in a polynomial ring, Proc. Amer. Math. Soc. 34(1972), 81-82.

78. _____, Rings of invariants of tori, Cohen–Macaulay rings generated by monomials, and polytopes, Annals Math. 96(1972), 318-337.

79. J. R. Isbell, On the multiplicative semigroup of a commutative ring, Proc. Amer. Math. Soc. 10(1959), 908-909.

80. P. Jaffard, Contribution a l'etude des groupes ordonnes, J. Math. Pures Appl. 32(1953), 203-280.

81. B. D. Janeway, Zero divisors in commutative semigroup rings, University of Houston Dissertation, 1981.

82. C. U. Jensen, Arithmetical rings, Acta Sci. Acad. Sci. Hungar. 17(1966), 115-123.

83. I. Kaplansky, Commutative Rings, Allyn & Bacon, Boston, 1969.

84. J.-H. Kim, Power invariant rings, Pacific J. Math. 51(1974), 207-213.

85. I. B. Kozhukhov, Chain semigroup rings (in Russian), Uspekhi Matem. Nauk 29(1974), 169-170.

86. J. Krempa, Isomorphic group rings with nonisomorphic commutative coefficients, Proc. Amer. Math. Soc. 83(1981), 459-460.

87. _____, Isomorphic group rings of free abelian groups, Canad. J. Math. 34(1982), 8-16.

88. J. Lambek, Lectures on Rings and Modules, Blaisdell, Toronto, 1966.

89. D. C. Lantz, R–automorphisms of R[G] for G abelian torsion–free, Proc. Amer. Math. Soc. 61(1976), 1-6.

90. D. C. Lantz, Preservation of local properties and chain conditions in commutative group rings, Pacific J. Math. 63(1976), 193-199.

91. M. D. Larsen and P. J. McCarthy, Multiplicative Theory of Ideals, Academic Press, New York, 1971.

92. F. Levi, Arithmetische Gesetze im Gebiete discreter Gruppen, Rend. Circ. Mat. Palermo 35(1913), 225-236.

93. Ja. B. Livchak, On orderable groups (in Russian), Uchen. Zap. Ural. Gos. Univ. 23(1959), 11-12.

94. A. I. Malcev, On the immersion of an algebraic ring into a field, Math. Ann. 113(1937), 686-691.

95. R. Matsuda, On algebraic properties of infinite group rings, Bull. Fac. Sci. Ibaraki Univ. 7(1975), 29-37.

96. _____, Infinite group rings II, Bull. Fac. Sci. Ibaraki Univ. 8(1976), 43-66.

97. _____, Torsion-free abelian group rings III, Bull. Fac. Sci. Ibaraki Univ. 9(1977), 1-49.

98. _____, Torsion-free abelian semigroup rings IV, Bull. Fac. Sci. Ibaraki Univ. 10(1978), 1-27.

99. _____, Torsion-free abelian semigroup rings V, Bull. Fac. Sci. Ibaraki Univ. 11(1979), 1-37.

100. _____, Krull properties of semigroup rings, Bull. Fac. Sci. Ibaraki Univ. 14(1982), 1-12.

101. W. May, Commutative group algebras, Trans. Amer. Math. Soc. 136(1969), 139-149.

102. _____, Group algebras over finitely generated rings, J. Algebra 39(1976), 483-511.

103. _____, Isomorphism of group algebras, J. Algebra 40(1976), 10-18.

104. N. H. McCoy, Remarks on divisors of zero, Amer. Math. Monthly 49(1942), 286-295.

105. T.-T. Moh, On a normalization lemma for integers and an application of four colors theorem, Houston J. Math. 5(1979), 119-123.

106. S. Mori, Axiomatische Begründung des Multiplikationsringes, J. Sci. Hiroshima Univ. Ser. A. 3(1932), 43-59.

107. _____, Allgemeine Z.P.I.-Ringe, J. Sci. Hiroshima Univ. Ser. A 10(1940), 117-136.

108. J. L. Mott, Equivalent conditions for a ring to be a multiplication ring, Canad. J. Math. 16(1964), 429-434.

109. M. Nagata, Local Rings, Wiley (Interscience), New York, 1962.

110. W. K. Nicholson, Local group rings, Canad. Math. B. 15(1972), 137-138.

111. D. G. Northcott, A generalization of a theorem on the contents of polynomials, Proc. Cambridge Philos. Soc. 55(1959), 282-288.

112. J. Ohm, Integral closure and $(x,y)^n = (x^n,y^n)$, Monatsh. Math. 71(1967), 32-39.

113. J. Ohm and P. Vicknair, Monoid rings as valuation rings, Commun. Algebra 11(1983), 1355-1368.

114. M. J. O'Malley, Isomorphic power series rings, Pacific J. Math. 41(1972), 503-512.

115. M. Orzech, Onto endomorphisms are isomorphisms, Amer. Math. Monthly 78(1971), 357-362.

116. T. Parker, The semigroup ring, Florida State University Dissertation, 1973.

117. T. Parker and R. Gilmer, Nilpotent elements of commutative semigroup rings, Mich. Math. J. 22(1975), 97-108.

118. M. M. Parmenter, Isomorphic group rings, Canad. Math. Bull. 18(1975), 567-576.

119. _____, Coefficient rings of isomorphic group rings, Bol. Soc. Bras. Mat. 7(1976), 59-63.

120. M. M. Parmenter and S. Sehgal, Uniqueness of the coefficient ring in some group rings, Canad. Math. Bull. 16(1973), 551-555.

121. D. S. Passman, Algebraic Structure of Group Rings, Wiley (Interscience), New York, 1977.

122. L. Redei, The theory of finitely generated commutative semigroups, Pergamon, Oxford-Edinburgh-New York, 1965.

123. S. K. Sehgal, On the isomorphism of integral group rings. I, Canad. J. Math. 21(1969), 410-413.

124. _____, Units in commutative integral group rings, Math. J. Okayama Univ. 14(1970), 135-138.

125. _____, Topics in Group Rings, Marcel Dekker, New York, 1978.

126. T. S. Shores, On generalized valuation rings, Mich. Math. J. 21(1974), 405-409.

127. E. Snapper, Completely primary rings. I, Annals Math. 52(1950), 666-693.

128. M. L. Teply, Semiprime semigroup rings and a problem of J. Weissglass, Glasgow Math. J. 21(1980), 131-134.

129. M. L. Teply, E. G. Turman, and A. Quesada, On semisimple semigroup rings, Proc. Amer. Math. Soc. 79(1980), 157-163.

130. S. G. Vleduc, On the coefficient ring of a semigroup ring, Math. USSR Izv. 10(1976), 899-911.

131. J. Weissglass, Regularity of semigroup rings, Proc. Amer. Soc. 25(1970), 499-503.

132. _____, Semigroup rings of semilattice sums of rings, Proc. Amer. Math. Soc. 39(1973), 471-478.

133. D. Whitman, Chain conditions on congruence relations in commutative semigroups (preprint).

134. K. Yoshida, On the coefficient ring of a torus extension, Osaka J. Math. 17(1980), 769-782.

135. A. Zaks, Atomic rings without a.c.c. on principal ideals, J. Algebra 74(1982), 223-231.

136. O. Zariski and P. Samuel, Commutative Algebra, Vol. I, Van Nostrand, Princeton, N. J., 1958.

137. _____, Commutative Algebra, Vol. II, Van Nostrand, Princeton, N. J., 1960.

TOPIC INDEX

The following abbreviations are used in some of the listings in the index: cong. for congruence, dom. for domain, elt. for element, gp. for group, id. for ideal, mon. for monoid, and smgp. for semigroup.

D

E

R

S

T

U

V

Z

INDEX OF MAIN NOTATION

SYMBOL	PAGE	MEANING
l.u.b.	29	Least upper bound
g.l.b.	29	Greatest lower bound
$\tilde{p}$	34	Congruence of p-equivalence
Δ	35	The minimal congruence with factor semigroup torsion-free and cancellative
$\rho(s)$	42	Congruence associated with ρ and s
$[\rho]^\circ$	42	An ideal of S associated with ρ
$[\rho]$	42	The radical of $[\rho]^\circ$
$a\|b$	52	a divides b
l.c.m.	53	Least common multiple
a.c.c.p.	57	Ascending chain condition for principal ideals
$\|A\|$	62	Cardinality of A
R[X;S] or R[S]	64	The semigroup ring of S over R
Supp(f)	68	The support of f
c(f)	68	Content of f
ker ϕ^*	69	The kernel of ϕ^*
$\vee$	72	least upper bound of congruences
$\wedge$	72	greatest lower bound of congruences
Z_0^n	77	Direct sum of n copies of Z_0
Char D	90	Characteristic of D
PID	165	Principal ideal domain
$r_0(M)$	165	The torsion-free rank of M
(g)	180	Cyclic subgroup generated by g
Z^w	185	Direct sum of w copies of Z
F_v	208	Divisorial ideal associated with F
div(A)	208	The divisor class of A
$\mathcal{D}(D)$	208	The set of divisor classes of D
P(D)	209	Principal divisor classes of D

SYMBOL	PAGE	MEANING
$C(D)$	209	Divisor class monoid of D
I_v	215	Divisorial ideal of $I \in F(S)$
div(I)	217	Divisor class of I
$\mathcal{D}(S)$	217	Divisors classes of S
$P(S)$	217	Principal divisor classes of S
$C(S)$	217	Divisor class monoid of S
SPIR	249	Special principal ideal ring
RS	317	Semigroup ring of S over R,S noncommutative
$\Sigma_1^n r_i s_i$	317	Element of RS
deg ϕ	339	Degree of an elementary isomorphism

INDEX OF SOME MAIN RESULTS

RESULT	PAGE	DESCRIPTION
10.6	118	Elements of Supp(f) , for f idempotent, are periodic
10.9	120	Conditions under which idempotents of R[X;S] are in R
10.14	123	Equivalent conditions for each idempotent of R[X;G] to be in R
10.17	126	R[X;S] is indecomposable iff R is indecomposable, the set G of periodic elements of S form a group,and the order of each nonzero element of G is a nonunit of R
11.1	129	D[X;S] has only trivial units if S is torsion-free and cancellative
11.3	131	Equivalent conditions for $f \in R[X;S]$ to be a unit
11.6	133	Conditions under which R[X;S] has only trivial units
11.13	140	If S is torsion-free and aperiodic, then $f \in R[X;S]$ is a unit iff the G-component of f is a unit of R[X;G] and each coefficient of the $(S\setminus G)$-component of f is nilpotent
11.14	140	For S torsion-free and aperiodic, the Jacobson radical and nilradical of R[X;S] coincide
11.18	143	If H is a finitely generated group of units of R[X;S] , then there exists a decomposition of 1 that splits each element of H
12.2	148	Relating the structure of T to a distinguished subring R
12.4	152	The concepts of integrality and almost integrality in cancellative monoids
12.5	153	Complete integral closure of R[S] in T[U]
12.7	155	D[S] is completely integrally closed iff D and S are
12.8	156	$\cap G_{V_\alpha}$ is the integral closure of S in G
12.10	158	Integral closure of R[S] in T[U]

RESULT	PAGE	DESCRIPTION
12.11	159	Integral closure of D[S]
13.6	167	Conditions for D[S] to be a Prüfer domain or a Bezout domain
13.8	168	D[S] is a Dedekind domain iff D is a field and S is isomorphic to Z or to Z_0
14.1	171	D[S] a GCD-domain implies D is a GCD-domain and S is a GCD-monoid
14.5	176	The converse of Theorem 14.1
14.6	177	a.c.c.p. in D[S]
14.8	178	D[S] is factorial iff D and K[S] are
14.16	187	Equivalent conditions for D[S] to be factorial
14.17	188	D[G] satisfies a.c.c.p. iff D satisfies a.c.c.p. and each nonzero element of G is of type (0,0,...)
15.1	191	Necessary conditions for D[S] to be a Krull domain
15.2	192	Characterization of Krull monoids
15.6	197	Equivalent conditions for D[S] to be a Krull domain
15.8	201	Uniqueness of a defining family of essential valuation monoids for a Krull monoid
15.9	202	Determination of the defining family for a Krull domain D[S]
16.2	211	C(D) imbeds in C(D[S])
16.3	213	$C(D) \oplus C(K[S]) \simeq C(D[S])$
16.4	215	Basic properties of the v-operation on a monoid
16.7	219	$C(S) \simeq C(K[S])$
16.10	223	A determination of C(D[S]) for a Krull domain D[S]
17.1	226	$\dim R[G] = \dim R[\{X_\lambda\}_{\lambda\in\Lambda}]$, where $\|\Lambda\| = r_0(G)$

RESULT	PAGE	DESCRIPTION
17.3	227	dim R[S] = dim R iff S is periodic
17.4	228	Equivalent conditions for R[S] to be (von Neumann) regular
17.9	234	Archimedean decomposition of S
17.10	236	The Archimedean components are cancellative iff S is free of asymptotic torsion
18.5	242	If R[S] is a P-ring, then R is regular
18.6	244	R is arithmetical iff each R_M is chained
18.7	246	R regular implies R[X] is a Bezout ring
18.9	248	Equivalent conditions for R[S] to be a Prüfer ring, a Bezout ring, or an arithmetical ring
18.10	250	The conditions PIR , general ZPI-ring, and multiplication ring in R[S]
19.1	254	R[U] is quasi-local iff R is quasi-local, Char(R/M) = p ≠ 0 and U is a p-group
19.4	256	Conditions for R[U] to be chained
19.7	259	Conditions for R[S] to be arithmetical, where S is neither torsion-free nor periodic
19.13	265	R[S] as a ZPI-ring
19.16	267	D[G] as a PID
20.5	273	If S satisfies a.c.c. and d.c.c. on congruences, then S is finite
20.6	274	R[S] is Artinian iff R is Artinian and S is finite
20.11	280	Necessary conditions for R[G] to be locally Noetherian
20.14	284	Converse of Theorem 20.11
21.4	290	dim R[S] = dim R[G] if S is cancellative

RESULT	PAGE	DESCRIPTION
21.9	298	Special Chain Theorem holds in R[S] for S torsion-free and finitely generated
22.1	303	Determination of the R-automorphisms of R[X]
22.3	306	R-automorphisms of $R[X,X^{-1}]$
22.4	309	R-automorphisms of $R[Z^n]$
23.12	329	R[S*] is the maximal regular subring of R[S]
23.16	331	$F[S] \simeq K[T]$ implies $F \simeq K$ if S is periodic
23.17	333	$F \simeq K$ if $F[S] \simeq K[T]$ and S is not aperiodic
24.2	338	Reduction to the case of elementary isomorphisms
24.6	342	$R[Z] \simeq T[Z]$ implies R and T are subisomorphic
24.9	346	$R[Z] \simeq T[Z]$ iff $R[Z^n] \simeq T[Z^n]$
24.11	347	Sufficient conditions for isomorphism of R[Z] and T[Z] to imply that of R and T

Printed and bound by CPI Group (UK) Ltd, Croydon, CR0 4YY

27/10/2024

14580398-0001